T0342368

Wireless Power Transfer

IEEE Press
445 Hoes Lane
Piscataway, NJ 08854

Wireless Power Transfer: Principles and Applications

Zhen Zhang and Hongliang Pang
Tianjin University, China

Published by John Wiley & Sons, Inc., Hoboken, New Jersey.
Published simultaneously in Canada.

Library of Congress Cataloging-in-Publication Data Applied for:

Hardback ISBN: 9781119654063

Cover Design: Wiley
Cover Image: © MuchMania/Shutterstock

Set in 9.5/12.5pt STIXTwoText by Straive, Chennai, India

To my baby girl, Olivia Yiqing Zhang.

Zhen Zhang, Ph.D.

Contents

Author Biographies *xv*
Preface *xvii*
Acknowledgments *xix*

Part I Introduction *1*

1 The Era of Wireless Power Transfer *3*
1.1 The Father of Wireless Power Transfer – Nikola Tesla *3*
1.2 Wireless Power Transfer *5*
1.2.1 Acoustic *6*
1.2.2 Optical *6*
1.2.3 Microwave *8*
1.2.4 Capacitive *9*
1.2.5 Inductive *10*
1.3 About This Book *13*
References *15*

2 Inductive Power Transfer *19*
2.1 Inductive Power Transfer *19*
2.1.1 Principle *19*
2.1.2 1-to-1 Transmission *21*
2.1.2.1 Primary Power Source *22*
2.1.2.2 Primary/Secondary Compensation Network *22*
2.1.2.3 Magnetic Coupling *23*
2.1.2.4 Pickup Unit *24*
2.1.3 1-to-*N* Transmission *24*
2.1.3.1 Single-Frequency Excitation *25*
2.1.3.2 Multifrequency Excitation *26*
2.2 1-to-1 Transmission *28*
2.2.1 Coupled Modeling *28*
2.2.1.1 Loosely Coupled Transformer Model *28*

2.2.1.2 T-model *30*
2.2.1.3 M-model *31*
2.2.1.4 Discussion *31*
2.2.2 Compensation *31*
2.2.2.1 Series Type *32*
2.2.2.2 Parallel Type *33*
2.2.3 Power Transmission *34*
2.2.3.1 Load Power *34*
2.2.3.2 Efficiency *36*
2.2.3.3 Relationship Between Power and Efficiency *36*
2.2.3.4 Considerations *36*
2.3 1-to-n Transmission *37*
2.3.1 General Configuration *37*
2.3.2 Two Pickups System Analysis *38*
2.3.2.1 Modeling *38*
2.3.2.2 Load Power *40*
2.3.2.3 Efficiency *41*
2.3.3 Multiple Pickups System Analysis *42*
2.3.3.1 Modeling *42*
2.3.3.2 Load Power *44*
2.3.3.3 Efficiency *44*
2.3.3.4 Discussion *45*
2.3.4 Cross-Coupling *45*
2.3.4.1 Cross-Coupling Effect *45*
2.3.4.2 Frequency Shifting *48*
2.3.4.3 Compensation of Cross-Coupling *49*
2.4 What Are the Differences Between 1-to-1 and 1-to-n Transmission *50*
2.4.1 Power Distribution *50*
2.4.2 Transmission Control *51*
2.4.3 Cross-Coupling Effects *51*
2.4.4 Energy Security *52*
2.A Appendix *52*
References *53*

Part II Design *55*

3 Design and Optimization for Coupled Coils *57*
3.1 Introduction *57*
3.2 Design Considerations *58*
3.2.1 Analysis of Power Transmission *58*

3.2.2 Coil Parameters *62*
3.2.2.1 Electrical Parameters *62*
3.2.2.2 Structure Parameters *64*
3.2.3 Shielding Methods *68*
3.3 Optimal Design *70*
3.3.1 Quality Factor *70*
3.3.1.1 Hollow Winding with Track-Width Ratio *70*
3.3.1.2 Double-Layer Printed Spiral Coil *74*
3.3.2 Coupling Effect *79*
3.4 Summary *89*
References *91*

4 Design and Optimization for Power Circuits *95*
4.1 Impedance Matching *95*
4.1.1 Compensation Networks *96*
4.1.1.1 Basic Topologies: SS/SP/PS/PP *96*
4.1.1.2 Hybrid Topologies *104*
4.1.2 Tunable Impedance Matching Networks *111*
4.1.2.1 Discontinuous Adjustment-Capacitor Array *111*
4.1.2.2 Continuous Adjustment-Virtual Impedance *114*
4.1.2.3 Hybrid Adjustment *122*
4.2 DC/AC Inverters *123*
4.2.1 Introduction *123*
4.2.2 Wide-Bandgap Semiconductor Devices *123*
4.2.3 Architectures *126*
4.2.3.1 Single-Phase Bridge Inverters *126*
4.2.3.2 Class-E Inverters *129*
4.2.4 Soft Switching *131*
4.2.4.1 Zero-Current Switching (ZCS) *132*
4.2.4.2 Zero-Voltage Switching (ZVS) *133*
4.2.5 Control Schemes *135*
4.2.5.1 Pulse-Width-Modulation Control *135*
4.2.5.2 Phase-Shift Control *135*
References *136*

Part III Control *141*

5 Control for Single Pickup *143*
5.1 Review of Control Schemes *143*
5.1.1 Factors Affecting Transmission Performances *143*

5.1.1.1 Effects of Magnetic Resonant State 143
5.1.1.2 Effects of Magnetic Coupling Coefficient and Load Resistance 144
5.1.2 Controls Ensuring Transmission Performances 147
5.2 Maximizing Efficiency Control Schemes 149
5.2.1 Resonant Control Schemes 149
5.2.1.1 Frequency Tracking 150
5.2.1.2 Controllable Impedance Matching 153
5.2.2 Maximizing Efficiency Control Schemes Based on Equivalent Load Resistance Adjustment 154
5.2.2.1 Equivalent Load Resistance Adjustment Schemes 155
5.2.2.2 Maximizing Efficiency Tracking Schemes 160
5.2.2.3 Maximizing Efficiency Control Schemes – Design Examples 166
References 173

6 Control Scheme for Multiple-pickup WPT System 179
6.1 Introduction 179
6.2 Transmission Strategy 179
6.2.1 Single-frequency Time-sharing Transmission 180
6.2.1.1 Modeling and Analysis 180
6.2.1.2 Verification 182
6.2.2 Multifrequency Simultaneous Transmission 184
6.2.2.1 Modeling and Analysis 184
6.2.2.2 Method of Multifrequency Excitation 185
6.2.2.3 Discussion 195
6.3 Impedance Matching Strategy for Multifrequency Transmission 196
6.3.1 Compensation Network for Multifrequency 196
6.3.1.1 Dual-frequency Compensation Network 196
6.3.1.2 Analysis for Multifrequency Compensation Network 199
6.3.2 Compensation for Cross-coupling on the Pickup Side 202
6.4 Others 204
6.4.1 Power Allocation 204
6.4.2 Maximum Efficiency for Multitransmitter 205
6.4.3 Constant Voltage Control 205
References 206

7 Energy Security of Wireless Power Transfer 209
7.1 Introduction 209
7.2 Characteristics of Frequency 210
7.2.1 Frequency Sensitivity 210
7.2.2 Frequency Splitting 211
7.3 Energy Encryption 215

7.3.1 Cryptography *216*
7.3.2 Energy Encryption Scheme *217*
7.4 Verifications *222*
7.4.1 Simulation *223*
7.4.1.1 Case 1 – One Single Transmitter with Authorized Pickups *224*
7.4.1.2 Case 2 – One Single Transmitter with Authorized Pickup and Unauthorized Pickup *224*
7.4.2 Experimentation *229*
7.5 Opportunities *232*
References *232*

8 Omnidirectional Wireless Power Transfer *235*
8.1 Introduction *235*
8.2 Mathematical Analysis *237*
8.2.1 2-Dimensional WPT with Multiple Pickups *237*
8.2.1.1 Load Current Calculation *241*
8.2.1.2 Output Power Calculation *241*
8.2.1.3 Input Power Calculation *243*
8.2.1.4 Efficiency Calculation *244*
8.2.1.5 Physical Implications of the Input Power in the Form of the Lemniscate of Bernoulli *245*
8.2.1.6 Electromagnetic Position *245*
8.2.2 3-Dimensional WPT with Multiple Pickups *247*
8.2.2.1 Load Current Calculation *251*
8.2.2.2 Output Power Calculation *251*
8.2.2.3 Input Power Calculation *253*
8.2.2.4 Efficiency Calculation *254*
8.3 Design of Transmitting Coils for Synthetic Magnetic Field *255*
8.4 Design and Control Considerations for Pickup Coils *262*
8.5 Load Detection *269*
8.6 Discussion *271*
References *272*

Part IV Application *275*

9 WPT for High-power Application – Electric Vehicles *277*
9.1 Introduction *277*
9.1.1 Origination of WPT for EVs *277*
9.1.2 Development of WPT for EVs *278*
9.1.2.1 Static Wireless Charging *278*

9.1.2.2 Dynamic Wireless Charging *279*
9.1.3 Regulations *280*
9.1.3.1 IEC *280*
9.1.3.2 SAE *280*
9.1.3.3 Other Works *281*
9.2 EV Wireless Charging *282*
9.2.1 Introduction *282*
9.2.2 Static Wireless Charging *282*
9.2.2.1 Introduction *282*
9.2.2.2 Typical Prototypes and Demonstration Projects *284*
9.2.3 Dynamic Wireless Charging *290*
9.2.3.1 Introduction *290*
9.2.3.2 Power Track *292*
9.2.3.3 Typical Demonstration Projects *294*
9.2.4 Market *300*
9.2.5 Patent *301*
9.2.5.1 Previous Development *302*
9.2.5.2 Patents from Enterprises *302*
9.3 Electromagnetic Field Reduction *303*
9.3.1 Standard *303*
9.3.1.1 ICNIRP *303*
9.3.1.2 IEC *305*
9.3.1.3 SAE *306*
9.3.2 Mitigation Schemes *308*
9.3.2.1 Passive Methods *308*
9.3.2.2 Active Methods *309*
9.4 Key Technologies *309*
9.4.1 Foreign Object Detection *310*
9.4.2 Wireless Vehicle-to-Grid *311*
9.4.3 Supercapacitor *312*
9.5 Summary *314*
9.5.1 Improvement of the Charging Power *315*
9.5.2 Enhancement of Misalignment Tolerance *315*
9.5.3 Foreign Object Detection *316*
9.5.4 Reduction of Cost *316*
9.5.5 Impact on Power Grid *317*
9.5.6 Promotion of Its Commercialization *317*
References *317*

10 WPT for Low-Power Applications *327*
10.1 Portable Consumer Electronics *327*
10.1.1 Introduction *327*
10.1.2 Wireless Charging Alliance *328*

10.1.2.1 Wireless Power Consortium *328*
10.1.2.2 Power Matters Alliance *328*
10.1.2.3 Alliance for Wireless Power *329*
10.1.2.4 Others *329*
10.1.3 Wireless Charging Standard *330*
10.1.3.1 Introduction *330*
10.1.3.2 Qi Wireless Charging Standard *331*
10.1.4 Wireless Charging for Mobile Phones *332*
10.1.4.1 Transmission Performance *336*
10.1.4.2 Transmission Stability *337*
10.1.4.3 User Experience (Practicality) *337*
10.1.5 Discussion *338*
10.2 Implantable Medical Devices *339*
10.2.1 Introduction *339*
10.2.2 Wireless Transfer for Implantable Medical Devices *340*
10.2.2.1 Inductive *340*
10.2.2.2 Capacitive *341*
10.2.2.3 Ultrasonic *341*
10.2.3 Various Applications *342*
10.2.3.1 Cochlear Implants *342*
10.2.3.2 Retinal Implants *343*
10.2.3.3 Cortical Implants *343*
10.2.3.4 Peripheral Nerve Implants *344*
10.2.4 Safety Consideration *345*
10.2.4.1 EM Safety *345*
10.2.4.2 Physical Safety *346*
10.2.4.3 Cyber Safety *346*
10.2.5 Future Challenges *346*
10.3 Drones *347*
10.3.1 Introduction *347*
10.3.2 Challenges *348*
10.3.3 Wireless In-flight Charging of Drones *350*
10.3.4 Discussion *352*
10.4 Underwater Wireless Charging *352*
10.4.1 Introduction *353*
10.4.2 Analysis of UWPT *353*
10.4.2.1 Challenges *353*
10.4.2.2 Analysis *354*
10.4.3 Applications *356*
10.4.4 Discussion *358*
References *359*

Index *361*

Author Biographies

Zhen Zhang, Ph.D., is a full professor with the School of Electrical and Information Engineering at Tianjin University. He has authored and co-authored numerous internationally refereed papers as well as two books published by Wiley-IEEE Press and Cambridge University Press. Prof. Zhang is currently the Chair of IEEE Beijing Section IES Chapter (Tianjin) and an Associate Editor for the *IEEE TRANSACTIONS ON INDUSTRIAL ELECTRONICS*, *IEEE TRANSACTIONS ON INDUSTRIAL INFORMATICS*, and *IEEE INDUSTRIAL ELECTRONICS MAGAZINE*. He is the recipient of the Humboldt Research Fellowship, Carl Friedrich von Siemens Research Fellowship, Japan Society for the Promotion of Science Visiting Fellowship, 2020 Outstanding Paper Award for IEEE TRANSACTIONS ON INDUSTRIAL ELECTRONICS, and IEEE J. David Irwin Early Career Award.

Hongliang Pang received the B.Eng. and M.Eng. degrees from Tianjin University, Tianjin, China, in 2017 and 2020, respectively. He is currently working toward the Ph.D. degree in electrical and electronic engineering at the Department of Electrical and Electronic Engineering, the University of Hong Kong, Hong Kong. He has published several technical papers and industrial reports in these areas. His current research interests include electric vehicle technologies, wireless power transfer, and power-electronic-based impedance matching.

Preface

Electrically isolated inductive magnetic coupling provides the wireless power transfer (WPT) with flexibility, noninvasivity, cleanliness, and security, enabling to deliver the power without any physical contact. This epoch-making technique has been favored by various electric-driven applications, thus fueling the industry with a new breed of technology. It is broadly expected that the WPT industry will grow persistently in the coming decades. Accompanied by the increasing penetration trend of practical niches as well as the rising development of WPT techniques, various scenarios pose complexity, diversity, and challenging issues on this emerging technique. Hence, a book that covers wide areas for WPT technologies, which concurrently takes theoretical analysis, optimal design, intelligent control, and emerging applications into consideration, is highly desirable. This book aims to achieve this mission in the research of WPT, especially for multiple-pickup WPT.

The purpose of this book is to offer readers a panoramic view of WPT technologies with emphasis on the multiple-pickup WPT which is different from single-pickup WPT. Along with the in-depth research on the WPT, the practical-demand-based optimal design and intelligent energy transmission control have attracted increasing attention from the academia and the industry. Hence, this book also aims to offer readers with hot-pot research topics in the current and near future. Most importantly, a whole chapter regarding the latest control scheme for multiple-pickup WPT is presented, which addresses the first systematic elaboration on this topic. Furthermore, this book discusses several typical applications for the wireless charging technique, including the high-power EV wireless charging, and low-power applications, e.g. portable consumer electronics, in-flight charging for unmanned aerial vehicles, and underwater wireless charging, wherein, WPT shows significant, profound, and irreplaceable roles. Thus, this book explores and provides the current methodologies, approaches, and foresight of the emerging technologies of multiple-objective WPT which nearly have not been published before.

This book covers the multidisciplinary aspects and is organized into the following four parts:

- Part I presents an introduction and addresses the differences between 1-to-1 and 1-to-*N* transmission. It consists of Chapters 1 and 2, which introduce the basic theory of inductive power transmission and conclude the system analysis.
- Part II focuses on the design and optimization of coupled coils and power circuits. It comprises two chapters, Chapters 3 and 4, respectively, which discuss the design consideration for coupled coils, coil topology, impedance matching compensation, and DC/AC inverters.
- Part III is the core section of this book, namely, the control scheme for WPT. It contains four chapters, Chapters 5–8, in which the maximum efficiency/power control, excitation modulation as well as power allocation for multiple-pickup WPT, energy-encryption-based security consideration control, and omnidirectional vector control are elaborated.
- Part IV unveils promising applications adopting the WPT technique. It includes Chapters 9 and 10, which emphasize electric vehicles (EVs), portable consumer electronics, implantable medical devices, drones, and underwater devices.

It is anticipated that WPT will experience explosive development. Hope this book to be a key reference for researchers, engineers, and administrators who need to make such progress.

My baby girl, Olivia Yiqing Zhang, was born in Hong Kong in the year 2011, which is right at the beginning of the place and the date for my research on WPT technologies. Today, more than 10 years have passed; my girl gradually realizes the importance of self-learning and independent thinking, while I also gradually realize the challenges and opportunities of WPT technologies. As a similarity as my girl is finishing her primary stage of education, my understanding on WPT technologies is shaping which seems a wonderful beginning for me. Hence, it is very necessary for me to summarize past work, re-understand basic concepts, and think about the future of WPT technologies, which is the reason why I decided to write this book. After the next 10 years, I firmly believe WPT technologies will embrace rapid development stage and draw increasing attention from various applications, similar to my baby girl who is growing up and has her own wonderful life. Lastly, this book is dedicated to my girl, as well as my wife who has been accompanying me to appreciate life all the time.

Hereby, I would therefore like to take this opportunity to express my heartfelt gratitude to my girl, Olivia Yiqing Zhang, and my wife, Zhenyan Liang, for their presence in my life.

PEIYANG Campus, Tianjin University *Written by Zhen Zhang, Ph.D.*

Acknowledgments

The authors wish to acknowledge many exceptional contributions toward the content of this book from our research team, namely Tianjin University Laboratory of Embedded Computing and Control (TJU-ECC), especially Mr. Lin Yang, Mr. Xingyu Li, Mr. Cong Xie, Mr. Shen Shen, Mr. Zhichao Wang, Mr. Yu Gu, Mr. Yitong Wu, and Mr. Yantian Gong. In particular, a large portion of the supporting materials presented are the results of our research groups.

The authors are deeply indebted to our tutors, colleagues, and friends worldwide for their continuous support and encouragement during the years. A note of gratitude to the editors and staff at Wiley who were instrumental in undertaking a diligent review of the text and editing the book through the production process.

The authors would like to express our gratitude to the National Natural Science Foundation of China (Grant No. 51977138), the International Teaching Project for Postgraduates of Tianjin University (Project No. ENT19021), and the Hum-boldt Research Fellowship (Ref. 3.5-CHN-1201512-HFST-P) for their financial support.

Last but not least, the authors also owe debts of gratitude to their families, who gave tremendous support during the process of writing this book.

Part I

Introduction

1

The Era of Wireless Power Transfer

As one of the most epoch-making technologies, the wireless power transfer (WPT) can realize the energy transmission in a cordless way [1, 2], which is obviously changing our traditional usage pattern of the energy, thus promoting the pervasive application of sustainable energies into our daily life. Surprisingly, such a miracle technique is not anything new, yet the original concept can date back more than 100 years ago. The story should begin with a great man, namely Nikola Tesla, as shown in Figure 1.1.

1.1 The Father of Wireless Power Transfer – Nikola Tesla

About 130 years ago, the wirelessly transmitting power was successfully demonstrated by Nikola Tesla's series of experiments, where the Geissler tubes and incandescent light bulbs can be lighted from across a stage based on near-field inductive and capacitive coupling, as depicted in Figure 1.2. The key technique of such amazing experiments is to use Tesla coils, which are spark-excited radio-frequency resonant transformers, to generate a high alternating current (AC) voltage [3, 4]. More importantly, Tesla found that the transmission distance could be increased if the LC circuit of receivers can be tuned to resonance with the LC circuit of transmitters [5], namely resonant inductive coupling [4]. Nevertheless, such an imaginative technique failed to proceed with commercialization due to various limitations at that time, such as semiconductor materials, power electronic technologies, and manufacturing.

Tesla's attempting efforts on WPT technologies have never stopped since the beginning of this story. Then, his focus moved to a wireless power distribution system, which can wirelessly deliver the power directly to everywhere in the world. Borrowing from the idea of Mahlon Loomis [3], he developed a demonstrating system composed of balloons to suspend transmitting and receiving electrodes in

Wireless Power Transfer: Principles and Applications, First Edition. Zhen Zhang and Hongliang Pang.

Figure 1.1 Nikola Tesla [3]. Source: Wikimedia Commons.

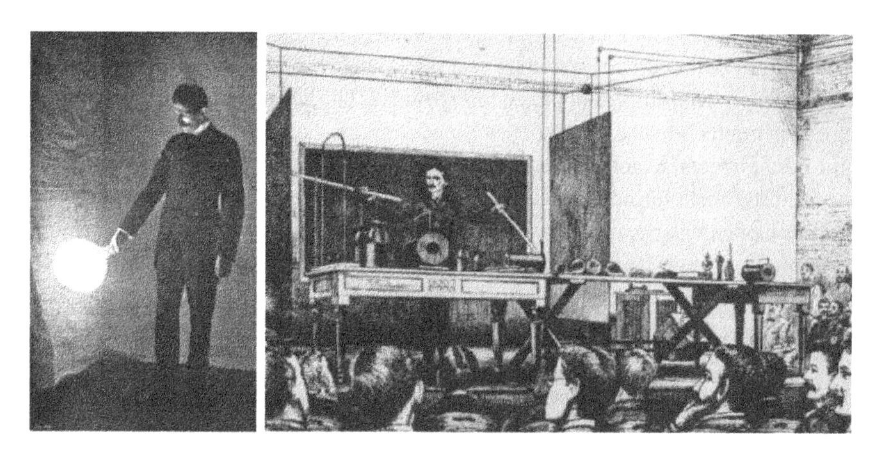

Figure 1.2 Demonstrations of wireless lighting by Tesla. Source: Ref. [2] Nikola Tesla/Wikimedia Commons/Public Domain.

Figure 1.3 Tesla's unsuccessful Wardenclyffe power station. Source: Ref. [2] Nikola Tesla/Wikimedia Commons/Public Domain.

the air above 9100-m altitude, because he believed that the low-pressure air at high altitude would allow higher voltage for long-distance power transmission. In 1899, Tesla build up a test facility at high altitude in Colorado Springs to further study the conductive characteristics of low-pressure air [3]. It is this famous experiment that led Tesla to incorrectly conclude that the entire Earth can be utilized to conduct electrical energy [3]. In addition, the potential of the Earth can be oscillated by driving AC pulses into the Earth at its resonant frequency using a grounded Tesla coil. In such a way, he believed that the AC power can be picked up at everywhere around the world using a similar capacitive antenna tuned to resonance with the Earth [3, 6].

Then, Tesla boldly proposed a "World Wireless System" to deliver both the information and the power around the world [3]. In 1901, a large high-voltage wireless power station, namely Wardenclyffe Tower, was built at Shoreham in New York, as depicted in Figure 1.3. Unfortunately, this project had to be halt due to dried-up investment by the year 1904. Despite all this, Nikola Tesla really redefined the energy transmission and opened up a brand-new research field, who is well deserved to be called the *Father of Wireless Power Transfer* [7].

1.2 Wireless Power Transfer

As one of the revolutionary technologies, WPT can realize the transmission of electric energy from the transmitting end to the desired pickup device in a contactless manner. From the perspective of the transmission distance, the WPT can be mainly divided to near-field transmission and far-field transmission. On the one hand, the near-field power transfer is based on the electromagnetic field coupling theory, including the inductive and capacitive coupling mechanisms. On the other hand, the far-field power transfer can be realized by means of acoustic, optical, and microwave methods, which are applied in low-power sensor networks and military fields. The different wireless power technologies are shown in Table 1.1.

1.2.1 Acoustic

The acoustic power transfer can achieve the wireless power transmission in the form of acoustic waves or mechanical vibrations. As shown in Figure 1.4, the system of the acoustic power transfer mainly includes four constituent units, namely the primary AC power supply, the primary and pickup sensors that realize the conversion of electrical energy and mechanical acoustic energy, and the energy pickup side.

The far-field power transfer can be realized in the acoustic mechanism with the help of the ultrasonic frequency of vibration. Meanwhile, a variety of transmission media, including living tissue, metal materials, and air, are suitable for the acoustic mechanisms. Nevertheless, the acoustic power transfer faces the following three challenges. Firstly, the principle of spatial resonance puts forward special requirements for the placement position of the pickup coil, which limits the application scenario of this technology. Secondly, the technology lacks complete and clear theoretical analysis, which hinders the further development and wide application of the technology. Finally, the design of sensor is an important part of this technology which needs to comprehensively consider the power, efficiency of the system, and the impact of reflections [8].

1.2.2 Optical

As shown in Figure 1.5, the optical WPT uses the laser as the medium to transmit energy to the pickup coil. This technology is mainly used in military or aerospace fields that require long-distance energy transmission. Compared with other energy transmission mechanisms, this technology has the following characteristics: the realization of the ultralong-distance transmission, the realization of the centralized and directional energy transmission, and zero interference to radio-frequency applications. Meanwhile, the optical WPT faces the challenges including the low conversion efficiency between light and electricity and the danger of laser radiation. Since the twenty-first century, optical WPT has also been used in industrial consumer electronics or low-power sensors [9].

Table 1.1 Different wireless power technologies.

Technology	Range	Directivity	Frequency	Antenna devices	Current and/or possible future applications
Inductive coupling	Short	Low	Hz to MHz	Wire coils	Electric tooth brush and razor battery charging, induction stovetops, and industrial heaters
Resonant inductive coupling	Mid	Low	kHz to GHz	Tuned wire coils and lumped element resonators	Charging portable devices (Qi), biomedical implants, electric vehicles, powering buses, trains, MAGLEV, Radio frequency identification (RFID), and smartcards
Capacitive coupling	Short	Low	kHz to GHz	Metal plate electrodes	Charging portable devices, power routing in large-scale integrated circuits, smartcards, and biomedical implants
Magneto-dynamic coupling	Short	N.A. (Not applicable)	Hz	Rotating magnets	Charging electric vehicles and biomedical implants
Microwaves	Long	High	GHz	Parabolic dishes, phased arrays, and rectennas	Solar power satellite, powering drone aircraft, and charging wireless devices
Light waves	Long	High	≥THz	Lasers, photocells, and lenses	Charging portable devices, powering drone aircraft, and powering space elevator climbers

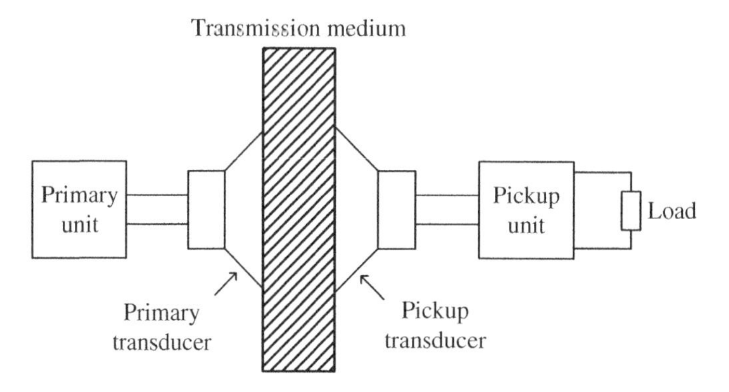

Figure 1.4 Acoustic wireless power transfer.

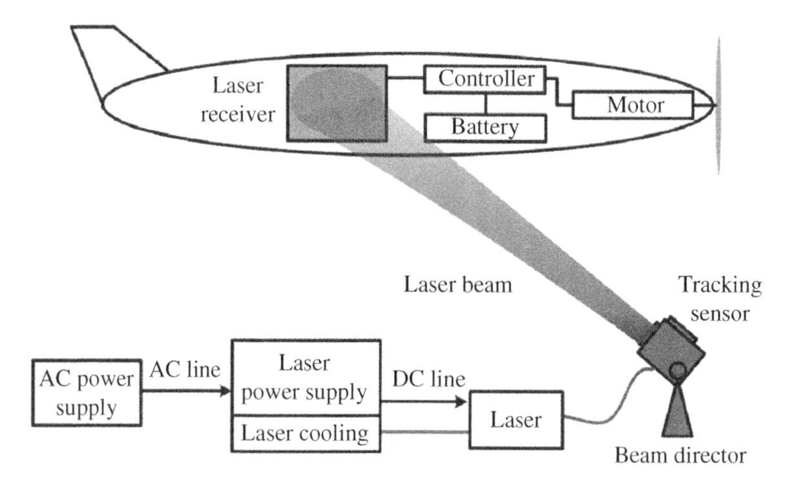

Figure 1.5 Optical wireless power transfer.

1.2.3 Microwave

Microwave power transfer technology is a typical far-field wireless power transmission mechanism, which is applied in the low-power sensor networks, space, and military fields. The principle and components of the microwave power transfer system are depicted in Figure 1.6. At the transmitting part, the microwave is generated by the microwave generator and transmitted through the coax-waveguide-adapted and waveguide circulator, which reduces the external radiation caused by microwave. Then, the tuner and directional coupler device are used to realize the separation of radiation signals according to different propagation directions, ensuring the propagation of radiation in the air. In the pickup part, the microwave radiation is received through the receiving antenna and then

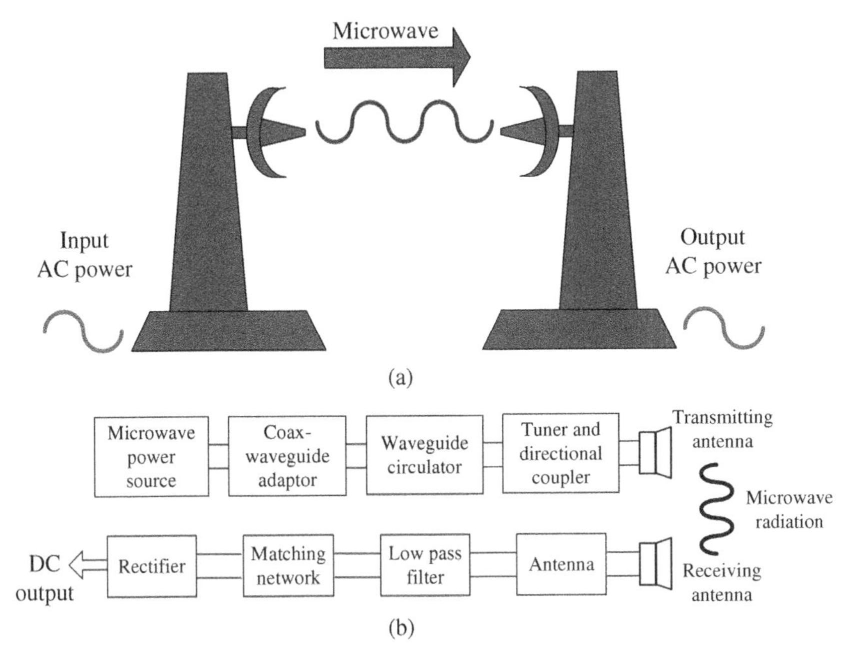

Figure 1.6 Microwave wireless power transfer: (a) schematic; (b) block diagram of working principle.

converted into DC power through the low-pass filter, matching network, and rectifier [10]. To obtain the best energy conversion efficiency in the wide-range input power levels, a novel rectifier antenna architecture was proposed in the microwave WPT system [11].

1.2.4 Capacitive

The capacitive WPT system mainly consists of the transmitting and receiving electrodes, where the transmitting plate voltage generates an induced AC electromotive force on the receiving plate through electrostatic induction. In this technology, the transmitting power is related to the switching frequency of the system and the capacitance between the plates. According to the number of plates, the capacitive WPT can be divided into unipolar and bipolar systems. As shown in Figure 1.7a, the bipolar system includes two sets of transmitting and pickup plates. Different transmitting plates have 180° voltage phase difference, and AC potentials with opposite phase are induced at the pickup plates to realize wireless power transmission in the bipolar method. In addition, as depicted in Figure 1.7b, the unipolar system can achieve the energy transmission with a set of plates, where the passive plates form the return path at the same time [12].

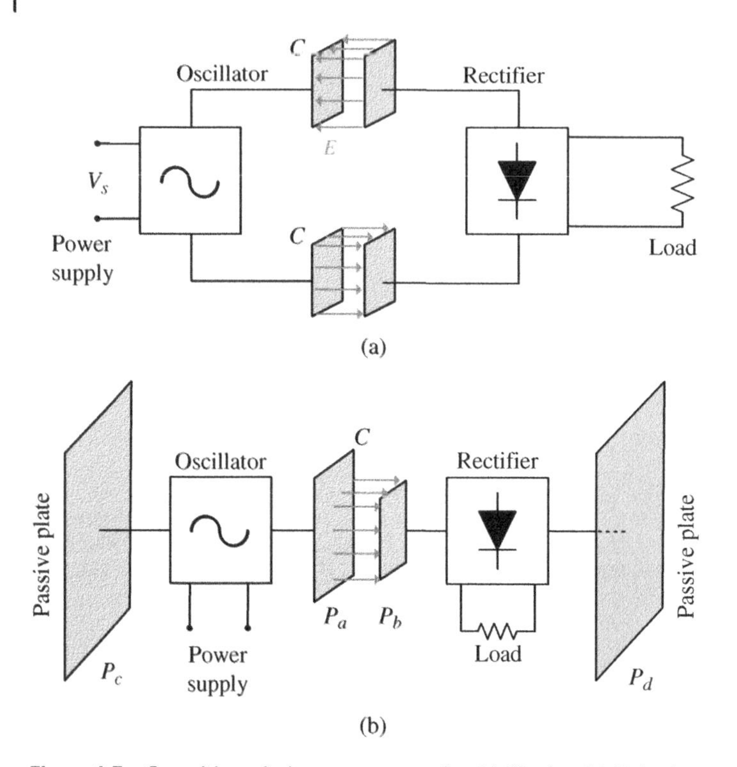

Figure 1.7 Capacitive wireless power transfer: (a) Bipolar; (b) Unipolar.

Compared with other energy transmission mechanisms, this technology reduces the need for alignment between the transmitter and the pickup coils and provides a closed energy field to reduce the external interferences. Nevertheless, since the high voltage on the electrode plate will lead to the generation of harmful gases, the technology can only be applied to some low-power wireless charging scenarios.

1.2.5 Inductive

As depicted in Figure 1.8, the common inductive WPT system is composed of the primary and secondary parts. In the primary part, the high-frequency AC is generated by DC power supply and inverter. Then, based on the electromagnetic induction mechanism, the high-frequency current is wirelessly transmitted from the primary coil to the pickup coil. In fact, the induced power transfer system is similar to the transformer system with weak coupling strength. Besides, the magnetic materials such as the ferrite can enhance the coupling strength between the primary and pickup coils [13].

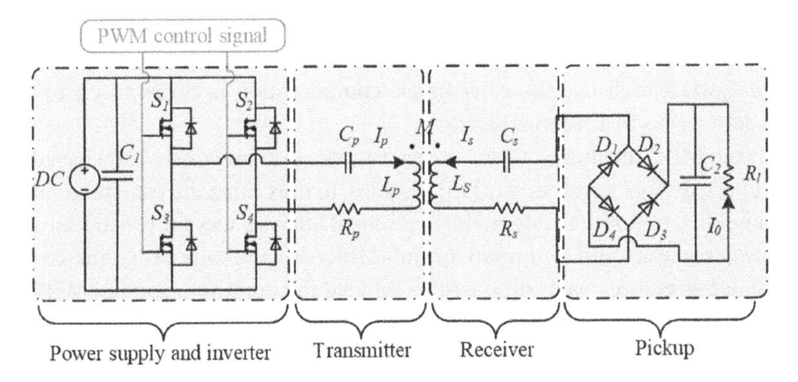

Figure 1.8 Inductive power transfer.

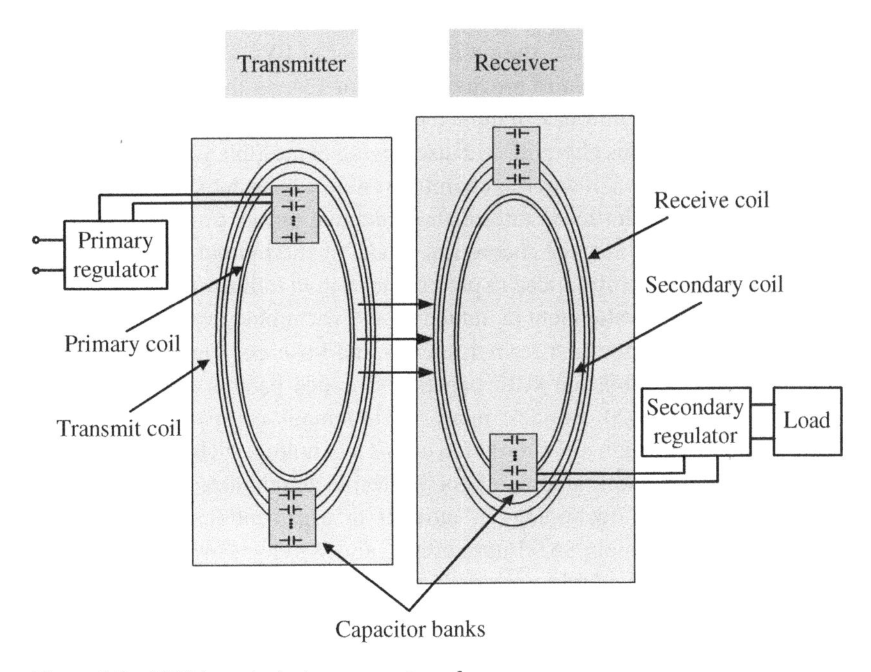

Figure 1.9 MRC-based wireless power transfer.

Moreover, the magnetic resonant coupling (MRC) mechanism can help to increase the transmission performance of inductive power transmission, which has broad application prospects in medium-distance and low-power applications. In Ref. [14], the effectiveness of MRC mechanism was proved by the WPT experiment with a transmission distance of 2 m and a power of 60 W. Figure 1.9 describes the basic components of four-coil MRC-based WPT system, including high-frequency AC power supply, compensation network of the transmitting and receiving coils, and the load side. Different from the inductive power transfer, the MRC-based system must use the capacitance compensation network to ensure that the system works in a resonant state.

Based on the MRC mechanism, the energy can be simultaneously transferred to different pickups in a wireless way [15]. Besides, to maximize the transmission efficiency and distance of the system, the theoretical analysis was proposed based on the equivalent circuit and Neumann formula [16]. Also, the coils with noncoaxial and circular structures were proposed to achieve the domino-resonator WPT system [17]. Furthermore, to avoid the adverse effects of system detuning, an adaptive impedance matching system was implemented by matching the primary and secondary compensation networks to the resonant frequency of the system [18].

As a typical application scenario of inductive power transfer technologies, electric vehicle (EV) wireless charging system has obtained abundant achievements. For instance, to realize the online charging of EVs, an inductive power transfer system was proposed to provide energy for EVs on the whole road [19]. Besides, a bidirectional interface of inductive power transmission was designed to achieve the simultaneous charging and discharging of multiple EVs [20]. Also, to ensure the safe operation of wireless charging system, a load detection model was proposed to monitor the load state of the inductive power transfer system in real time [21]. In addition, the theoretical model of the mutual coupling effect between inductors was introduced to predict the mutual inductance [22].

In fact, due to the development of inductive WPT technology, wireless charging has attracted a lot of attention from academia and business. Figure 1.10 shows the statistics of the published SCIE papers and issued patents about inductive WPT from 2011 to 2020. Besides, many world-famous companies have been engaged in the application and promotion of this technology, such as BMW, Audi, Tesla, Apple, and Huawei, among others. Hence, to standardize the application of the inductive WPT technology, a number of organizations (International Telecommunication Union, SAE International, and Wireless Power Consortium, among others) have issued relevant standards as shown in Table 1.2.

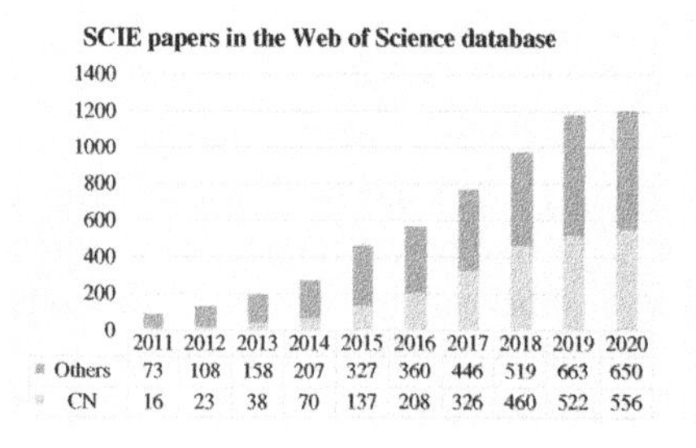

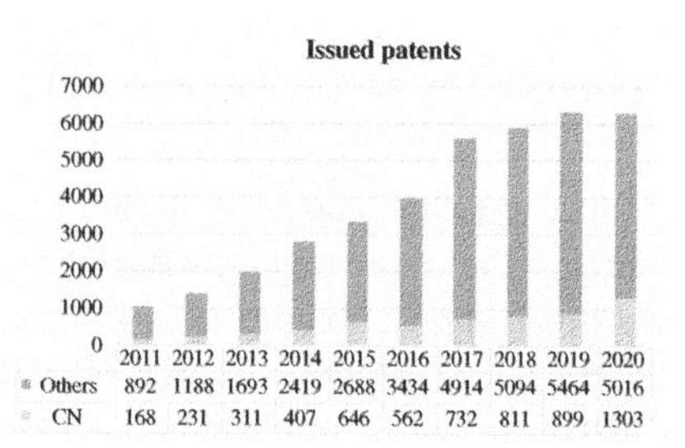

Figure 1.10 Statistics of studies on IPT: the numbers of SCIE papers from 2011 to 2020 and patents from 2011 to 2020. CN: China

Table 1.2 Different standard of WPT.

Standard	Title (Content)	Application
Recommendation ITU-R SM.2110	Guidance for the use of frequency ranges for operation of non-beam wireless power transmission for electric vehicle	EV
Report ITU-R SM.2303-2	Wireless power transmission using technologies other than radio frequency beam	All
Recommendation ITU-R SM.2110	Frequency ranges for operation of non-beam Wireless Power Transmission systems	All
IEC61980-1	Electric vehicle wireless power transfer (WPT) systems - Part 1: General requirements	EV
IEC61980-2	Electric vehicle wireless power transfer (WPT) systems - Part 2: Specific requirements for communication between electric road vehicle (EV) and infrastructure	EV
IEC61980-3	Electric vehicle wireless power transfer (WPT) systems - Part 3: Specific requirements for the magnetic field wireless power transfer systems	EV
ISO 19363:2017	Electrically propelled road vehicles? Magnetic field wireless power transfer? Safety and interoperability requirements	EV
ISO 19363:2020	Electrically propelled road vehicles? Magnetic field wireless power transfer? Safety and interoperability requirements	EV
SAE J2954	Wireless Power Transfer for Light-Duty Plug-in/Electric Vehicles and Alignment Methodology	EV
SAE J2847/6	Communication for Wireless Power Transfer Between Light-Duty Plug-in Electric Vehicles and Wireless EV Charging Stations	EV
Qi standard	The world's de facto wireless charging standard for providing 5-15 watts of power to small personal electronics	Phone

1.3 About This Book

As mentioned in the preface, this book is trying to introduce working mechanisms, summarize recent research works, and discuss about classic applications of WPT technologies, especially during the past 20 years. It is expected to provide a big picture of WPT technologies for academic researchers, industrial engineers, postgraduate students, and readers who are interested in this research topic. Figure 1.11 shows the organization and the relationship among basic concepts, key technique,

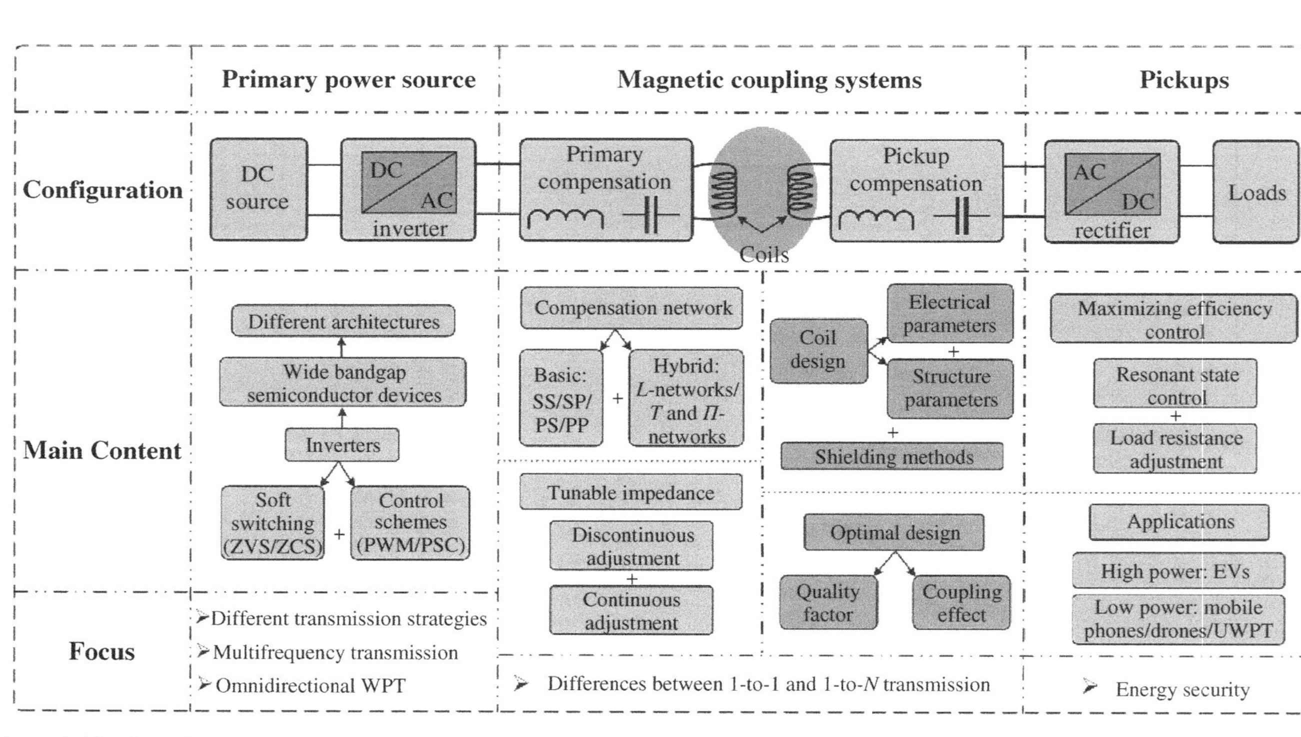

Figure 1.11 Organization of key concepts in this book.

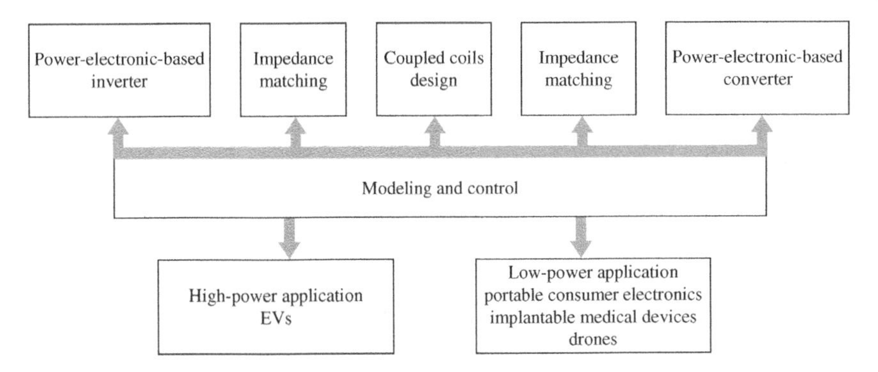

Figure 1.12 Structure of this book.

and main contents of this book, which is utilized to help readers understand WPT technologies more readily, more intuitively, and more in-depth. As depicted in Figure 1.12, this book covers the multidisciplinary aspects and is organized into the following four parts as:

- Part I presents an introduction and addresses the differences between 1-to-1 and 1-to-N transmission. It consists of Chapters 1 and 2, which introduce the basic theory of inductive power transmission and conclude the system analysis.
- Part II focuses on the design and optimization of coupled coils and power circuits. It comprises two chapters, Chapters 3 and 4, respectively, which discuss the design consideration for coupled coils, coil topology, impedance matching compensation, and DC/AC inverters.
- Part III is the core section of this book, namely the control scheme for WPT. It contains four chapters, Chapters 5–8, in which the maximum efficiency/power control, excitation modulation as well as power allocation for multiple-pickup WPT, energy-encryption-based security consideration control, and omnidirectional vector control are elaborated.
- Part IV unveils promising applications adopting the WPT technique. It includes Chapters 9 and 10, which emphasize EVs, portable consumer electronics, implantable medical devices, drones, and underwater devices.

References

1 Zhang, Z., Pang, H., Georgiadis, A., and Cecati, C. (2019). Wireless power transfer – an overview. *IEEE Transactions on Industrial Electronics* 66 (2): 1044–1058.

2 Wireless Power Transfer. https://en.wikipedia.org/wiki/Wireless_power_transfer.

3 Carlson, W.B. (2013). *Tesla: Inventor of the Electrical Age*. Princeton University Press.

4 Lee, C.K., Zhong, W.X., and Hui, S.Y.R. (2012). Recent progress in mid-range wireless power transfer. *Proceedings of 2012 IEEE Energy Conversion Congress and Exposition*, Raleigh, NC, USA (15-20 September 2012) https://ieeexplore.ieee.org/document/6472081.

5 Wheeler, L.P. (1943). II – Tesla's contribution to high frequency. *Electrical Engineering* 62 (8): 355–357.

6 Tesla, N. (1904). The transmission of electric energy without wires. *Electrical World and Engineer* 43: 23760–23761.

7 Tesla, N. (1900). System of transmission of electrical energy. US Patent 645,576, filed 02 September 1897 and issued 20 March 1900. https://patents.google.com/patent/US645576A/en.

8 Roes, M.G.L., Duarte, J.L., Hendrix, M.A.M., and Lomonova, E.A. (2013). Acoustic energy transfer: a review. *IEEE Transactions on Industrial Electronics* 60 (1): 242–248.

9 Sahai, A. and Graham, D. (2011). Optical wireless power transmission at long wavelengths. In: *Proceedings of 2011 International Conference on Space Optical Systems and Applications*, Santa Monica, CA, USA, 146–170. https://ieeexplore.ieee.org/document/5783662.

10 Reddy, M.V., Hemanth, K.S., and Venkat Mohan, C.H. (2013). Microwave power transmission – a next generation power transmission system. *IOSR Journal of Electrical and Electronics Engineering* 4: 24–28.

11 Marian, V., Allard, B., Vollaire, C., and Verdier, J. (2012). Strategy for microwave energy harvesting from ambient field or a feeding source. *IEEE Transactions on Power Electronics* 27 (11): 4481–4491.

12 Liu, C., Hu, A.P., and Nair, N.K.C. (2011). Modeling and analysis of a capacitively coupled contactless power transfer system. *IET Power Electronics* 4: 808–815.

13 Govic, G.A. and Boys, J.T. (2013). Inductive power transfer. *Proceedings of the IEEE* 101 (6): 1276–1289.

14 Kurs, A., Karalis, A., Moffatt, R. et al. (2007). Wireless power transfer via strongly coupled magnetic resonances. *Science* 317 (5834): 83–86.

15 Cannon, B.L., Hoburg, J.F., Stancil, D.D., and Goldstein, S.C. (2009). Magnetic resonant coupling as a potential means for wireless power transfer to multiple small receivers. *IEEE Transactions on Power Electronics* 24 (7): 1819–1825.

16 Imura, T. and Hori, Y. (2011). Maximizing air gap and efficiency of magnetic resonant coupling for wireless power transfer using equivalent circuit

and Neumann formula. *IEEE Transactions on Industrial Electronics* 58 (10): 4746–4752.

17 Zhong, W.X., Lee, C.K., and Hui, S.Y.R. (2012). Wireless power domino-resonator systems with noncoaxial axes and circular structures. *IEEE Transactions on Power Electronics* 27 (11): 4750–4762.

18 Beh, T.C., Kata, M., Imura, T. et al. (2013). Automated impedance matching system for robust wireless power transfer via magnetic resonance coupling. *IEEE Transactions on Industrial Electronics* 60 (9): 3689–3698.

19 Huh, J., Lee, S.W., Lee, W.Y. et al. (2011). Narrow-width inductive power transfer system for online electric vehicles. *IEEE Transactions on Power Electronics* 26 (12): 3666–3679.

20 Madawala, U.K. and Thrimawithana, D.J. (2011). A bidirectional inductive power interface for electric vehicle in V2G systems. *IEEE Transactions on Industrial Electronics* 58 (10): 4789–4796.

21 Wang, Z., Li, Y., Sun, Y. et al. (2013). Load detection model of voltage-fed inductive power transfer system. *IEEE Transactions on Power Electronics* 28 (11): 5233–5243.

22 Raju, S., Wu, R., Chan, M., and Yue, C.P. (2014). Modeling of mutual coupling between planar inductors in wireless power applications. *IEEE Transactions on Power Electronics* 29 (1): 481–490.

2

Inductive Power Transfer

2.1 Inductive Power Transfer

In this chapter, the basic theories and fundamental principles of the inductive power transfer (IPT) are first presented, where no distinction between the definition of IPT and that of magnetic resonant coupling (MRC) is made. From the perspective of the quantity for pickups, the IPT system is divided into 1-to-1 transmission and 1-to-n transmission. Then, the general configuration, coupled modeling, and compensation and power transfer characteristics of the aforementioned two transmission forms are illustrated in detail. Finally, the differences between 1-to-1 and 1-to-n transmission are explained, with emphasis on the unique characteristics of 1-to-N transmission.

Specific discussions on design, control, and application are extended in the remaining chapters of the book.

2.1.1 Principle

Some other literature reviews may divide the magnetic induction-based wireless power transfer (WPT) into IPT and MRC. However, there only remain slight differences between IPT which requires actual capacitors to resonate with the inductance coils and MRC which utilizes the designed parasitic capacitance to compensate the inductance reactance. Consequently, the fundamental principles and operation of these two classifications are inherently the same, although various quality factors (Q) of the winding coils may cause the theoretical analysis differences between IPT and MRC, where the extremely high Q is often adopted in the MRC system [1, 2]. Hence, this chapter makes no distinction between them and focuses on the IPT which possesses more representatives compared with MRC to elaborate the magnetic induction-based WPT.

The principle of the IPT system is shown in Figure 2.1 where power is transmitted wirelessly through two loosely coupled coils, namely the transmitting coils and

Wireless Power Transfer: Principles and Applications, First Edition. Zhen Zhang and Hongliang Pang.

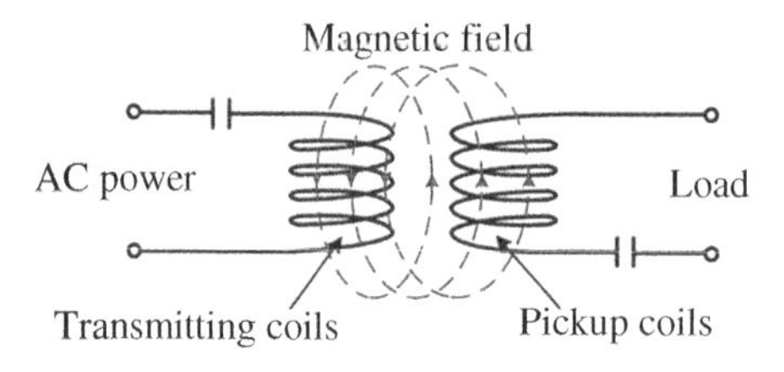

Figure 2.1 Principle of IPT.

the pickup coils. The specific power transmission procedures will be elaborated step by step as follows:

(1) Firstly, the resonant inverter/converter is adopted to modulate the DC power source or utility AC power source at the power frequency (50 or 60 Hz) into high-frequency AC power which is utilized in WPT system with the scope of about 20 kHz [3–5] to a few MHz (6.78 or 13.56 MHz) [6].
(2) Then, the time-varying AC current flows through the transmitting coils, accordingly, producing an alternatingly induced magnetic field nearby.
(3) Meantime, based on the Faraday's law, the alternatingly induced voltage with the same frequency is excited in the pickup coils through a weak magnetic coupling.
(4) Finally, after power electronics conversion, the power will be delivered directly to the load that is connected to each pickup coil.

Similar to the conventional transformer theory, the aforementioned IPT principle is well illustrated; however, where are the differences between transformer theory and IPT principle? Four slight but noticeable features for IPT compared with transformers will be discussed as listed below:

- ***Low mutual inductance and large leakage inductance*** – Unlike the traditional transformers, due to the large winding separation, the IPT system has a relatively large leakage inductance, where the leakage inductance holds 10 times larger than the magnetizing inductance in IPT system compared with the magnetizing inductance that holds about 50 times larger than the leakage inductance in conventional transformers. Moreover, the relatively long distance also leads to a significant reduction in magnetic flux, which results in the weaker coupling coefficient $\kappa < 0.2$ [7] and lower mutual inductance M; thus, the IPT is also called loosely coupled power transmission. Hence, both the reasonable selection of compensation circuit so as to eliminate the large leakage inductance reactance and the elaborate design procedure for the transmitting and pickup coils with emphasis on high Q [8] prove to be the critical technical challenges in the IPT system.
- ***The hollow inductor*** – The transmitting and pickup devices are both inductors in the IPT system, while most of them are hollow inductors to eliminate

core loss (except the EV charging pad and EV track coils). However, regarding the transformers, the iron cores around the windings form a magnetic path and avoid magnetic leakage losses. Consequently, whether adopting the iron core as the magnetic circuit or not turns into a key factor for the differences between traditional transformers and the IPT system.

- ***The phase between primary- and secondary-side current*** – Ignoring the loss and hysteresis, the currents of the primary and secondary sides are in phase or of opposite phase in conventional ideal transformers; however, for IPT system, there often exists approximately 90° phase difference between the primary and the secondary currents, which is completely different from that of the transformers.
- ***The parameters perturbation and parasitic parameters*** – With the changing transmission distance, misalignment of coupled coils, variable load values, and uncertainty quantity for pickups, the inherent or reflected parameters of the IPT system suffer violent fluctuations, which result in the extreme degradation of transmission efficiency and power. Besides, IPT systems employ the frequency band within 6.78 MHz ± 15 kHz [6] and a low band of several kHz [3–5]. Hence, the parasitic resistance and capacitance become severe through the high-frequency IPT system, thus impeding the normal operation. As a result, the maintenance of high efficiency and stable output current/voltage under the circumstance of complex parameters perturbation falls into the focal concerns of the IPT system.

After conducting an overall discussion aiming at the principles of IPT, as well as revealing the differences between conventional transformers and IPT principle, the detailed introduction with emphasis on the essential theories for IPT system will follow up for the readers.

From the perspective of the quantity for pickups, the IPT system is divided into 1-to-1 (single transmitter single pickup) transmission and 1-to-n (single transmitter multiple pickups) transmission. As for the n-to-n (multiple transmitters multiple pickups) transmission, it could be considered as the cascaded connected operation mode for 1-to-1 transmission system with varying frequencies. Hence, this chapter will not regard n-to-n transmission as a separate classification and incorporate it into a special circumstance of 1-to-1 transmission.

2.1.2 1-to-1 Transmission

As shown in Figure 2.2, the typical configuration of the 1-to-1 IPT system is divided into five parts, which include the primary power source, primary compensation circuit, magnetic coupling, secondary compensation circuit, and pickup unit.

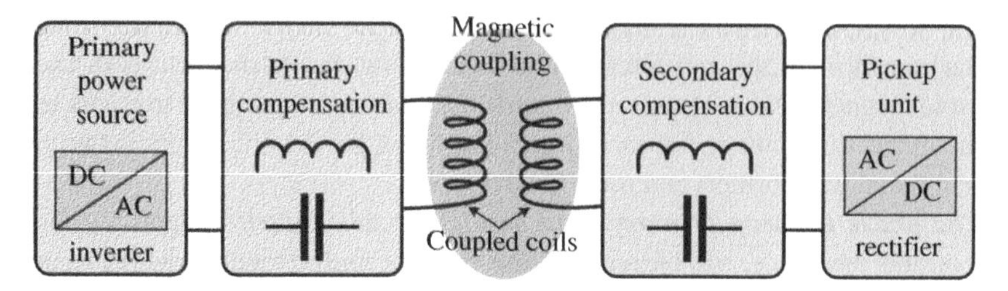

Figure 2.2 Configuration of 1-to-1 IPT system.

2.1.2.1 Primary Power Source

Commonly, the high-frequency low-loss inverter which operates at 20 kHz to a few MHz is adopted to modulate the DC power source. Due to the clear parameters design and low control difficulty, the half- or full-bridge Class D inverter prevails in the practical low- or medium-power IPT system. However, it should be noted that the spike voltage across the switching devices of the full-bridge Class D inverter is twice as high as the DC supply voltage. Hence, the parallel full-bridge and cascaded full-bridge Class D inverters are applied to provide higher and more flexible output power [9]. Besides, apart from the conventional resonant inverters, signal generator combined with power amplifier is sometimes utilized as the AC power supply to feed the circuit, which can reduce the total harmonic distortion (THD) in the transmitting coils, although the efficiency performance is poor to some extent. The behaviors and the parameters design procedure of the inverters will be addressed in Chapter 4.

2.1.2.2 Primary/Secondary Compensation Network

In order to nullify the large leakage inductance reactance of the loosely coupled coils, the well-designed compensation circuit which consists of one or several inductance-capacitance (*LC*) resonant tanks offers a feasible solution to minimize the voltage–ampere (VA) rating of the converter/power supply in the primary side circuit, while maximizing the efficiency of the wirelessly transmitted power in the secondary sides [10, 11].

Both the transmitting and the pickup coils provide the inductance L in either the primary or secondary compensation network. Besides, the *LC* resonant tanks also own the characteristics of filtering out higher or lower switching harmonics of the high-frequency inverter to regulate the transmitting current into the sinusoidal wave.

In addition, some high-order compensation circuits can not only achieve the zero-voltage switching (ZVS) or zero-current switching (ZCS) to promote the maximum transfer efficiency [12] but also realize the load-independent constant output under various parameters perturbation in practical IPT systems [13]. If choosing the metal–oxide–semiconductor field-effect transistors (MOSFETs) as the switching device, the realization of soft switching is able to decrease the switching power loss, thus promoting the overall transfer efficiency prominently, especially in the high-frequency IPT systems [14].

To sum up, an appropriate selection of the compensation networks as well as the careful design of the parameters of LC resonant tanks contributes significantly to the transmission performance of the IPT. The characteristics of the traditional compensation circuits, the operating principle of the constant output current/voltage, and the realization of soft switching will be analyzed in detail in Chapter 4.

2.1.2.3 Magnetic Coupling

Figure 2.3 exhibits a simplified magnetic coupling system, where r represents the radius of the transmitting coils, and d means the distance from the transmitting coils to a certain location. Based on the Biot–Savart law, the magnetic field intensity at the random distance d can be deduced as [15]

$$H(d) = \frac{IN}{2}\frac{r^2}{(r^2 + d^2)^{\frac{3}{2}}} \tag{2.1}$$

As revealed in Eq. (2.1), both promoting the transmitting current I and increasing the number of turns N of the transmitting coils contribute to the larger magnetic field intensity. Theoretically speaking, the larger the magnetic field intensity is, the more power the pickup coils can gain. However, there remain complex relationships between the radius r and distance d; hence, the normalized magnetic field strength versus d/r is plotted intuitively in Figure 2.4, where the normalized magnetic field strength experiences the process of increasing slowly and then decreasing dramatically with the rising ratio of d/r. During this period, the peak value of the magnetic field is gained at $d = \frac{\sqrt{2}}{2}r$, which can be derived from

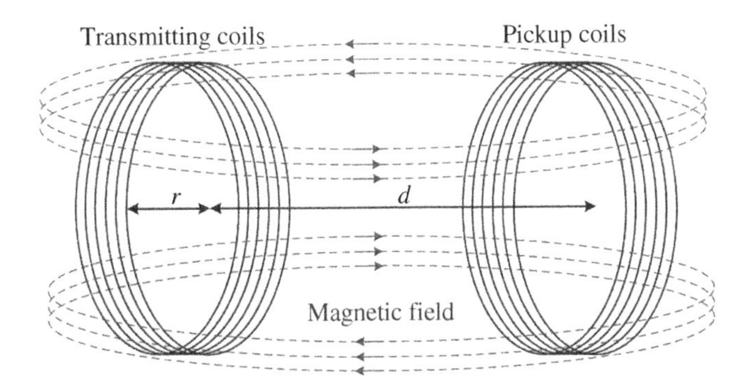

Figure 2.3 Simplified magnetic coupling system.

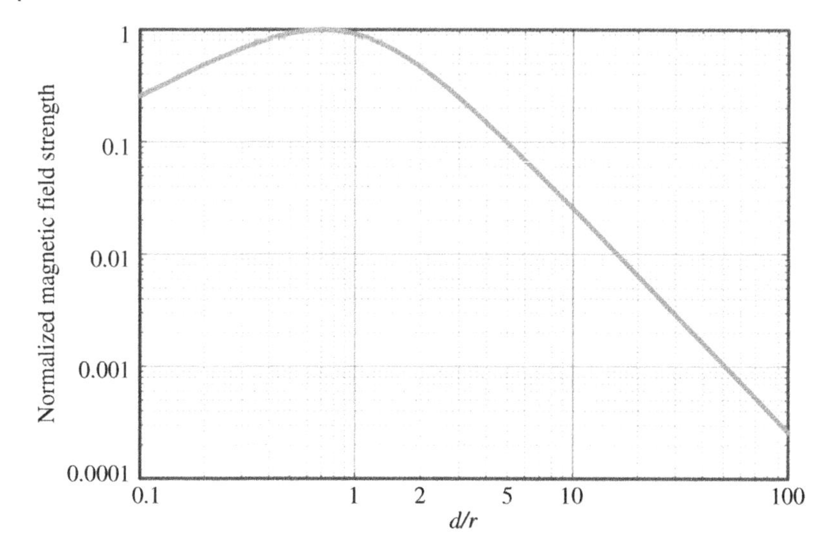

Figure 2.4 Normalized magnetic field strength versus *d*/*r*.

Eq. (2.1). Accordingly, the optimal transmission distance is achieved at approximately the radius length of the transmitting coils. Besides, the various Q values and different shapes of the transmitting coils, for example the round, square, and rectangle transmitting coils also pose significant impacts on the transmission performance. Those readers who intend to understand the principle, structure, design, and optimization of the transmitting coils may refer Chapter 3.

2.1.2.4 Pickup Unit

Commonly, a bridge rectifier is adopted to convert the AC current into DC current, which comes with the filtering circuit to minimize the pulsation. According to the different requirements of various loads, sometimes the DC–DC converter is connected to stabilize the output current/voltage. Besides, the DC–DC converter may also be treated as an impedance transformer via adjusting the duty ratio of the switches. The detailed information and control strategy will be elaborated in Chapter 5.

2.1.3 1-to-*N* Transmission

Literally, the definition of 1-to-n IPT system indicates that only single transmitter is responsible for simultaneously or time-sharingly transmitting the power to multiple pickups in a cordless way, which is shown in Figure 2.5. However, it is noteworthy that the phrase "single transmitter" does not mean "only one inverter" or "only one transmitting coil" or "only one operating frequency." Actually, it

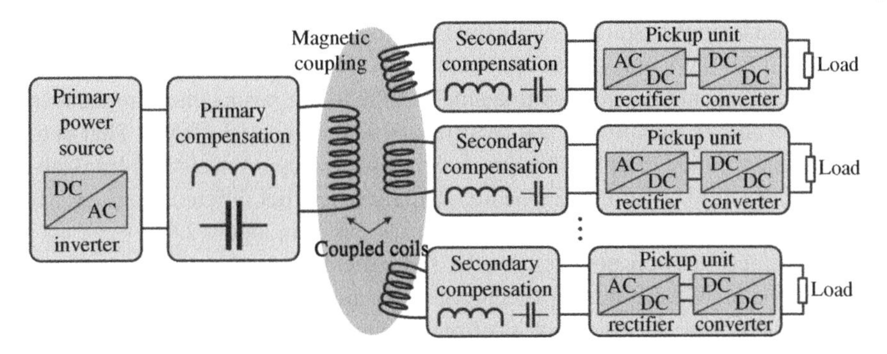

Figure 2.5 Configuration of 1-to-*n* IPT system.

means "only one induced electromagnetic field" which is fed by one transmitting coil or multiple transmitting coils connected in series or in parallel. Accordingly, the concept of 1-to-*n* IPT system in this chapter is clear, namely multiple pickups embrace within one open induced electromagnetic field which is excited by one power supply. From the perspective of the operating frequency, this chapter classifies 1-to-*n* IPT system into two different working modes, that is single-frequency excitation and multifrequency excitation.

2.1.3.1 Single-Frequency Excitation

Literally, single frequency means only one resonant frequency existing in the IPT system, where the power can be wirelessly transmitted to multiple pickups simultaneously, as shown in Figure 2.6a. However, unlike the 1-to-1 IPT system, where only one resonant frequency is elaborately designed, different operating frequencies are sometimes adopted to excite the induced electromagnetic field in 1-to-*n* IPT system, for example the selective frequency WPT [16], the energy encryption strategy-based WPT [17], and smart power distribution in multiple-receiver IPT system [18]. Nevertheless, the abovementioned power transmission mode is

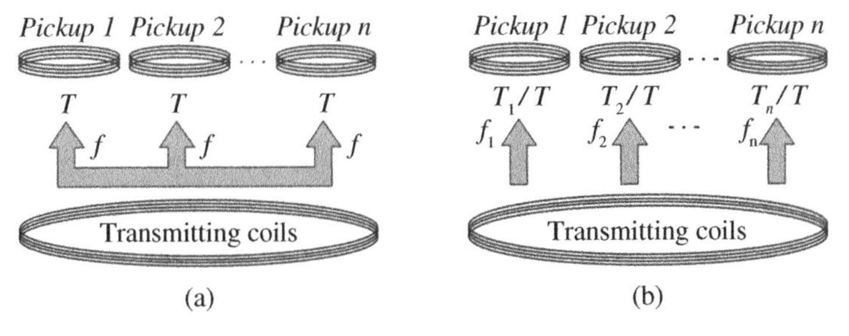

Figure 2.6 Different power distribution modes for single-frequency 1-to-*n* transmission: (a) Simultaneous transmission; (b) Time-sharing transmission.

more like a various-frequency-based time-sharing strategy. Actually, at one specific point of time, there only exists one operating frequency, although the entire IPT system can resonant at various frequencies. Due to the mismatch of impedance in the secondary side, not all the pickups can acquire enough power. That is to say, during a period of time, only one designed frequency is selected for transmitting the power to an authorized pickup, and then, another different frequency is utilized for feeding another different pickup, as shown in Figure 2.6b.

Accordingly, this chapter defines the single-frequency excitation as: at each specific point of time during the power transmission, only one frequency is adopted to excite the induced electromagnetic field. Figure 2.6 depicts two different power distribution modes, namely simultaneous mode or time-sharing mode, for single-frequency excitation in 1-to-n transmission.

2.1.3.2 Multifrequency Excitation

Compared to the single-frequency excitation, the multifrequency excitation is defined as: at each specific point of time during the power transmission, multiple frequencies are superposed to excite the induced electromagnetic field. That is, the operating frequency is a composition of different frequencies. Accordingly, varied schemes have been proposed to achieve the multifrequency excitation. For example, a scheme that produces dual frequencies generated by novel PWM control-based full-bridge inverter was proposed to implement the double-frequency excitation, as shown in Figure 2.7 [19]. Reference [20]

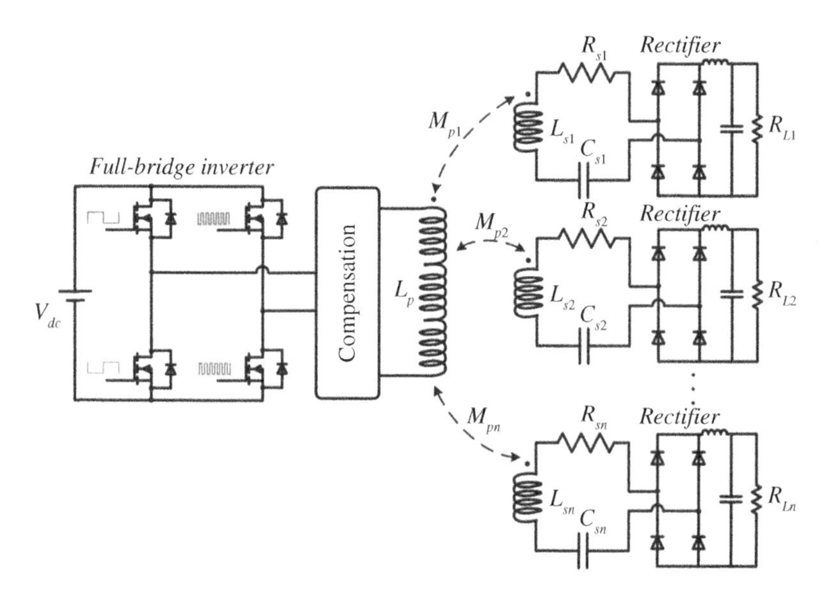

Figure 2.7 Configuration of PWM-based multifrequency 1-to-n transmission.

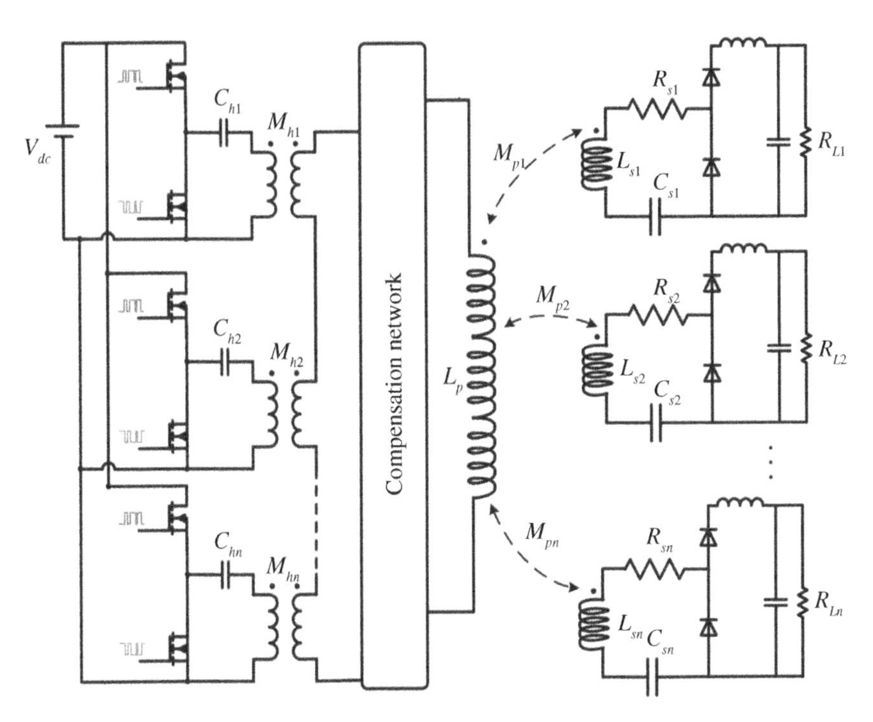

Figure 2.8 Configuration of multifrequency superposition-based 1-to-n transmission.

proposed a method that amplifies both the fundamental and third-harmonic frequency to implement the double-frequency excitation. Besides, an ingenious inverter topology was presented to achieve multifrequency in Ref. [21], and a multifrequency superposition methodology-based high-efficiency IPT system was verified, as shown in Figure 2.8 [22].

Differing from the conventional 1-to-1 IPT systems, the 1-to-n transmission holds the characteristics of simultaneously energizing various pickups with a more flexible and intelligent solution. By comparing with adopting multiple 1-to-1 IPT systems, most of the 1-to-n transmission can reduce the number of inverters, thus saving the cost and increasing the power density. In addition, due to the 1-to-n structures consisting only one transmitter, the mutual interference among multiple induced electromagnetic fields within a limited working area can be avoided. Lastly, the charging area can be dramatically extended via increasing the number of transmitting coils, which shows significant meanings for some specific applications, such as portable charging tables, sensor networks, and dynamic wireless charging for electric vehicles (EVs). However, for its salient advantages as aforementioned, the 1-to-n transmission have had mixed success; there are still critical technical challenges ahead, especially for the power energization and distribution, cross-coupling effects among multiple pickups, and energy securities. The detailed discussion and control strategy will be addressed in Section 2.4 and Chapter 6.

2.2 1-to-1 Transmission

For IPT system, 1-to-1 transmission is regarded as the most common structure, which contains a single transmitter and a single pickup. This section elaborates 1-to-1 transmission, with emphasis on the coupled model, compensation topology, and transmission performance.

2.2.1 Coupled Modeling

The configuration of the 1-to-1 IPT system has already been shown in Figure 2.2. Without mechanical and electrical connection between the primary and the secondary sides, there are merely loosely coupled coils utilizing induced magnetic field as the carrier to transfer power from the transmitting coil to the pickup coil. Consequently, the magnetic coupling serves an important role with regard to the power transmission. Similar to the 1-to-1 IPT system, the basic principle of the loosely coupled transformer is firstly introduced in Section 2.2.1.1.

2.2.1.1 Loosely Coupled Transformer Model

Practically, in loosely coupled transformer system, due to the relatively large air gap between the transmitter and the pickups, the coupling coefficient is low, while the leakage inductance is large. Accordingly, the leakage magnetic flux is inevitable. A coupled transformer containing leakage magnetic flux is depicted in Figure 2.9, which describes the distribution of the magnetic flux between the transmitting and the pickup coils.

Some of the magnetic flux generated by the transmitting coil intersects the pickup coil, denoted as ϕ_{ps}, and the rest circulates, denoted as ϕ_{pl}. Similarly, denote the generated magnetic flux of the pickup coil that crosses the transmitting coil as ϕ_{sp}, and the others as ϕ_{sl}. Meanwhile, U_p, U_s, I_p, and I_s are defined as the primary exciting voltage, secondary pickup voltage, primary exciting current, and secondary current, respectively. Define ϕ_m as the sharing magnetic flux, which is given by

$$\phi_m = \phi_{ps} - \phi_{sp} \tag{2.2}$$

Then, the equivalent flux generated in the transmitting coils can be obtained as

$$\phi_p' = \phi_{ps} - \phi_{sp} + \phi_{pl} \tag{2.3}$$

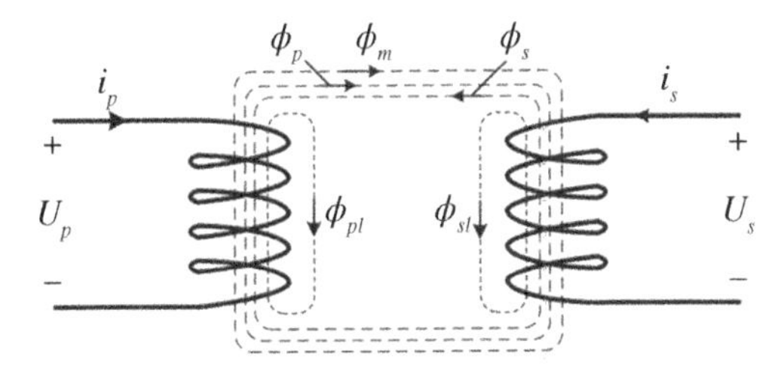

Figure 2.9 Schematic diagram of loosely coupled transformer.

Similarly, the equivalent flux excited in the pickup coils can be deduced as

$$\phi_s' = \phi_{sp} - \phi_{ps} + \phi_{sl} \tag{2.4}$$

Accordingly, the primary exciting voltage and secondary pickup voltage can be, respectively, calculated as

$$\begin{cases} U_p = N_p \dfrac{d\phi_p'}{dt} = N_p \dfrac{d(\phi_{pl} + \phi_{ps} - \phi_{sp})}{dt} = L_{pl}\vec{i}_p + \left(L_{ps}\vec{i}_p - \dfrac{L_{sp}\vec{i}_s}{n} \right) \\ U_s = N_s \dfrac{d\phi_s'}{dt} = N_s \dfrac{d(\phi_{sl} + \phi_{sp} - \phi_{ps})}{dt} = L_{sl}\vec{i}_s + (L_{sp}\vec{i}_s - nL_{ps}\vec{i}_p) \end{cases} \tag{2.5}$$

where N_p represents the turns of the primary windings, and N_s represents the turns of the secondary windings. Besides, the relative inductances and turn ratio n are expressed as below:

$$L_{pl} = \frac{N_p\phi_{pl}}{i_p}, \quad L_{sl} = \frac{N_s\phi_{sl}}{i_s}, \quad L_{ps} = \frac{N_p\phi_{ps}}{i_p}, \quad L_{sp}\frac{N_s\phi_{sp}}{i_s}, \quad n = \frac{N_s}{N_p} \tag{2.6}$$

Afterward, the magnetic circuit formula of this loosely coupled transformer can be indicated as

$$N_p i_p = N_s i_s = \Re\phi_m \tag{2.7}$$

where $\Re$ represents the magnetic resistance of the loosely coupled coils.

Subsequently, L_{ps} can be recalculated as

$$L_{ps} = \frac{N_p\phi_{ps}}{i_p} = \left.\frac{N_p\phi_m}{i_p}\right|_{i_s=0} \tag{2.8}$$

Substituting (2.7) into (2.8) gives

$$L_{ps} = \frac{N_p^2}{\Re} \tag{2.9}$$

Similarly, L_{sp} can be derived as

$$L_{sp} = \frac{N_s^2}{\Re} \tag{2.10}$$

Comparing (2.9) with (2.10), the relationship can be gained as

$$L_{sp} = n^2 L_{ps} \tag{2.11}$$

Substituting (2.11) into (2.5) gives

$$\begin{cases} U_p = L_{pl}\vec{i}_p + \left(L_{ps}\vec{i}_p - \frac{L_{sp}\vec{i}_s}{n}\right) = L_{pl}\vec{i}_p + L_{ps}(\vec{i}_p - n\vec{i}_s) \\ U_s = L_{sl}\vec{i}_s + (L_{sp}\vec{i}_s - nL_{ps}\vec{i}_p) = L_{sl}\vec{i}_s - nL_{ps}(\vec{i}_p - n\vec{i}_s) \\ U_m = L_{ps}(\vec{i}_p - n\vec{i}_s) \end{cases} \tag{2.12}$$

Accordingly, as shown in Figure 2.10, the specific circuit can be constructed based on Eq. (2.12), which consists of an ideal transformer with the turn ratio of n.

2.2.1.2 T-model

Another commonly used model, namely T-model, which is shown in Figure 2.11, is proposed, whose name originates from its topology, that is three inductances resembling the letter "T."

Let $M = nL_{ps}$, $L_p = L_{pl} + L_{ps}$, and $L_s = L_{sl} + L_{sp}$. By substituting (2.11) into (2.12), the expression can be arranged as follows:

$$\begin{cases} U_p = L_{pl}\vec{i}_p + L_{ps}(\vec{i}_p - n\vec{i}_s) = (L_{pl} + L_{ps} - nL_{ps})\vec{i}_p \\ \quad + nL_{ps}(\vec{i}_p - \vec{i}_s) = (L_p - M)\vec{i}_p + M(\vec{i}_p - \vec{i}_s) \\ U_s = L_{sl}\vec{i}_s + (L_{sp}\vec{i}_s - nL_{ps}\vec{i}_p) = (L_{sl} + L_{sp} - nL_{ps})\vec{i}_s \\ \quad - nL_{ps}(\vec{i}_p - \vec{i}_s) = (L_s - M)\vec{i}_s - M(\vec{i}_p - \vec{i}_s) \end{cases} \tag{2.13}$$

The circuit reconstruction for Eq. (2.13) is shown in Figure 2.11, where the polarity of the secondary pickup voltage is reversed compared with Figure 2.10.

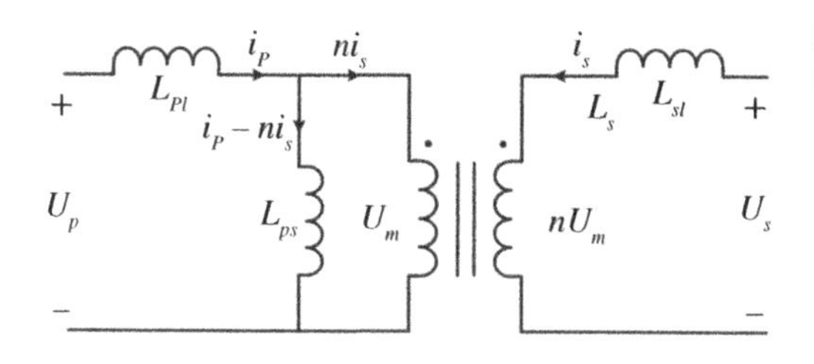

Figure 2.10 Loosely coupled transformer model.

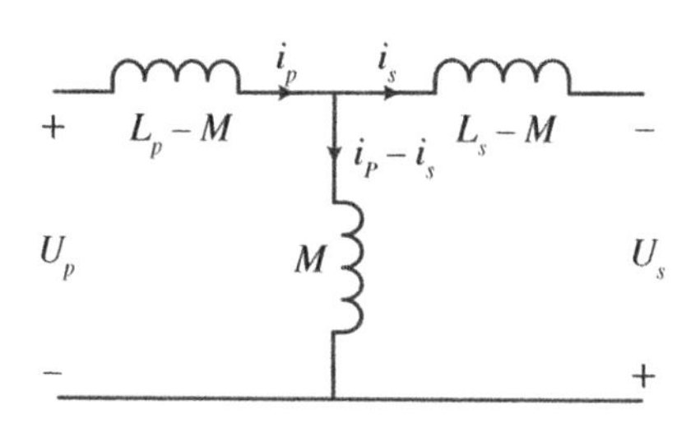

Figure 2.11 T-model for 1-to-1 transmission.

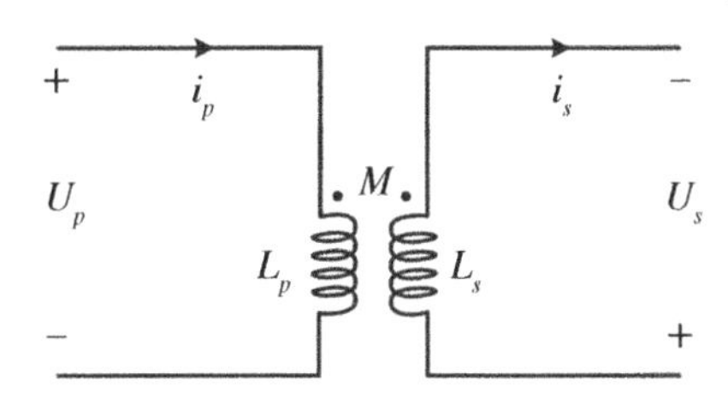

Figure 2.12 M-model for 1-to-1 transmission.

2.2.1.3 M-model

In addition to the T-model, M-model is widely used in 1-to-1 transmission, as shown in Figure 2.12. Meanwhile, the M-model can be deduced from the T-model by rearranging (2.13) as follows:

$$\begin{cases} U_p = (L_p - M)\vec{i}_p + M(\vec{i}_p - \vec{i}_s) = L_p\vec{i}_p - M\vec{i}_s \\ U_s = (L_s - M)\vec{i}_s - M(\vec{i}_p - \vec{i}_s) = L_s\vec{i}_s - M\vec{i}_p \end{cases} \tag{2.14}$$

where M is defined as the mutual inductance, which can be derived as

$$M = nL_{ps} \tag{2.15}$$

The specific calculation procedure for M will be given in the Appendix 2.A.

Accordingly, as shown in Figure 2.12, the specific circuit can be reconstructed based on Eq. (2.14). It should be noted that the polarity of the secondary pickup voltage in Figure 2.12 is identical with that in Figure 2.11 but inverse with that compared with Figure 2.10. The readers should be cautious when defining the polarity of the proposed models.

2.2.1.4 Discussion

According to the aforementioned analysis and numerical calculation, it can be concluded that the loosely coupled transformer model, T-model, and M-model can actually be interconverted into each other, while the most significant difference lies in the division of magnetic flux. It is feasible for scholars to choose any model for conducting the research. Additionally, their merits and demerits are summarized in Table 2.1.

2.2.2 Compensation

Due to the large leakage inductive reactance, if there exist no compensation capacitors matching with it, both the primary and secondary sides will consume massive reactive power, thus reducing the power factor of the power supply. Accordingly, reactive power compensation plays a vital role in minimizing the VA rating of the converter/power supply in the primary side circuit, while maximizing the efficiency of the wirelessly transmitted power in the secondary sides.

Table 2.1 Comparison of three models.

Model	Loosely coupled transformer	T-model	M-model
Advantage	Comprehensive description[a)]	Simple	Simple
Disadvantage	Sophisticated	No current isolation	Incomplete
Scope of application	Coil design[b)] and magnetic-based analysis	Circuit-based analysis	Equation-based analysis

a) It consists of two leakage inductances, one magnetizing inductance, turn ratio n, and an ideal transformer.
b) The relatively large air gap, misalignment, and location between the transmitting and the pickup coils contribute significantly to the leakage inductance along with the magnetizing inductance. Hence, the design procedure of the transmitting or pickup coils should take all the factors into consideration.

Right now, capacitance compensation is considered as the most effective approach to eliminate the inductive reactance in IPT system, where the compensation network contains a number of capacitors whose values are well designed to resonate with the inductance of coils. The ultimate goal of adopting the compensation network is to turn the primary equivalent input impendence and the secondary impendence into pure resistive, thus nullify the reactive power.

According to the types of connections between capacitance and inductance, compensation circuits are divided into two modes, namely series type and parallel type.

2.2.2.1 Series Type

In this RLC series circuit, as shown in Figure 2.13a, the total impedance Z_S can be deduced as

$$Z_S = R + j\omega L + \frac{1}{j\omega C} = R + j\left(\omega L - \frac{1}{\omega C}\right) = R + jX = |Z|\angle\phi \tag{2.16}$$

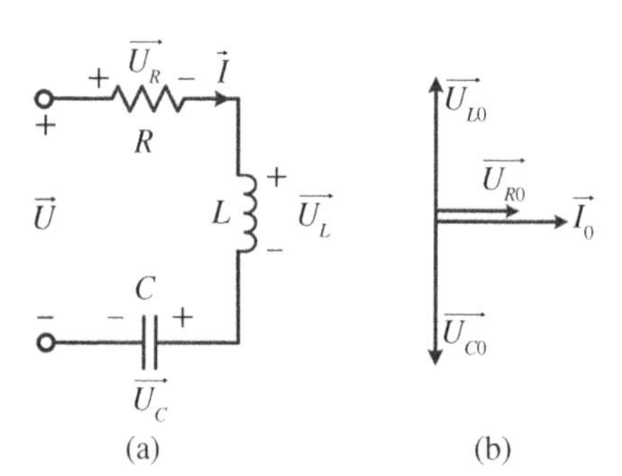

Figure 2.13 (a) Circuit diagram of series compensation circuit; (b) Vector diagram of series resonance.

where X represents the imaginary part of the total equivalent impedance. Let $X = 0$, the resonance angular frequency ω_0 can be obtained as

$$\omega_0 = \frac{1}{\sqrt{LC}} \tag{2.17}$$

Then, the resonance frequency is derived as

$$f_0 = \frac{1}{2\pi\sqrt{LC}} \tag{2.18}$$

Based on Eqs. (2.17) and (2.18), the characteristics of series resonance are given:

(1) Impedance angle $\phi = 0$, power factor $\cos\phi = 1$, and the total impedance Z_s is pure resistive.
(2) Under the constant voltage U, the resonance current $I_0 = \frac{U}{|Z|} = \frac{U}{R}$ reaches maximum.
(3) The amplitude of the inductance voltage $U_{L0} = j\omega_0 L\vec{I}_0$ and the capacitance voltage $U_{C0} = -j\omega_0 \frac{1}{C_0}\vec{I}_0$ is the same, while the phase of them is opposite. Thus, the LC resonance tank is equal to short circuit, and all the voltage is energized to the resistance, as depicted in Figure 2.13b. Accordingly, the series resonance is also named voltage resonance.

In addition, defining the ratio of resonance inductance voltage or resonance capacitance voltage to the voltage of power supply as quality factor (Q) of the resonance circuit can be obtained as

$$Q = \frac{U_{L0}}{U} = \frac{U_{C0}}{U} = \frac{\omega_0 L}{R} \tag{2.19}$$

In practical IPT system, due to the neglectable impact of the line resistance, the internal resistance of coils assumes the actual resistance of the resonance circuit. Thus, Q can also be defined as the quality factor of coils. The specific influence of Q on the transmission performance will be addressed in detail in Chapter 3.

2.2.2.2 Parallel Type

In practical IPT system, the compensation capacitance is parallelly connected with coils, where the coils hold the inductance L and internal resistance R, as shown in Figure 2.14a, and the total impedance Z_p can be deduced as

$$Z_p = \frac{(R + j\omega L)\frac{1}{j\omega C}}{R + j\omega L + \frac{1}{j\omega C}} \tag{2.20}$$

Let $Im\{Z_p\} = 0$, the resonance angular frequency ω_0 can be obtained as

$$\omega_0 = \sqrt{\frac{1}{LC} - \frac{R^2}{L^2}} \tag{2.21}$$

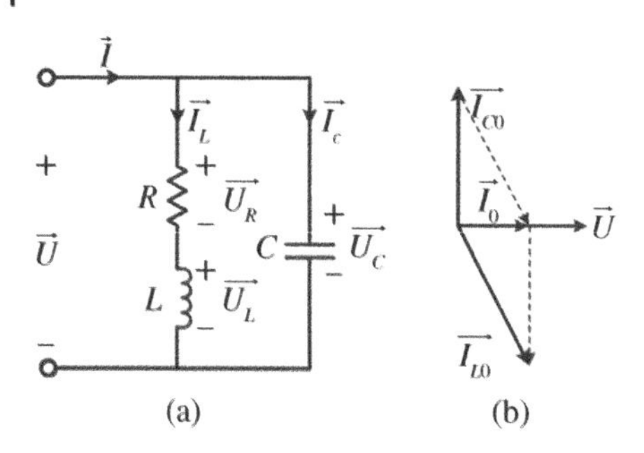

Figure 2.14 (a) Circuit diagram of parallel compensation circuit; (b) Vector diagram of parallel resonance.

Equation (2.21) can be derived when $R < \sqrt{\frac{L}{C}}$.
Then, the resonance frequency is derived as

$$f_0 = \frac{1}{2\pi}\sqrt{\frac{1}{LC} - \frac{R^2}{L^2}} \tag{2.22}$$

Accordingly, the vector diagram of the parallel resonance can be found in Figure 2.14b.

In practical IPT system, high-order topology, which consists of more than one series or parallel connected resonance LC tanks, is always adopted to realize load-independent constant current/voltage and soft switching of inverters. The detailed introduction will be given in Chapter 4.

However, due to the simple structure, independent parameters design process, and wide application, this chapter adopts the series-series (SS) compensation network in the following discussion. More information about the compensation network will be elaborated in Chapter 4.

2.2.3 Power Transmission

In this section, a simple and most common 1-to-1 IPT system with SS compensation network is adopted to analyze the power transmission performance, where the load power, total efficiency, and the relationship between them will be discussed in detail.

2.2.3.1 Load Power

Based on Kirchhoff's voltage law (KVL), as shown in Figure 2.15, the circuit equations of SS compensation topology can be obtained as

$$\begin{bmatrix} \vec{U} \\ 0 \end{bmatrix} = \begin{bmatrix} R_p + j\omega L_p + \frac{1}{j\omega C_p} & M_{ps} \\ M_{ps} & R_L + R_s + j\omega L_s + \frac{1}{j\omega C_s} \end{bmatrix} \begin{bmatrix} \vec{I}_p \\ \vec{I}_s \end{bmatrix} \tag{2.23}$$

Besides, L, C, and R with subscripts p or s denote the inductance, compensation capacitance, and coil internal resistance of the primary and secondary sides,

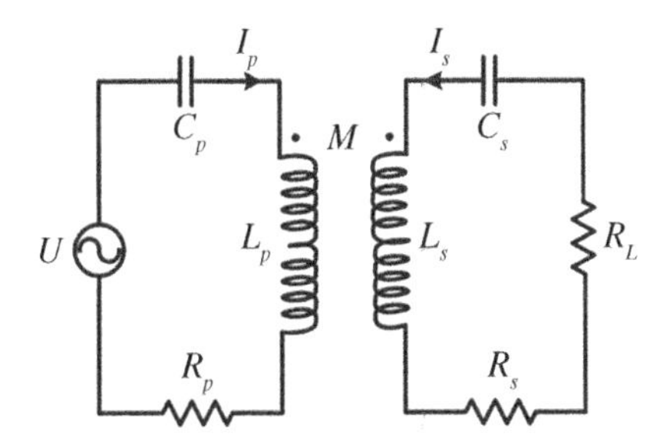

Figure 2.15 Simplified 1-to-1 IPT system with SS compensation.

respectively; R_L represents the equivalent resistance; and M_{ps} means the mutual inductance between the transmitting and the pickup coils.

When the system operates at the resonant frequency, as shown in Eq. (2.17), the inductance of each coil can be matched with the compensation capacitance under SS compensation, which can be obtained as

$$j\omega L_k + \frac{1}{j\omega C_k} = 0, \quad k \in [p, s] \tag{2.24}$$

By substituting (2.24) into (2.23), the primary current I_p and the secondary current $\overrightarrow{I_s}$ can be solved as

$$\begin{cases} I_p = \dfrac{U}{R_p + \dfrac{\omega^2 M_{ps}^2}{R_s + R_L}} = \dfrac{U}{Z_{in}} \\ \overrightarrow{I_{s1}} = \dfrac{-j\omega M_{ps} U}{R_p(R_s + R_L) + \omega^2 M_{ps}^2} \end{cases} \tag{2.25}$$

Defining Z_r as the reflected impedance from the secondary to the primary side gives

$$Z_r = \frac{\omega^2 M_{ps}^2}{R_s + R_L} \tag{2.26}$$

Based on (2.25), the input power P_{in} and the load power P_L can be expressed as

$$\begin{cases} P_{in} = \dfrac{(R_s + R_L)U^2}{R_p(R_s + R_L) + \omega^2 M_{ps}^2} \\ P_L = \dfrac{\omega^2 M_{ps}^2 U^2 R_L}{\left[R_p(R_s + R_L) + \omega^2 M_{ps}^2\right]^2} \end{cases} \tag{2.27}$$

Additionally, by solving the derivative equations $\frac{\partial P_L}{\partial R_L} = 0$, the optimal equivalent load values R_{Lpmax} for the maximum load power can be derived as

$$R_{Lpmax} = \frac{\omega^2 M_{ps}}{R_p} + R_s \tag{2.28}$$

By substituting (2.28) into (2.27), the maximum load power P_{Lmax} can be derived and simplified as

$$P_{Lmax}\frac{\omega^2 M_{ps}^2 U^2}{4\left(\omega^2 M_{ps}^2 + R_p R_{s1}\right) R_p} \tag{2.29}$$

2.2.3.2 Efficiency

There is no doubt that the efficiency is at the paramount priority with regard to the analysis of the IPT system.

Based on Eq. (2.27), the corresponding transmission efficiency η is obtained as

$$\eta = \frac{P_L}{P_{in}} = \frac{\omega^2 M_{ps}^2 R_L}{\left[R_p(R_s + R_L) + \omega^2 M_{ps}^2\right](R_s + R_L)} \tag{2.30}$$

Similarly, by solving the derivative equations $\frac{\partial \eta}{\partial R_L} = 0$, the optimal equivalent load values $R_{L\eta max}$ for the maximum transmission efficiency can be derived as

$$R_{L\eta max} = R_{s1}\sqrt{1 + \frac{\omega^2 M_{ps}^2}{R_p R_s}} \tag{2.31}$$

By substituting (2.31) into (2.30), the maximum transmission efficiency can be deduced and simplified as

$$\eta_{max} = 1 - \frac{2}{\sqrt{1 + \frac{\omega^2 M_{ps}^2}{R_p R_s}} + 1} \tag{2.32}$$

Consequently, based on Eq. (2.32), by reducing the coil internal resistances of the primary and secondary sides as well as increasing the mutual inductance between the transmitting and the pickup coils, a higher efficiency can be achieved in 1-to-1 transmission.

2.2.3.3 Relationship Between Power and Efficiency

To further explore the relationship between power and efficiency, the efficiency and normalized load power versus Z_r/R_p is plotted intuitively in Figure 2.16, where the normalized load power experiences the process of dramatically increasing and then slowly decreasing with the rising ratio of Z_r/R_p. It should be noted that the maximum load power is gained at approximately $\frac{Z_r}{R_p} = 1$. As for the efficiency, it increases in a nonlinear trend with the rising ratio of Z_r/R_p and finally close to 1.

2.2.3.4 Considerations

For an ideal situation, the transmission efficiency should equal 100%. Why the maximum efficiency only achieves approximately 95%? Three main factors that contribute to the power loss are given:

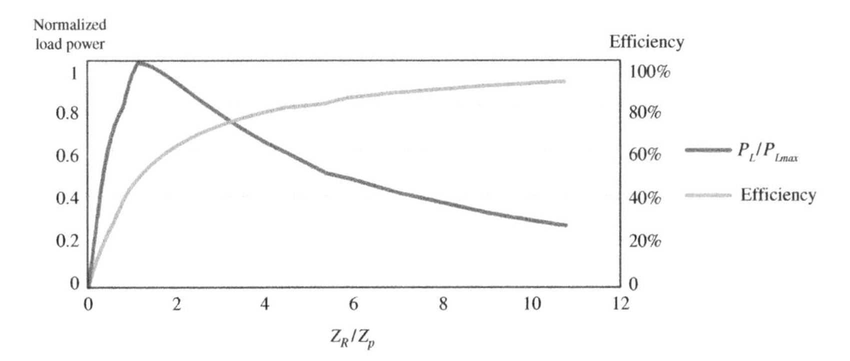

Figure 2.16 Efficiency and normalized load power versus Z_r/R_p.

(1) ***Power loss of the inverters*** – including the conduction loss of diodes, conduction loss of MOSFETs, turn-on loss of MOSFETs, and turn-off loss of MOSFETs, where the turn-on or -off loss can be eliminated by realizing soft switching, which will be discussed in detail in Chapter 4.
(2) ***Power loss of the rectifiers*** [13] – including forward/reverse loss and switch loss of diodes, which is mainly influenced by the equivalent on-state resistance and threshold voltage of the diode.
(3) ***Power loss of resonance circuit*** – mainly caused by the mismatching of the compensation and the power loss on the line resistance.

2.3 1-to-*n* Transmission

With the increasing commercialization and marketization of the integrated IPT system, utilizing single transmitter to energize multiple pickups has attracted more and more attentions [16–22]. In this section, modeling, power distribution, efficiency analysis, effects of cross-coupling, and frequency shifting of single-frequency excitation 1-to-*n* transmission are elaborated in detail. Notably, the analysis and control scheme for multifrequency excitation 1-to-*n* transmission will be given in Chapter 6.

2.3.1 General Configuration

A simplified multiple pickups IPT system configuration is shown in Figure 2.17, where the high-frequency full-bridge inverter is utilized to feed the transmitting coils, the SS compensation network is adopted to nullify the inductance reactance in both the primary and secondary sides, and the mutual inductance among different pickups is presented. Meanwhile, multiple pickups can absorb the energy simultaneously in the same induced electromagnetic field.

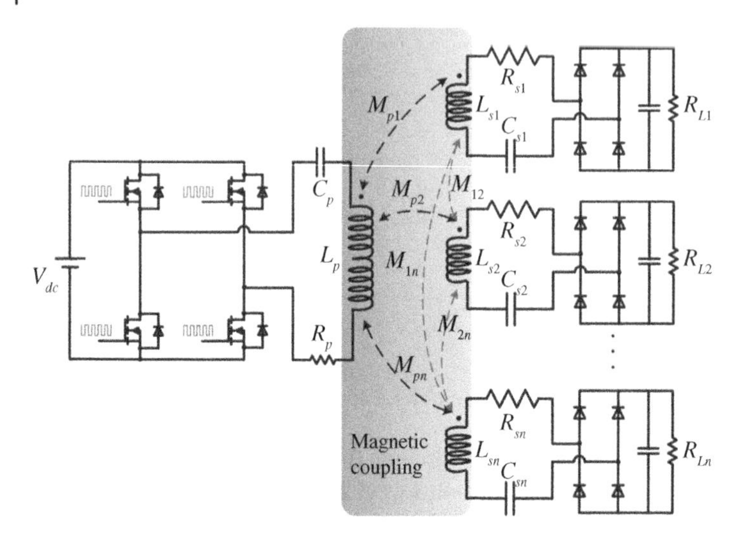

Figure 2.17 Simplified multiple pickups IPT system configuration.

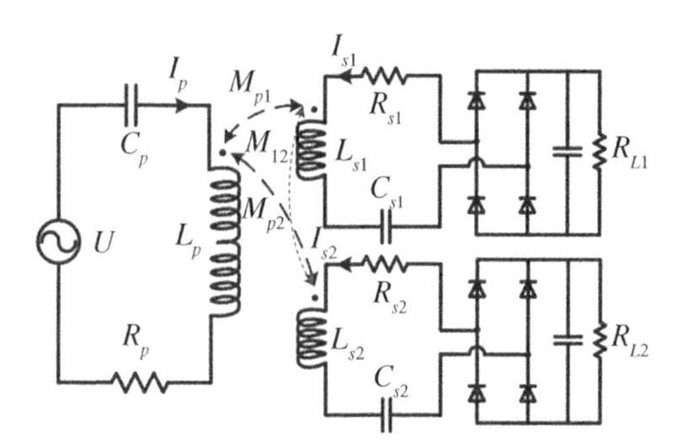

Figure 2.18 A general two pickups IPT system.

2.3.2 Two Pickups System Analysis

Aiming to facilitate the following theoretical analysis, a general two pickups IPT system, shown in Figure 2.18, is exhibited to further explore the related transmission characteristics. Besides, a more generalized multiple pickups IPT system will follow up with some discussions in Section 2.3.3.4.

2.3.2.1 Modeling

Comparing the distance between two pickups to the radius of the pickup coils is sufficiently large, and two pickups are located in the same plane; hence, the mutual inductance between two pickups, which is also named cross-coupling, is ignored in this section.

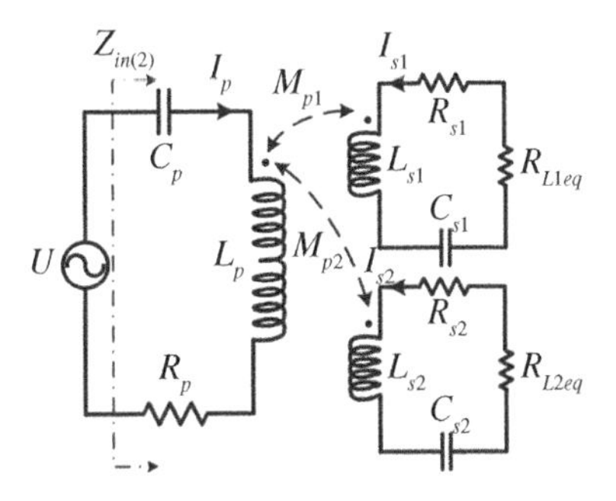

Figure 2.19 Simplified two pickups IPT system without cross-coupling.

As shown in Figure 2.19, there exist one transmitter and two pickups which are loosely coupled to the transmitting coils with the mutual inductance M_{p1} and M_{p2}, respectively. The coupling coefficient between the transmitting coils and each pickup coil can be expressed as

$$\kappa_{p1} = \frac{M_{p1}}{\sqrt{L_p L_{s1}}} \tag{2.33}$$

$$\kappa_{p2} = \frac{M_{p2}}{\sqrt{L_p L_{s2}}} \tag{2.34}$$

Besides, L, C, and R with subscripts p, $s1$, or $s2$ denote the inductance, compensation capacitance, and coil internal resistance of the primary side and two secondary sides, respectively. Considering the neglectable impact of the line resistance and parasitic resistance, the extra resistance is ignored in this section. By ignoring the loss of diodes, the equivalent resistances R_{L1eq} and R_{L2eq} can be deduced as

$$R_{L1eq} = \frac{8}{\pi^2} R_{L1} \tag{2.35}$$

$$R_{L2eq} = \frac{8}{\pi^2} R_{L2} \tag{2.36}$$

where R_{L1} and R_{L2} are defined as the load resistance of each pickup side, respectively, as shown in Figure 2.18.

Based on KVL and equivalent M-model circuit of the SS topology, the circuit as shown in Figure 2.19 can be expressed as

$$\begin{bmatrix} \overrightarrow{U} \\ 0 \\ 0 \end{bmatrix} = \begin{bmatrix} \overrightarrow{Z_p} & j\omega M_{p1} & j\omega M_{p2} \\ j\omega M_{p1} & \overrightarrow{Z_{s1}} + R_{L1eq} & 0 \\ j\omega M_{p2} & 0 & \overrightarrow{Z_{s2}} + R_{L2eq} \end{bmatrix} \begin{bmatrix} \overrightarrow{I_p} \\ \overrightarrow{I_{s1}} \\ \overrightarrow{I_{s2}} \end{bmatrix} \tag{2.37}$$

Denote ω, Z_p, Z_{s1}, and Z_{s2} as the angular frequency, impedance of the primary side, secondary pickup 1, and secondary pickup 2, respectively, which are given by

$$\begin{cases} \overrightarrow{Z_p} = R_p + j\omega L_p + \frac{1}{j\omega C_p} \\ \overrightarrow{Z_{s1}} = R_{s1} + j\omega L_{s1} + \frac{1}{j\omega C_{s1}} \\ \overrightarrow{Z_{s2}} = R_{s2} + j\omega L_{s2} + \frac{1}{j\omega C_{s2}} \end{cases} \tag{2.38}$$

When the system operates at the resonant frequency, as shown in Eq. (2.17), the inductance of each coil can be matched with the compensation capacitance under SS compensation, which can be obtained as

$$j\omega L_k + \frac{1}{j\omega C_k} = 0, \quad k \in [p, s1, s2] \tag{2.39}$$

By substituting (2.39) into (2.38), the expression (2.37) can be simplified and unfolded as

$$\begin{cases} R_p\overrightarrow{I_p} + j\omega M_{p1}\overrightarrow{I_{s1}} + j\omega M_{p2}\overrightarrow{I_{s2}} = U \\ j\omega M_{p1}\overrightarrow{I_p} + (R_{s1} + R_{L1eq})\overrightarrow{I_{s1}} = 0 \\ j\omega M_{p2}\overrightarrow{I_p} + (R_{s2} + R_{L2eq})\overrightarrow{I_{s2}} = 0 \end{cases} \tag{2.40}$$

Then, the primary current I_p and the secondary current $\overrightarrow{I_{s1}}$ and $\overrightarrow{I_{s2}}$ can be solved as

$$\begin{cases} I_p = \dfrac{(R_{s1} + R_{L1eq})(R_{s2} + R_{L2eq})U}{R_p(R_{s1} + R_{L1eq})(R_{s2} + R_{L2eq}) + \omega^2 M_{p1}^2 + \omega^2 M_{p2}^2} = \dfrac{U}{Z_{in(2)}} \\ \overrightarrow{I_{s1}} = \dfrac{-j\omega M_{p1}}{R_{s1} + R_{L1eq}} I_p = \dfrac{-j\omega M_{p1}(R_{s2} + R_{L2eq})U}{R_p(R_{s1} + R_{L1eq})(R_{s2} + R_{L2eq}) + \omega^2 M_{p1}^2 + \omega^2 M_{p2}^2} \\ \overrightarrow{I_{s2}} = \dfrac{-j\omega M_{p2}}{R_{s2} + R_{L2eq}} I_p = \dfrac{-j\omega M_{p2}(R_{s1} + R_{L1eq})U}{R_p(R_{s1} + R_{L1eq})(R_{s2} + R_{L2eq}) + \omega^2 M_{p1}^2 + \omega^2 M_{p2}^2} \end{cases} \tag{2.41}$$

Defining $Z_{in(2)no}$ as the total equivalent impedance seen from the power supply, it can be deduced as

$$Z_{in(2)no} = R_p + Z_{re(2)no} = R_p + \frac{\omega^2 M_{p1}^2}{R_{s1} + R_{L1eq}} + \frac{\omega^2 M_{p2}^2}{R_{s2} + R_{L2eq}} \tag{2.42}$$

where $Z_{re(2)no}$ is defined as the reflected impedance from the pickups to the primary side, which is given by

$$Z_{re(2)no} = \frac{\omega^2 M_{p1}^2}{R_{s1} + R_{L1eq}} + \frac{\omega^2 M_{p2}^2}{R_{s2} + R_{L2eq}}$$

2.3.2.2 Load Power

Accordingly, the total input power from the DC power supply $P_{in(2)}$, load power of pickup 1 P_{L1}, and load power of pickup 2 P_{L2} can be, respectively, calculated as

$$\begin{cases} P_{in(2)} = I_p U = \dfrac{(R_{s1}+R_{L1eq})(R_{s2}+R_{L2eq})U^2}{R_p(R_{s1}+R_{L1eq})(R_{s2}+R_{L2eq})+\omega^2 M_{p1}^2+\omega^2 M_{p2}^2} \\ P_{L1} = \vec{I_{s1}}^2 R_{L1eq} = \dfrac{\omega^2 M_{p1}^2 (R_{s2}+R_{L2eq})^2 U^2 R_{L1eq}}{\left[R_p(R_{s1}+R_{L1eq})(R_{s2}+R_{L2eq})+\omega^2 M_{p1}^2+\omega^2 M_{p2}^2\right]^2} \\ P_{L2} = \vec{I_{s2}}^2 R_{L2eq} = \dfrac{\omega^2 M_{p2}^2 (R_{s1}+R_{L1eq})^2 U^2 R_{L2eq}}{\left[R_p(R_{s1}+R_{L1eq})(R_{s2}+R_{L2eq})+\omega^2 M_{p1}^2+\omega^2 M_{p2}^2\right]^2} \end{cases} \tag{2.43}$$

Hence, the power distribution ratio between two pickups can be derived as

$$P_{L1} : P_{L2} = \frac{M_{p1}^2 R_{L1eq}}{(R_{s1}+R_{L1eq})^2} : \frac{M_{p2}^2 R_{L2eq}}{(R_{s2}+R_{L2eq})^2} \tag{2.44}$$

To sum up, the power distribution among two pickups can be regulated by changing the equivalent load values and mutual inductance.

2.3.2.3 Efficiency

Based on Eq. (2.43), the corresponding transmission efficiencies η_1, η_2 and the total transmission efficiency η can be, respectively, obtained as

$$\begin{cases} \eta_1 = \dfrac{P_{L1}}{P_{in(2)}} = \dfrac{\dfrac{\omega^2 M_{p1}^2 R_{L1eq}}{(R_{s1}+R_{L1eq})^2}}{R_p + \dfrac{\omega^2 M_{p1}^2}{R_{s1}+R_{L1eq}} + \dfrac{\omega^2 M_{p2}^2}{R_{s2}+R_{L2eq}}} \\ \eta_2 = \dfrac{P_{L2}}{P_{in(2)}} = \dfrac{\dfrac{\omega^2 M_{p2}^2 R_{L2eq}}{(R_{s2}+R_{L2eq})^2}}{R_p + \dfrac{\omega^2 M_{p1}^2}{R_{s1}+R_{L1eq}} + \dfrac{\omega^2 M_{p2}^2}{R_{s2}+R_{L2eq}}} \\ \eta = \dfrac{P_{L1}+P_{L2}}{P_{in(2)}} = \dfrac{\dfrac{\omega^2 M_{p1}^2 R_{L1eq}}{(R_{s1}+R_{L1eq})^2} + \dfrac{\omega^2 M_{p2}^2 R_{L2eq}}{(R_{s2}+R_{L2eq})^2}}{R_p + \dfrac{\omega^2 M_{p1}^2}{R_{s1}+R_{L1eq}} + \dfrac{\omega^2 M_{p2}^2}{R_{s2}+R_{L2eq}}} \end{cases} \tag{2.45}$$

Thus, the transmission efficiency ratio between two pickups can be expressed as

$$\eta_1 : \eta_2 = P_{L1} : P_{L2} = M_{p1}^2 R_{L1eq}(R_{s2}+R_{L2eq})^2 : M_{p2}^2 R_{L2eq}(R_{s1}+R_{L1eq})^2 \tag{2.46}$$

Additionally, by solving the following two partial derivative equations (2.47), the optimal equivalent load values $R_{L1eq\eta max}$ and $R_{L2eq\eta max}$ for the maximum total transmission efficiency η can be derived:

$$\begin{cases} \dfrac{\partial \eta}{\partial R_{L1eq}} = 0 \\ \dfrac{\partial \eta}{\partial R_{L2eq}} = 0 \end{cases} \tag{2.47}$$

The result can be deduced as

$$\begin{cases} R_{L1eq\eta max} = R_{s1}\sqrt{1 + \dfrac{\omega^2 M_{p1}^2}{R_p R_{s1}} + \dfrac{\omega^2 M_{p2}^2}{R_p R_{s2}}} \\ R_{L2eq\eta max} = R_{s2}\sqrt{1 + \dfrac{\omega^2 M_{p1}^2}{R_p R_{s1}} + \dfrac{\omega^2 M_{p2}^2}{R_p R_{s2}}} \end{cases} \tag{2.48}$$

By substituting (2.48) into (2.45), the maximum efficiency can be deduced and simplified as

$$\eta_{max} = 1 - \frac{2}{\sqrt{1 + \dfrac{\omega^2 M_{p1}^2}{R_p R_{s1}} + \dfrac{\omega^2 M_{p2}^2}{R_p R_{s2}}} + 1} \tag{2.49}$$

Consequently, based on Eq. (2.49), by reducing the coil internal resistances or line resistances as well as increasing the coupling coefficient between the transmitting coil and each pickup coil, a higher efficiency can be achieved.

2.3.3 Multiple Pickups System Analysis

2.3.3.1 Modeling

Similarly, the aforementioned conclusions can be extended to general multiple pickups IPT systems, where the identical procedures of circuit analysis can be applied once again. As shown in Figure 2.20, the coupling coefficient between any transmitting and pickup coils can be expressed as

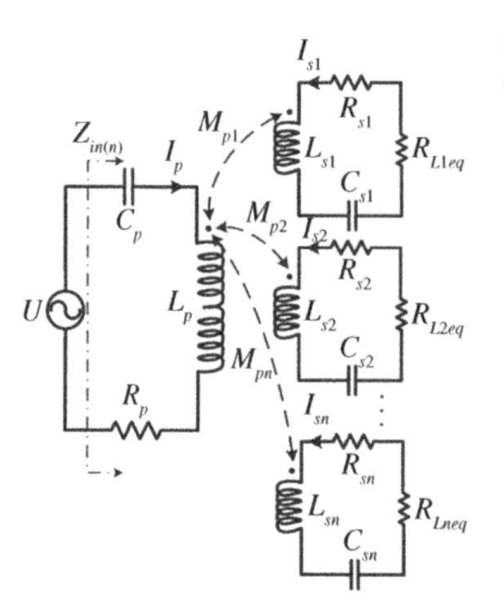

Figure 2.20 Simplified multiple pickups IPT system without cross-coupling.

$$k_{pi} = \frac{M_{pi}}{\sqrt{L_p L_{si}}}, \quad i = [1,2, \dots, n] \tag{2.50}$$

By ignoring the loss of diodes, the equivalent resistance R_{Lneq} can be deduced as

$$R_{Lieq} = \frac{8}{\pi^2} R_{Li}, \quad i = [1,2, \dots, n] \tag{2.51}$$

where R_{Li} is defined as the load resistance of any pickup side.

Based on KVL, the circuit as shown in Figure 2.20 can be expressed as

$$\begin{bmatrix} \vec{U} \\ 0 \\ 0 \\ \vdots \\ 0 \end{bmatrix} = \begin{bmatrix} \vec{Z_p} & j\omega M_{p1} & j\omega M_{p2} & \cdots & j\omega M_{pn} \\ j\omega M_{p1} & \vec{Z_{s1}} + R_{L1eq} & 0 & \cdots & 0 \\ j\omega M_{p2} & 0 & \vec{Z_{s2}} + R_{L2eq} & \cdots & 0 \\ \vdots & \vdots & \vdots & \ddots & \vdots \\ j\omega M_{pn} & 0 & 0 & \cdots & \vec{Z_{sn}} + R_{Lneq} \end{bmatrix} \begin{bmatrix} \vec{I_p} \\ \vec{I_{s1}} \\ \vec{I_{s2}} \\ \vdots \\ \vec{I_{sn}} \end{bmatrix} \tag{2.52}$$

Denoting Z_{si}, $i \in [p, s1, s2, \dots, sn]$, as the impedance of the primary side or the arbitrary one of secondary pickup, respectively,

$$\vec{Z_k} = R_k + j\omega L_k + \frac{1}{j\omega C_k}, \quad k \in [p, s1, s2, \dots, sn] \tag{2.53}$$

Then, Eq. (2.39) can be redefined as

$$j\omega L_i + \frac{1}{j\omega C_i} = 0, \quad i \in [p, s1, s2, \dots, sn] \tag{2.54}$$

By substituting (2.54) into (2.53), the expression (2.52) can be simplified and unfolded as

$$\begin{cases} R_p\vec{I_p} + j\omega M_{p1}\vec{I_{s1}} + j\omega M_{p2}\vec{I_{s2}} + \cdots + j\omega M_{pn}\vec{I_{sn}} = U \\ j\omega M_{pi}\vec{I_p} + \vec{I_{si}}(R_{si} + R_{Lieq}) = 0, \quad i \in [1,2, \dots, n] \end{cases} \tag{2.55}$$

Solving Eq. (2.55), the primary current I_p and the secondary current $\vec{I_{si}}$, $i = [1,2, \dots, n]$, are obtained as

$$\begin{cases} I_p = \dfrac{\prod_{j=1}^{n}(R_{sj}+R_{Ljeq})U}{R_p\prod_{j=1}^{n}(R_{sj}+R_{Ljeq})+\sum_{j=1}^{n}\omega^2 M_{pj}^2} = \dfrac{U}{Z_{in(n)}} \\ \vec{I_{si}} = \dfrac{-j\omega M_{pi}}{R_{si}+R_{Lieq}} I_p = \dfrac{-j\omega M_{pi}}{R_{si}+R_{Lieq}} \cdot \dfrac{\prod_{j=1}^{n}(R_{si}+R_{Ljeq})U}{R_p\prod_{j=1}^{n}(R_{sj}+R_{Ljeq})+\sum_{j=1}^{n}\omega^2 M_{pj}^2}, \quad i = [1,2, \dots, n] \end{cases} \tag{2.56}$$

Defining $Z_{in(n)}$ as the total equivalent impedance seen from the power supply, it can be deduced as

$$Z_{in(n)} = R_p + Z_{re(n)no} = R_p + \sum_{j=1}^{n} \frac{\omega^2 M_{pj}^2}{R_{sj} + R_{Ljeq}} \tag{2.57}$$

where $Z_{re(n)no} = \sum_{j=1}^{n} \frac{\omega^2 M_{pj}^2}{R_{sj}+R_{Ljeq}}$ is defined as the reflected impedance from the pickups to the primary side.

2.3.3.2 Load Power

Accordingly, the total input power from the DC power supply $P_{in(n)}$ and the load power of any pickup iP_{Li}, $i = [1, 2, \ldots, n]$, can be, respectively, calculated as

$$\begin{cases} P_{in(n)} = I_p U = \dfrac{\prod\limits_{j=1}^{n}(R_{sj}+R_{Ljeq})U^2}{R_p\prod\limits_{i=1}^{n}(R_{sj}+R_{Ljeq})+\sum\limits_{j=1}^{n}\omega^2 M_{pj}^2} \\ P_{Li} = \vec{I}_{si}^{\,2} R_{Lieq} = \dfrac{\omega^2 M_{pi}^2}{(R_{si}+R_{Lieq})^2}\cdot\dfrac{\prod\limits_{j=1}^{n}(R_{sj}+R_{Ljeq})^2U^2}{\left[R_p\prod\limits_{j=1}^{n}(R_{sj}+R_{Ljeq})+\sum\limits_{j=1}^{n}\omega^2 M_{pj}^2\right]^2}, \quad i = [1,2,\ldots,n] \end{cases} \tag{2.58}$$

Hence, the power distribution ratio between any pickups can be derived as

$$P_{L1} : P_{L2} : \cdots : P_{Ln} = \frac{M_{p1}^2 R_{L1eq}}{(R_{s1}+R_{L1eq})^2} : \frac{M_{p2}^2 R_{L2eq}}{(R_{s2}+R_{L2eq})^2} : \cdots : \frac{M_{pn}^2 R_{Lneq}}{(R_{sn}+R_{Lneq})^2} \tag{2.59}$$

2.3.3.3 Efficiency

Based on Eq. (2.58), the corresponding transmission efficiency $\eta_i, i = [1,2,\ldots,n]$, and the total transmission efficiency η can be, respectively, obtained as

$$\begin{cases} \eta_i = \dfrac{P_{Li}}{P_{out(n)}} = \dfrac{\frac{\omega^2 M_{pi}^2 R_{Lieq}}{(R_{si}+R_{Lieq})^2}}{R_p+\sum\limits_{j=1}^{n}\frac{\omega^2 M_{pj}^2}{R_{sj}+R_{Ljeq}}}, \quad i = [1,2,\ldots,n] \\ \eta = \dfrac{\sum\limits_{j=1}^{n} P_{Lj}}{P_{out(n)}} = \dfrac{\sum\limits_{j=1}^{n}\frac{\omega^2 M_{pj}^2}{R_{sj}+R_{Ljeq}}}{R_p+\sum\limits_{j=1}^{n}\frac{\omega^2 M_{pj}^2}{R_{sj}+R_{Ljeq}}} \end{cases} \tag{2.60}$$

Thus, the transmission efficiency ratio between any pickups can be expressed as

$$\eta_1 : \eta_2 : \cdots : \eta_n = P_{L1} : P_{L2} : \cdots : P_{Ln} = \frac{M_{p1}^2 R_{L1eq}}{(R_{s1}+R_{L1eq})^2} : \frac{M_{p2}^2 R_{L2eq}}{(R_{s2}+R_{L2eq})^2} : \cdots : \frac{M_{pn}^2 R_{Lneq}}{(R_{sn}+R_{Lneq})^2} \tag{2.61}$$

Additionally, by solving the following partial derivative equation (2.62), the optimal equivalent load value $R_{Lieq\eta max}$ for the maximum total transmission efficiency η can be derived

$$\frac{\partial \eta}{\partial R_{Lieq}} = 0, \quad i = [1,2,\ldots,n] \tag{2.62}$$

The result can be deduced as

$$R_{Lieq\eta max} = R_{si}\sqrt{1+\sum_{j=1}^{n}\frac{\omega^2 M_{pj}^2}{R_p R_{sj}}}, \quad i = [1,2,\ldots,n] \tag{2.63}$$

By substituting (2.63) into (2.60), the maximum efficiency can be deduced and simplified as

$$\eta_{max} = 1 - \frac{2}{\sqrt{1 + \sum_{j=1}^{n} \frac{\omega^2 M_{pj}^2}{R_p R_{sj}}} + 1} \tag{2.64}$$

2.3.3.4 Discussion

From the aforementioned calculation and analysis, it should be noted that multiple pickups IPT system without cross-coupling has the following characteristics:

(1) For SS topology, the selection for the compensation capacitance of both the primary and secondary sides only depends on its own inductance, holding no relevance to the mutual inductance and load resistance.
(2) Based on Eq. (2.42), the total equivalent impedance seen from the power supply $Z_{in(n)}$ is purely resistive under the resonant state, which is different from the equivalent impedance in multiple pickups IPT system with cross-coupling effect, as shown in Eq. (2.70). The specific discussion will be elaborated in Section 2.3.4.
(3) According to Eq. (2.60), the mutual inductance and the related resistances contribute to the maximum efficiency. By reducing the coil internal resistances or line resistances as well as increasing the coupling coefficient between the transmitting coil and each pickup coil, a higher efficiency can be significantly achieved.
(4) The power acquired by any one of the pickups can be regulated by changing the equivalent load values and the corresponding mutual inductance. Theoretically, arbitrary power distribution ratio can be achieved according to Eq. (2.59). Additionally, if adopting a constant current source, the power received by any one of the pickups is not affected by the others.
(5) Just as (4) addresses, the decoupling of multiple pickups IPT system, which simplifies the analysis of transmission performance, enables the process of parameters selection, design, and control more straightforwardly. Hence, the many characteristics of 1-to-1 IPT transmission can be applied in the multiple pickups IPT system.

2.3.4 Cross-Coupling

2.3.4.1 Cross-Coupling Effect

Comparing the distance between two pickups to the radius of pickup coils is close enough; while two pickups are located in different planes, the cross-coupling effect is obvious, which will affect the corresponding transmission performance

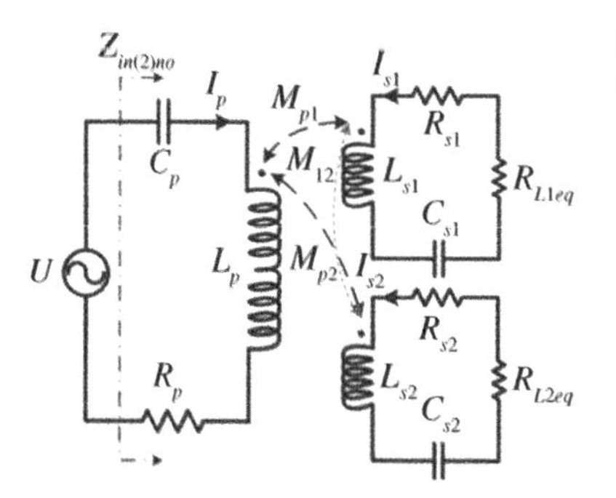

Figure 2.21 Simplified two pickups IPT system with cross-coupling.

and should not be ignored in 1-to-n IPT systems. Aiming to facilitate the following theoretical analysis, a general two pickups IPT system with cross-coupling, shown in Figure 2.21, is depicted to further explore the related transmission characteristics.

As shown in Figure 2.21, the coupling coefficient κ_{12} between pickup coil one and pickup coil two can be expressed as

$$\kappa_{12} = \frac{M_{12}}{\sqrt{L_{s1}L_{s2}}} \tag{2.65}$$

where M_{12} is defined as the mutual inductance between them. Due to nearly the same structure compared with Figure 2.19 without cross-coupling, the validity of Eqs. (2.33)–(2.36), (2.38) and (2.39) can also be proven in two pickups IPT system with cross-coupling.

Accordingly, based on KVL, the circuit as shown in Figure 2.20 can be expressed as

$$\begin{bmatrix} \vec{U} \\ 0 \\ 0 \end{bmatrix} = \begin{bmatrix} \vec{Z_p} & j\omega M_{p1} & j\omega M_{p2} \\ j\omega M_{p1} & \vec{Z_{s1}} + R_{L1eq} & j\omega M_{12} \\ j\omega M_{p2} & j\omega M_{12} & \vec{Z_{s2}} + R_{L2eq} \end{bmatrix} \begin{bmatrix} \vec{I_p} \\ \vec{I_{s1}} \\ \vec{I_{s2}} \end{bmatrix} \tag{2.66}$$

By substituting (2.39) into (2.38), the expression (2.66) can be simplified and unfolded as

$$\begin{cases} R_p\vec{I_p} + j\omega M_{p1}\vec{I_{s1}} + j\omega M_{p2}\vec{I_{s2}} = U \\ j\omega M_{p1}\vec{I_p} + (R_{s1} + R_{L1eq})\vec{I_{s1}} + j\omega M_{12}\vec{I_{s2}} = 0 \\ j\omega M_{p2}\vec{I_p} + j\omega M_{12}\vec{I_{s1}} + (R_{s2} + R_{L2eq})\vec{I_{s2}} = 0 \end{cases} \tag{2.67}$$

Then, the primary current I_p and the secondary currents $\vec{I_{s1}}$ and $\vec{I_{s2}}$ can be, respectively, deduced as

$$\begin{cases} \overrightarrow{I_p} = \dfrac{U}{R_p + \dfrac{\omega^2 M_{p1}^2 (R_{s2} + R_{L2eq}) + \omega^2 M_{p1}^2 (R_{s2} + R_{L2eq}) + 2j\omega^3 M_{p1} M_{p2} M_{12}}{\omega^2 M_{12}^2 + (R_{s1} + R_{L1eq})(R_{s2} + R_{L2eq})}} = \dfrac{U}{Z_{in(2)}} \\ \overrightarrow{I_{s1}} = -\dfrac{\omega^2 M_{p2} M_{12} + j\omega M_{p1}(R_{s2} + R_{l2eq})}{\omega^2 M_{12}^2 + (R_{s1} + R_{L1eq})(R_{s2} + R_{L2eq})} \overrightarrow{I_p} \\ \overrightarrow{I_{s2}} = -\dfrac{\omega^2 M_{p1} M_{12} + j\omega M_{p2}(R_{s1} + R_{l1eq})}{\omega^2 M_{12}^2 + (R_{s1} + R_{L1eq})(R_{s2} + R_{L2eq})} \overrightarrow{I_p} \end{cases} \tag{2.68}$$

Defining $\overrightarrow{Z_{in(2)}}$ as the total equivalent impedance seen from the power supply, it can be deduced as

$$\overrightarrow{Z_{in(2)}} = R_p + Z_{re(2)} = R_p + \frac{\omega^2 M_{p1}^2 (R_{s2} + R_{L2eq}) + \omega^2 M_{p1}^2 (R_{s2} + R_{L2eq}) - 2j\omega^3 M_{p1} M_{p2} M_{12}}{\omega^2 M_{12}^2 + (R_{s1} + R_{L1eq})(R_{s2} + R_{L2eq})} \tag{2.69}$$

where $Z_{re(n)no}$ is defined as the reflected impedance from the pickups to the primary side, which is given by

$$\overrightarrow{Z_{re(2)no}} = \frac{\omega^2 M_{p1}^2 (R_{s2} + R_{L2eq}) + \omega^2 M_{p1}^2 (R_{s2} + R_{L2eq}) - 2j\omega^3 M_{p1} M_{p2} M_{12}}{\omega^2 M_{12}^2 + (R_{s1} + R_{L1eq})(R_{s2} + R_{L2eq})} \tag{2.70}$$

With regard to Eq. (2.70), the $\overrightarrow{Z_{re(2)no}}$ owns the capacitive part, although each pickup circuit is pure resistance.

Then, the total input power from the DC power supply $P_{in(2)}$, the load power of pickup 1 P_{L1}, and the load power of pickup 2 P_{L2} can be, respectively, obtained as

$$\begin{cases} P_{in(2)} = \overrightarrow{I_p} U = \dfrac{U^2}{R_p + \frac{\omega^2 M_{p1}^2 (R_{s2} + R_{L2eq}) + \omega^2 M_{p1}^2 (R_{s2} + R_{L2eq})}{\omega^2 M_{12}^2 + (R_{s1} + R_{L1eq})(R_{s2} + R_{L2eq})}} \\ P_{L1} = \overrightarrow{I_{s1}}^2 R_{L1eq} = \dfrac{\left[\omega^4 M_{p2}^2 M_{12}^2 + \omega^2 M_{p1}^2 (R_{s2} + R_{l2eq})^2\right] R_{L1eq} U^2}{\left[\omega^2 M_{12}^2 + (R_{s1} + R_{L1eq})(R_{s2} + R_{L2eq})\right]^2 \left[R_p + \frac{\omega^2 M_{p1}^2 (R_{s2} + R_{L2eq}) + \omega^2 M_{p1}^2 (R_{s2} + R_{L2eq})}{\omega^2 M_{12}^2 + (R_{s1} + R_{L1eq})(R_{s2} + R_{L2eq})}\right]^2} \\ P_{L2} = \overrightarrow{I_{s2}}^2 R_{L2eq} = \dfrac{\left[\omega^4 M_{p1}^2 M_{12}^2 + \omega^2 M_{p2}^2 (R_{s1} + R_{l1eq})^2\right] R_{L2eq} U^2}{\left[\omega^2 M_{12}^2 + (R_{s1} + R_{L1eq})(R_{s2} + R_{L2eq})\right]^2 \left[R_p + \frac{\omega^2 M_{p1}^2 (R_{s2} + R_{L2eq}) + \omega^2 M_{p1}^2 (R_{s2} + R_{L2eq})}{\omega^2 M_{12}^2 + (R_{s1} + R_{L1eq})(R_{s2} + R_{L2eq})}\right]^2} \end{cases} \tag{2.71}$$

Hence, the power distribution ratio between two pickups can be derived as

$$P_{L1} : P_{L2} = \omega^2 M_{p2}^2 M_{12}^2 + M_{p1}^2 (R_{s2} + R_{l2eq})^2 : \omega^2 M_{p1}^2 M_{12}^2 + M_{p2}^2 (R_{s1} + R_{l1eq})^2 \tag{2.72}$$

To sum up, the power distribution among two pickups can be regulated by three factors, which consist of the equivalent load values, mutual inductance, and cross-coupling.

Based on Eq. (2.71), the corresponding transmission efficiencies η_1, η_2 and the total transmission efficiency η can be, respectively, obtained as

$$\eta_1 = \frac{P_{L1}}{P_{in(2)}};\ \eta_2 = \frac{P_{L2}}{P_{in(2)}};\ \eta = \frac{P_{L1} + P_{L2}}{P_{in(2)}} \tag{2.73}$$

Given space limitations, the final expression is ignored, which can be easily obtained by substituting (2.71) into (2.73).

Thus, the transmission efficiency ratio between any pickups can be expressed as

$$\begin{aligned}\eta_1 : \eta_2 = P_{L1} : P_{L2} &= \omega^2 M_{p2}^2 M_{12}^2 + M_{p1}^2 (R_{s2} + R_{l2eq})^2 \\ &: \omega^2 M_{p1}^2 M_{12}^2 + M_{p2}^2 (R_{s1} + R_{l1eq})^2\end{aligned} \tag{2.74}$$

where the cross-coupling effect affects the transmission efficiency and the ratio between them.

2.3.4.2 Frequency Shifting

Obviously seen from Eq. (2.70), the total equivalent impedance $\overrightarrow{Z_{in(2)}}$ holds the capacitive parts. Accordingly, the actual resonant frequency $f_o = \frac{\omega_o}{2\pi}$ should be recalculated by solving

$$\begin{aligned}Im\left\{Z_{in(2)} + j\omega_o L_p + \frac{1}{j\omega_o C_p}\right\} &= \omega L_p - \frac{1}{\omega_o C_p} \\ &- \frac{2\omega_o^3 M_{p1} M_{p2} M_{12}}{\omega_o^2 M_{12}^2 + (R_{s1} + R_{L1eq})(R_{s2} + R_{L2eq})} = 0\end{aligned} \tag{2.75}$$

where $Im\{\}$ means the imaginary part of a complex number. However, the inherent resonant frequency $f = \frac{\omega}{2\pi}$ is given by solving

$$Im\left\{j\omega L_p + \frac{1}{j\omega C_p}\right\} = \omega L_p + \frac{1}{j\omega C_p} = 0 \tag{2.76}$$

By comparing Eq. (2.75) with Eq. (2.76), the actual resonant frequency is entirely different from the inherent resonant frequency. The reason for the difference lies in whether the reflected impedance $\overrightarrow{Z_{re(2)no}}$, as shown in equation, is pure resistive or not. If the reflected impedance is pure resistive, there exists no distinction between the actual resonant frequency and the inherent resonant frequency. However, if there remains inductive/capacitive part, the actual resonant frequency will stray away from the inherent resonant frequency.

Regarding the IPT system with cross-coupling, the aforementioned analysis reveals three main differences compared with the IPT system without the cross-coupling effect:

(1) Based on Eq. (2.70), the reflected impedance owns the capacitive parts, which will lead to the deviation between the actual operating frequency deviation and the inherent resonant frequency, thus causing the attenuation of the transmission performance.
(2) All the main features, for example the primary current, input or load power, efficiency, and the power/efficiency ratio, are influenced by the coupling coefficient κ_{12}. Hence, the general analysis approach regarding multiple pickups without the cross-coupling effect is invalid.
(3) Supposing the system parameters and the number of loads are identical, the power received by each pickup in the 1-to-n IPT system considering the cross-coupling effect is smaller than that of the pickup without cross-coupling.
(4) For IPT system considering the cross-coupling effect with SS topology, the selection for the compensation capacitance of both the primary and the secondary sides not only depends on the inductance of coils but also is affected by the mutual inductance, cross-coupling, and load resistance. Thus, the compensation of cross-coupling supposes to become the prerequisite to carry on the decoupling analysis of multiple pickups IPT systems, which is shown in Section 2.3.3. Accordingly, the detailed compensation method aiming to cancel out the cross-coupling effect is given in Section 2.3.4.3.

2.3.4.3 Compensation of Cross-Coupling

Compared with Eqs. (2.43)–(2.71) and (2.45)–(2.73), it should be noted that if the two pickups IPT system holds the same set of the primary and secondary currents as well as identical equivalent resistance, its important characteristics, total equivalent impedance seen from the power supply, transmission efficiency, input power, and load power will be equal whether there remains cross-coupling effect or not [23]. The abovementioned conditions can be realized by adopting properly designed extra reactance, as shown in Figure 2.22.

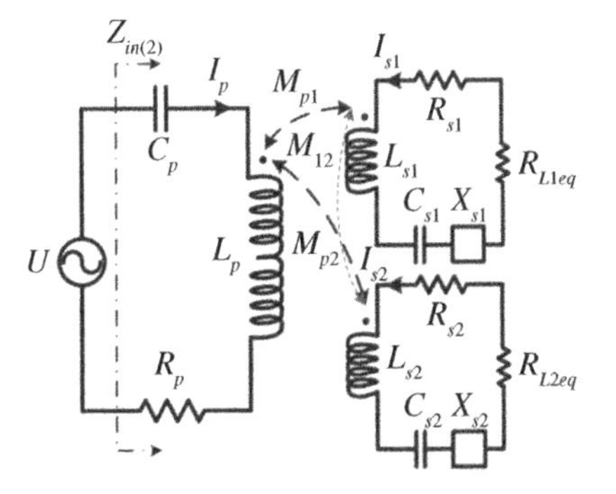

Figure 2.22 Simplified two pickups IPT system with cross-coupling and extra reactance.

Denote X_{s1} and X_{s2} as the extra reactance of pickup 1 and pickup 2, respectively. Accordingly, based on KVL, the circuit as shown in Figure 2.22 can be described as

$$\begin{bmatrix} \vec{U} \\ 0 \\ 0 \end{bmatrix} = \begin{bmatrix} \vec{Z_p} & j\omega M_{p1} & j\omega M_{p2} \\ j\omega M_{p1} & \vec{Z_{s1}} + R_{L1eq} + jX_{s1} & j\omega M_{12} \\ j\omega M_{p2} & j\omega M_{12} & \vec{Z_{s2}} + R_{L2eq} + jX_{s2} \end{bmatrix} \begin{bmatrix} \vec{I_p} \\ \vec{I_{s1}} \\ \vec{I_{s2}} \end{bmatrix} \tag{2.77}$$

Substituting (2.39) into (2.77) gives

$$\begin{cases} X_{s1} = -\dfrac{\omega M_{p2} M_{12} (R_{s1} + R_{L1eq})}{M_{p1} (R_{s2} + R_{L2eq})} \\ X_{s2} = -\dfrac{\omega M_{p1} M_{12} (R_{s2} + R_{L2eq})}{M_{p2} (R_{s1} + R_{L1eq})} \end{cases} \tag{2.78}$$

The negative value of X_{s1} and X_{s2} means the capacitive impedance, which signifies that this method can be implemented by adjusting the series-connected capacitors. Adopting this approach, the numerical and theoretical analysis of the total equivalent impedance, transmission efficiency, input power, and load power in Eqs. (2.42), (2.43), and (2.45), respectively, can be well preserved. Accordingly, all the significant conclusions addressed in Section 2.3.3.4 (under the circumstances without cross-coupling) remain valid.

2.4 What Are the Differences Between 1-to-1 and 1-to-*n* Transmission

Sections 2.2 and 2.3 introduced the modeling and power transmission performance of 1-to-1 and 1-to-*n* IPT system, respectively. Being different from the conventional 1-to-1 IPT system, the unique and key technical challenges of 1-to-*n* IPT systems can be summarized as follows:

2.4.1 Power Distribution

Compared with the traditional 1-to-1 IPT system, in which the number of pickups and value of resistance is fixed, the 1-to-*n* IPT holds the features of random access of loads and uncertain quantity of load quantity, thus causing large-scale load fluctuation. Based on Eq. (2.57), large-scale load fluctuation in the secondary sides eventually results in the wide-range variation in the reflected impedance and input impedance seen from the power supply. In IPT system with SS compensation, when adopting the constant voltage source, the input power tightly bonds with the input impedance, the power acquired by any pickup is closely connected to the reflected impedance, and the power distribution ratio strongly links to the relationship among the related reflected impedance.

More concretely, as the number of pickups accessing in 1-to-*n* IPT system increases, less power is received by each pickup, and as the corresponding reflected impedance enlarges, more power-related pickup is acquired. Accordingly, how to keep the power received by each pickup and the power distribution ratio unchanged under the circumstance of large-scale load fluctuation and random access of pickups becomes the first technical issue for 1-to-*n* IPT systems.

2.4.2 Transmission Control

The transmission performance of the IPT systems is partly determined by the relative position between the transmitting and the pickup coils. Generally, aiming to realize the maximum power or maximum transmission efficiency, transmitting and pickup coils are placed in the same coaxial direction without misalignment in the 1-to-1 IPT systems. However, the misalignment is the inevitable operating condition in the 1-to-*n* IPT system since multiple pickups embrace within one open induced electromagnetic field. It will unavoidably result in the shifting of the optimal transmission performance. Besides, the selective transmission control and multi-frequency-based excitation approaches come into the researchers' minds.

Hence, how to ensure the transmission performance in the case of the misalignment, how to implement the multifrequency excitation, how to track the maximum transmission power, and how to track the maximum transmission efficiency become the research focuses for 1-to-*n* IPT systems.

2.4.3 Cross-Coupling Effects

Due to multiple pickup coils locating in a limited induced electromagnetic field, the cross-coupling effect in the pickup side becomes the inevitable technical concern, which should not appear in the 1-to-1 IPT system. Based on Eq. (2.70), the reflected impedance owns the capacitive parts although the pickup impedance is pure resistance, which will lead to the deviation between the actual operating frequency deviation and inherent resonant frequency, thus causing the attenuation of transmission performance. Besides, the power acquired by each pickup in the 1-to-*n* IPT system considering the cross-coupling effect is smaller than that of the pickup without cross-coupling. Meanwhile, the power distribution ratio is also significantly affected by the cross-coupling effect. Accordingly, how to cancel out the cross-coupling effect in the pickup side and how to adjust the impedance precisely and rapidly to ensure the resonant state is another technical issue for 1-to-*n* systems.

2.4.4 Energy Security

In 1-to-n IPT system, due to the open characteristics of the induced electromagnetic field, all the involved pickups can gain the wirelessly transferred power. Consequently, the energy security issue becomes even more important than before. Specifically, the energy is expected to be delivered to authorized pickups while preventing unauthorized pickup from accessing the energy illegally. Accordingly, how to maintain the energy security is the final key technical issue for 1-to-n IPT systems.

2.A Appendix

The mutual inductance between two one-turn loops in any shapes is shown in Figure 2.A.1, which can be gained using the Neumann integral [24] as follows:

$$M_{12} = \frac{\psi_{12}}{\vec{I_1}} = \frac{\mu_0}{4\pi} \oint_{C_1} \oint_{C_2} \frac{\vec{dl_1} \cdot \vec{dl_2}}{\vec{R_{12}}}, \quad \mu_0 = 4\pi \times 10^{-7}$$

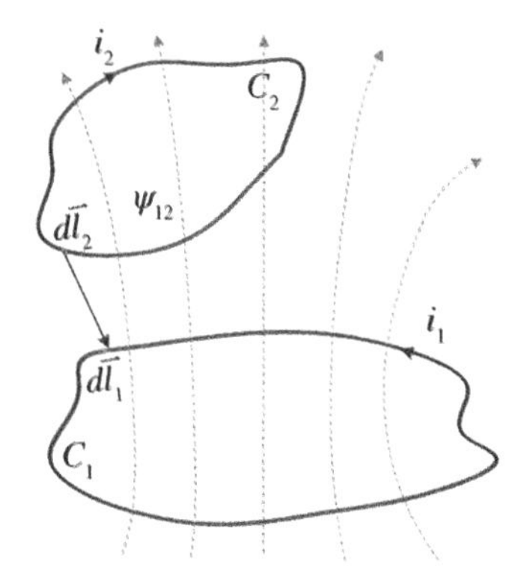

Figure 2.A.1 Mutual inductance between two one-turn loops in any shapes.

where

M_{12} the mutual inductance between one-turn loop 1 and one-turn loop 2
ψ_{12} the magnetic flux penetrating the surface of loop 2
$\vec{I_1}$ the current through loop 1
C_1 contour of loop 1
C_2 contour of loop 2
$\vec{dl_1}$ the infinitesimal length vector on the contour C_i
$\vec{dl_2}$ the infinitesimal length vector on the contour C_j
$\vec{R_{12}}$ the distance vector between two points at $\vec{dl_1}$ and $\vec{dl_2}$

References

1 Kurs, A., Karalis, A., Moffatt, R. et al. (2007). Wireless power transfer via strongly coupled magnetic resonances. *Science* 317 (5834): 83–86.

2 Chen, C.-J., Chu, T.-H., Lin, C.-L., and Jou, Z.-C. (2010). A study of loosely coupled coils for wireless power transfer. *IEEE Transactions on Circuits and Systems II: Express Briefs* 57 (7): 536–540.

3 Huh, J., Lee, S.W., Lee, W.Y. et al. (2011). Narrow-width inductive power transfer system for online electrical vehicles. *IEEE Transactions on Power Electronics* 26 (12): 3666–3679.

4 Wireless Power Consortium Specification. (2013). System description wireless power transfer. Volume 1: Low power.

5 Power Matter Alliance. (2014). PMA inductive wireless power and charging transmitter specification-system release 1. Standard PMA-TS-003-0 v1.00.

6 Alliance for Wireless Power. (2014). A4WP wireless power transfer system baseline system specification. BSS Standard A4WP-S-0001 v1.2.

7 Mur-Miranda, J.O., Fanti, G., Feng, Y. et al. (2010). Wireless power transfer using weakly coupled magnetostatic resonators. *IEEE Energy Conversion Congress and Exposition* 4179–4186.

8 Green, E.I. (1955). The story of Q. *American Scientist* 43 (4): 584–594.

9 Jiang, C., Chau, K.T., Liu, C., and Lee, C.H.T. (2017). An overview of resonant circuits for wireless power transfer. *Energies* 10 (7): 894–913.

10 Hui, S. (2016). Magnetic resonance for wireless power transfer [a look back]. *IEEE Power Electronics Magazine* 3 (1): 14–31.

11 Zhang, W. and Mi, C.C. (2016). Compensation topologies of high-power wireless power transfer systems. *IEEE Transactions on Vehicular Technology* 65 (6): 4768–4778.

12 Jiang, Y., Wang, L., Wang, Y. et al. (2019). Analysis, design, and implementation of accurate ZVS angle control for EV battery charging in wireless high-power transfer. *IEEE Transactions on Industrial Electronics* 66 (5): 4075–4085.

13 Li, Y., Hu, J., Li, X. et al. (2020). Analysis, design, and experimental verification of a mixed high-order compensations-based WPT system with constant current outputs for driving multistring LEDs. *IEEE Transactions on Industrial Electronics* 67 (1): 203–213.

14 Li, H., Wang, K., Fang, J., and Tang, Y. (2019). Pulse density modulated ZVS full-bridge converters for wireless power transfer systems. *IEEE Transactions on Power Electronics* 34 (1): 369–377.

15 Mayordomo, I., Dräger, T., Spies, P. et al. (2013). An overview of technical challenges and advances of inductive wireless power transmission. *Proceedings of IEEE* 101 (6): 1302–1311.

16 Zhang, Y., Lu, T., Zhao, Z. et al. (2015). Selective wireless power transfer to multiple loads using receivers of different resonant frequencies. *IEEE Transactions on Power Electronics* 30 (11): 6001–6005.

17 Zhang, Z., Chau, K.T., Qiu, C., and Liu, C. (2015). Energy encryption for wireless power transfer. *IEEE Transactions on Power Electronics* 30 (9): 5237–5246.

18 Kim, Y.-J., Ha, D., Chappell, W.J., and Irazoqui, P.P. (2016). Selective wireless power transfer for smart power distribution in a miniature-sized multiple-receiver system. *IEEE Transactions on Industrial Electronics* 63 (3): 1853–1862.

19 Papani, S.K., Neti, V., and Murthy, B.K. (2015). Dual frequency inverter configuration for multiple-load induction cooking application. *IET Power Electronics* 8 (4): 591–601.

20 Pantic, Z., Lee, K., and Lukic, S.M. (2014). Multifrequency inductive power transfer. *IEEE Transactions on Power Electronics* 29 (11): 5995–6005.

21 Liu, W., Chau, K.T., Lee, C.H.T. et al. (2019). Multi-frequency multi-power one-to-many wireless power transfer system. *IEEE Transactions on Magnetics* 55 (7): 8001609: 1–9.

22 Liu, F., Yang, Y., Ding, Z. et al. (2018). A multifrequency superposition methodology to achieve high efficiency and targeted power distribution for a multiload MCR WPT system. *IEEE Transactions on Power Electronics* 33 (10): 9005–9016.

23 Fu, M., Zhang, T., Zhu, X. et al. (2016). Compensation of cross coupling in multiple-receiver wireless power transfer systems. *IEEE Transactions on Industrial Informatics* 12 (2): 474–482.

24 Paul, C.R. (2010). *Inductance Loop and Partial.* Hoboken, NJ: Wiley.

Part II

Design

3

Design and Optimization for Coupled Coils

This chapter aims to discuss about the design and optimization of coupled coils. Firstly, the transmission efficiency is analyzed for wireless power transfer (WPT) systems with a series–series compensation topology. Then, two key indices affecting the transmission efficiency are deduced and confirmed, that is the quality factor and the coupling coefficient. In addition, the influence of coil parameters and the shielding plate is also discussed, with emphasis on its impact on the transmission performance. Lastly, based on the above discussion, optimal design schemes are introduced from two aspects, which are the quality factor and the coupling coefficient.

3.1 Introduction

In recent years, WPT technologies have been successfully incorporated with a great number of applications such as battery charging for consumer electronics devices, implantable bioelectronics, and static and dynamic charging for electric vehicles (EVs). The power transmission efficiency, transmitted power, transmission distance, electromagnetic radiation, and construction cost are some key indices regarding the WPT system design process. Meanwhile, the power transmission performance is also limited by the switching device, the topology of inverters, and the control strategies [1], wherein the transmission efficiency is determined by the power switching loss and the electromagnetic loss. According to the analysis of Eq. (3.5), two of the most concerning parameters of transmission efficiency involve the quality factor and the coupling coefficient. These two indicators can be optimized by the cautious design of the indispensable and key component, namely the coupled coil, to realize the wireless transmission and reception efficiently. In other words, the parameters, topology, and the size of the coupled coils play an imperative role to ensure the transmission performance of WPT systems.

Wireless Power Transfer: Principles and Applications, First Edition. Zhen Zhang and Hongliang Pang.

In terms of the first indicator, a higher quality factor means lower loss, which can be achieved by decreasing the AC resistance, reducing skin and proximity losses, and increasing the self-inductance of coils. Regarding another indicator, the coupling coefficient is determined by the misalignment and the distance between coupled coils. Unfortunately, the misalignment and the distance cannot be compromised due to the limitation of specific applications. In terms of the wireless charging for EVs, the distance between the transmitting and the pickup coils has strict requirements so that it is incapable to change easily. Consequently, the optimal design of coils is the most direct way to improve the transmission performance of WPT systems. The corresponding analysis of coil parameters and the optimization design of coils will be elaborated in this chapter.

3.2 Design Considerations

3.2.1 Analysis of Power Transmission

Figure 3.1 shows the typical structure of WPT systems, which mainly consists of three parts: the high-frequency power terminal, the resonant coupling circuit, and the load-receiving terminal. The coupling part is key to WPT systems, which can be divided into two types in terms of resonant coils, that is the self-resonance coil and the *LC* resonance coil. The self-resonance coil operates at the resonate state matching with the parasitic capacitance [2], while the *LC* resonator depends on the matching between the coil inductance and the additional lumped capacitance [3]. Regarding the self-resonance coils, the loss is relatively small; however, the parasitic capacitance is extremely low, thus leading to the deterioration of the implementation ability if the operating frequency is not high enough. Hence, the most common type is the *LC* resonance coil since it can offer the design and control flexibility for WPT systems.

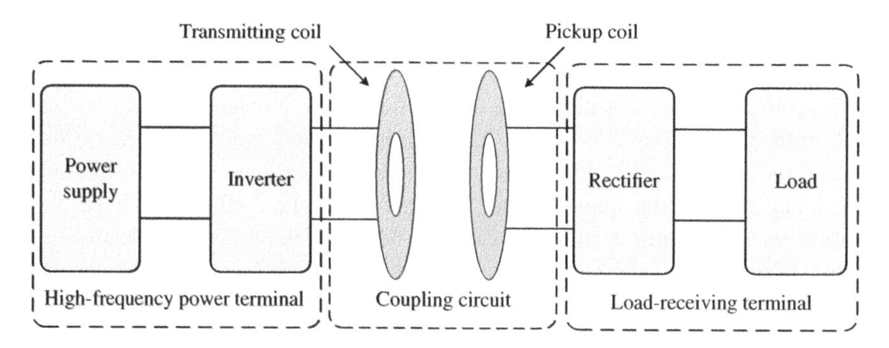

Figure 3.1 Schematic of the typical WPT systems.

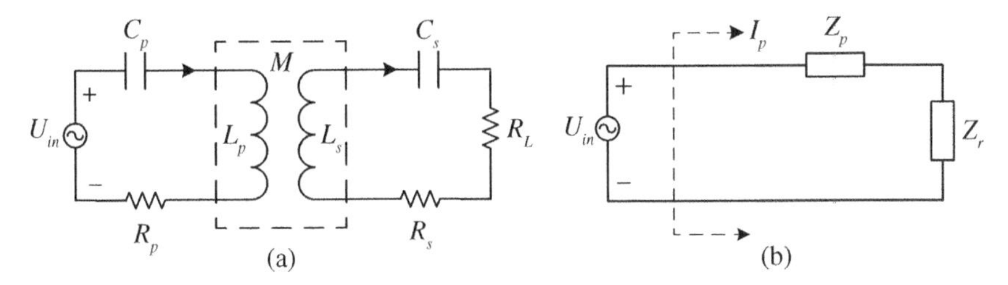

Figure 3.2 Circuit: (a) topology of SS compensation and (b) equivalent circuit model of SS compensation.

Regarding the LC resonance coil, there exist four fundamental compensation topologies, which are primary series–secondary series (SS), primary series–secondary parallel (SP), primary parallel–secondary series (PS), and primary parallel–secondary parallel (PP). The corresponding modeling and analysis will be elaborated in detail in Chapter 4. In order to reveal the impact of coil parameters on the transmission performance, this section takes the SS compensation topology as the exemplification for the following analysis. Figure 3.2 shows the SS compensation topology structure diagram and the equivalent M-model circuit diagram.

Based on the Kirchhoff's law, the corresponding circuit equations can be obtained:

$$\begin{cases} U_{in} = \left[R_p + j\left(\omega L_p - \frac{1}{\omega C_p}\right)\right] I_p + j\omega M I_s \\ 0 = j\omega M I_p + \left[R_s + R_L + j\left(\omega L_s - \frac{1}{\omega C_s}\right)\right] I_s \end{cases} \tag{3.1}$$

When the WPT system works at the resonant state, the impedance shows the characteristics of purely resistive. Then, the relationship between L and C can be expressed as

$$\begin{cases} \omega L_p = \frac{1}{\omega C_p} \\ \omega L_s = \frac{1}{\omega C_s} \end{cases} \tag{3.2}$$

In order to ensure the transmission performance, the operating frequency should equal the natural resonant frequency. In such a way, the input power P_{in} and the output power P_{out} can be obtained as

$$\begin{cases} P_{in} = \frac{(R_s + R_L)U_{in}^2}{R_p(R_s + R_L) + \omega^2 M^2} \\ P_{out} = \frac{\omega^2 M^2 (R_s + R_L) U_{in}^2}{[R_p(R_s + R_L) + \omega^2 M^2]^2} \end{cases} \tag{3.3}$$

Accordingly, the transmission efficiency η can be expressed as

$$\eta = \frac{P_{out}}{P_{in}} = \frac{\omega^2 M^2}{R_p(R_L + R_s) + \omega^2 M^2} \tag{3.4}$$

which shows that the efficiency and the output power are both determined by the load resistance and the mutual inductance. In the meanwhile, the output power and the transmission efficiency are the most direct indicators for WPT systems. However, the maximum efficiency does not mean the maximum output power and vice versa, which has been explained in Chapter 2. As a result, this chapter mainly focuses on the maximum transmission efficiency when adopting four fundamental compensation topologies. Regarding the SS compensation which is taken as the exemplification in this chapter, the maximum achievable η_{max} can be derived as [4]

$$\eta_{max} = \frac{\kappa^2 Q_1 Q_2}{(1 + \sqrt{1 + \kappa^2 Q_1 Q_2})^2} \tag{3.5}$$

where κ is the coupling coefficient between the transmitting and the pickup coils. Additionally, Q_1 and Q_2 are the quality factors of the transmitting and the pickup coils, respectively, which are given by:

$$\begin{cases} \kappa = \dfrac{M}{\sqrt{L_s L_p}} \\ Q_1 = \dfrac{\omega L_s}{R_s} \\ Q_2 = \dfrac{\omega L_p}{R_p} \end{cases} \tag{3.6}$$

It illustrates that the quality factor of the transmitting and the pickup coils and the corresponding coupling coefficient is key to improve the transmission performance.

Before the analysis of the maximum transmission efficiency, let us recall the meanings of the quality factor and the coupling coefficient for WPT system. The quality factor is a physical quantity used as a quality index in electricity and magnetism. It represents the ratio of the energy stored by the energy storage device to the energy consumed by the entire circuit in the same cycle. In the resonant circuit, the higher the quality factor generally means the higher frequency selectivity. Meanwhile, in WPT systems, the coupling coefficient is used to indicate the tightness or the strengthening of the coupling between the transmitting and the pickup coils. In the conventional transformer, the coupling coefficient can reach an extremely high level by adopting the ferrite core connecting the primary and the secondary windings, which means a strongly coupled system. However, the WPT system holds a totally different situation, where the coupling is realized over the air gap instead of the ferrite core. As a result, the coupling coefficient is far lower than that of the transformers, which means a loosely coupled system.

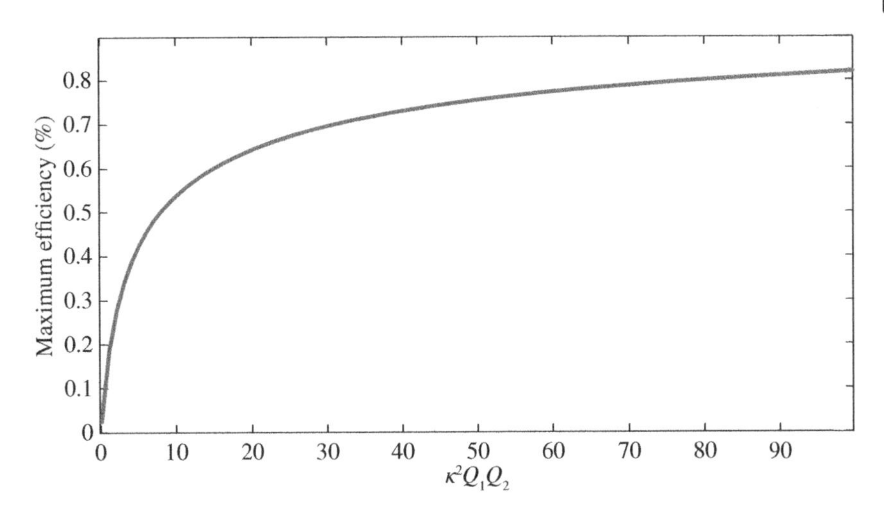

Figure 3.3 Relationship between maximum efficiency and $\kappa^2 Q_1 Q_2$.

After specifying the concept above, let us come back to the impact on the transmission efficiency of WPT systems. Figure 3.3 shows the quantitative relationship between the quality factor multiplied by the coupling coefficient and the maximum efficiency of the WPT systems. That is, the larger the quality factor multiplied by the coupling coefficient, the higher the maximum system efficiency. Consequently, the optimal design of the transmitting and the pickup coils should focus on the coupling coefficient and the quality factor.

From Eq. (3.6), it can be seen that the quality factor can be increased by reducing the internal resistance and increasing the inductance of the coil. In previous studies, a great number of efforts have been attempted to optimize the coil for achieving the goal of lower internal resistance, higher inductance, and strong coupling effect, for example the number of turns, the track width, the winding topology, and the core design [5]. The improving method of circular planer spirals has been proved by removing the inner turns [6, 7]. By this way, the quality factor Q is improved via the smaller high-frequency winding resistance. What's more, the strategy of track-width variations has also been proved to increase the Q effectively for radio frequency applications [8, 9]. Furthermore, the hollow approach combined with varying track width is proposed to enhance the performance of planer spirals [10]. Instead of removing the turns, this method reduces the resistance loss without compromising the inductance since the number of turns is constant. Thus, this combination technology provides a higher quality factor Q than removing turns or varying width independently. In order to enhance the magnetic coupling, the ferrite core or substrate is usually applied between the primary and secondary coils. A helical coil with ferrite cylinder core is applied to improve the magnetic flux and coupling in the volume of helix [11]. Besides, compound coil topologies such as the DD and the DDQ pads are proposed to increase the misalignment adaptability between the transmitting and the pickup coils [12].

3.2.2 Coil Parameters

For electromagnetic-conversion-based WPT systems, the transmitting and the pickup coils are the vital components to realize efficient power transmission. As aforementioned, the corresponding coil parameters determine the coupling effect and the quality factor [13]. Accordingly, this section gives a comprehensive analysis of the parameters for circular planar spiral coil, which is commonly utilized for WPT systems by means of its simple structure, perfect symmetry, and low manufacturing cost [14, 15]. In Sections 3.2.2.1 and 3.2.2.2, the relationship between the coil and the transmission performance is comprehensively elaborated from the perspectives of the electrical and the structure parameters.

3.2.2.1 Electrical Parameters

The electrical parameters directly determine the optimization objectives of coils, namely the quality factor and the coupling coefficient. Taking the circular planar spiral coil as the exemplification, consequently, it will be analyzed before taking consideration of the structure parameters that intuitively reflect the optimization for the transmitting and the pickup coils. In this section, the main electrical parameters of coupled coil, resistance, capacitance, and self-inductance, are discussed [16].

3.2.2.1.1 Resistance High-frequency current flows on the surface under the skin effect, where a higher AC frequency results in a more obvious skin effect. Generally, the skin depth is usually used to describe the skin level of current, which is defined as

$$\delta = \sqrt{\frac{2}{\omega \mu_0 \sigma}} \tag{3.7}$$

where ω is the current frequency, μ_0 is the absolute permeability ($\mu_0 = 4\pi \times 10^{-7}$ N/A^2), and σ is the conductivity. The effective area of current flowing through the conductor can be expressed by $S = 2\pi \alpha \sigma$. Then, the AC resistance R is calculated as

$$R = \frac{1}{S\delta} = \sqrt{\frac{\mu_0 \omega}{2\sigma}} \frac{Nr}{a} \tag{3.8}$$

where N is the number of turns, r is the radius of the coil, and a is the radius of the winding wire.

3.2.2.1.2 *Capacitance* The high operating frequency inevitably results in the distributed capacitance between adjacent turns. The interturn capacitance can be expressed by

$$C = \frac{2\pi^2 r\varepsilon_0}{\ln\left(\frac{p}{2a} + \sqrt{\frac{p}{2a}^{2} - 1}\right)} \tag{3.9}$$

where ε_0 is the dielectric coefficient of air ($\varepsilon_0 = 8.85 \times 10^{-12}$ F/m), and p is the line spacing, which is shown in Figure 3.4. The existence of interturn capacitance results in the increasing of the equivalent impedance of the coil, thus reducing the corresponding quality factor, which should be taken into account for the coil. Generally, the capacitance and the AC resistance should be reduced as much as possible to achieve a higher quality factor.

3.2.2.1.3 *Self-Inductance* The circular planar spiral coil is usually regarded as a concentric coil with N turns when calculating the self-inductance. The self-inductance of the single-turn coil can be given by

$$L = \mu_0 r\left(\ln\frac{8r}{a} - 1.75\right) \tag{3.10}$$

Then, the total self-inductance of the N-turns coil can be given by

$$L = \mu_0 N^2 r\left(\ln\frac{8r}{a} - 1.75\right) \tag{3.11}$$

where r is the radius of the coil, and a is the radius of the wire.

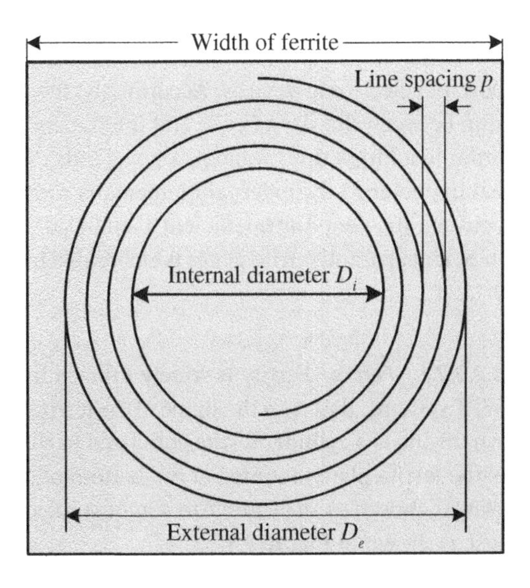

Figure 3.4 Schematic of circular planar spiral coils.

In order to achieve a higher quality factor, the self-inductance should be maintained as high as possible, which is determined by N and r according to (3.8) and (3.11). However, it also means that the increasing of the self-inductance leads to the increased usage of the wire, which means the inevitably increasing resistance. Eventually, the quality factor may be deduced caused by the excessive increasing of the resistance. It can be seen that it is difficult to obtain a satisfying quality factor when just taking the electrical parameters into consideration. Hence, it is necessary for the analysis of the influence of structure parameters on WPT systems.

3.2.2.2 Structure Parameters

As aforementioned, electrical parameters determine the quality factor of coupled coils, thus affecting the transmission efficiency. According to Eqs. (3.8)–(3.11), the resistance, capacitance, and self-inductance are determined by the structure parameters of the coils, including the wire radius, the coil radius, the line spacing, and the number of turns. This section focuses on the analysis of the structure parameters and the corresponding impact on the transmission performance of WPT systems.

Since the coupling coefficient κ indicates the coupling strengthen between the transmitting and the pickup coils, it is expected to be as high as possible. As for the circular planar spiral coils, the coupling coefficient κ significantly depends on the air gap, the line spacing, the external diameter, the internal diameter, and the ferrite plate, as shown in Figure 3.4. When the air gap between the coupled coils is fixed, κ is proportional to the size of the loosely coupled windings [17]. Hence, the simplest way to improve the coupling coefficient is to increase the effective coupling area of the coupled coils. However, the size cannot be expended arbitrarily due to the limitation of the practical applications such as biomedical implants and portable electronic devices. Accordingly, the excessive attention on the relationship between the size of coils and the transmission efficiency makes no sense. In order to address this problem, a novel index D_i/D_e, namely the ratio of the internal diameter to the external diameter, is proposed in Ref. [18]. The relationship between the coupling coefficient κ and D_i/D_e with respect to different line spacings, air gaps, and ferrite plates is elaborated in the Sections 3.2.2.2.1, 3.2.2.2.2, and 3.2.2.2.3.

3.2.2.2.1 Ferrite Ferrite is widely utilized to enhance the magnetic coupling of WPT systems. Based on the shape of the ferrites, there are mainly two applications, where one is a cylinder ferrite core fixed in the middle of the coil, while the other is the ferrite plate mounted at the bottom of coils [19]. In this section, the ferrite plate is chosen as an example to demonstrate the effectiveness of adopting ferrites, just as shown in Figure 3.4.

The ferrite plate is commonly made up by assembling a number of small square ferrite sheets, where the width of the ferrite plate is usually greater than the external diameter of the coil so as to acquire the expected effect of magnetic coupling. In order to investigate the influence of ferrite parameters, the change in self-inductance L and the coupling coefficient k will be analyzed with respect to different ratios of D_i/D_e and the thickness of the ferrite plate.

As depicted in Figure 3.5, the self-inductance of circular planar spiral coils significantly increases when adding the ferrite and meanwhile decreases along with the increasing of D_i/D_e. The additional ferrite can effectively increase the self-inductance of coils even though all design parameters remain the same. As a result, the usage of wire can be effectively reduced to meet the requirement of increased self-inductance. Additionally, there is an interesting phenomenon, that is the value of self-inductance has no significant improvement when increasing the thickness of the ferrite plate. The usage of ferrite is extremely essential to ensure the high-quality factor Q of coils. As shown in Figure 3.6, the coupling coefficient κ varies with the change in the thickness of the ferrite plate and the ratio D_i/D_e. Similar to self-inductance, the coupling coefficient can be increased by adding the ferrite plate to the circular planar spiral coil. Meanwhile, it should be noted that the thickness of the ferrite plate has little effect on the coupling coefficient.

To sum up, the ferrite can not only effectively improve the quality factor Q by increasing the self-inductance of the coil but also increase the coupling coefficient κ by enhancing the magnetic field. Undoubtedly, the involvement of ferrites is an ideal technical solution to improve the transmission performance of WPT systems.

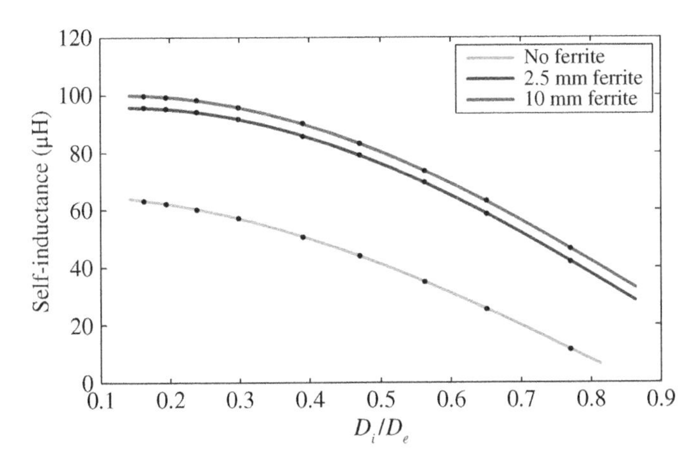

Figure 3.5 Self-inductance with respect to different ferrite thicknesses and D_i/D_e (line spacing: 0.1 mm; air gap: 100 mm).

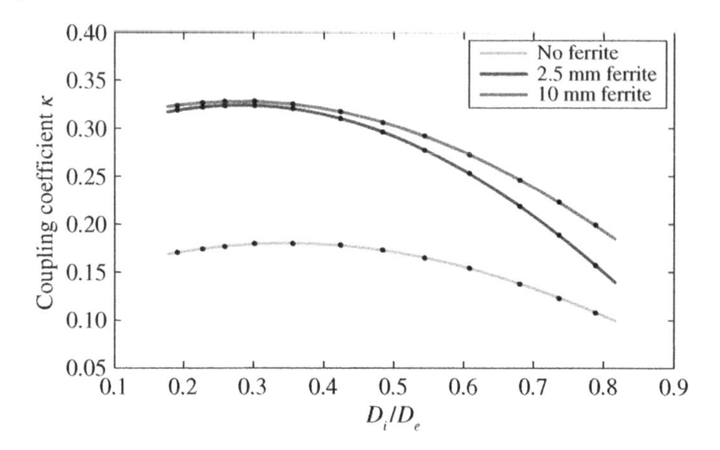

Figure 3.6 Coupling coefficient with respect to different ferrite thicknesses and D_i/D_e (line spacing: 0.1 mm; air gap: 100 mm).

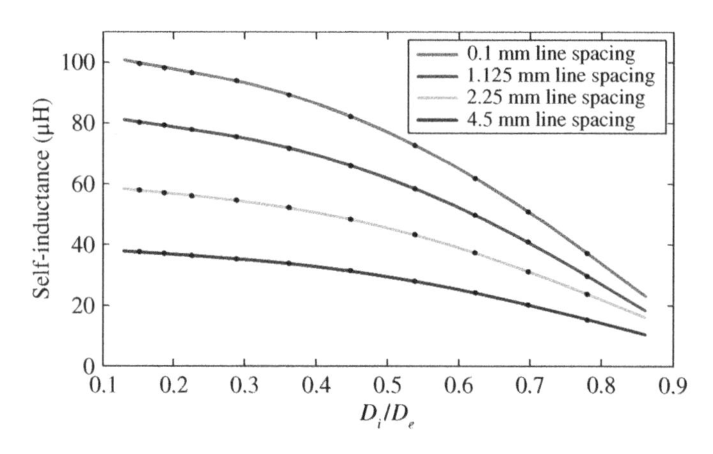

Figure 3.7 Self-inductance with respect to different line spacings and D_i/D_e (ferrite substrate: 2.5 mm; air gap: 100 mm).

However, it does not mean the more the better. In other words, the thickness of ferrites should be particularly paid attention since the excessive usage results in the additional power loss.

3.2.2.2.2 Line Spacing Figure 3.7 illustrates the impact of D_i/D_e on the self-inductance with respect to various values of line spacing. The relationship between self-inductance and the ratio D_i/D_e has been elaborated as before, which indicates that the key factor is the number of turns of the coil. Actually, the relationship between self-inductance and line spacing can also be explained as

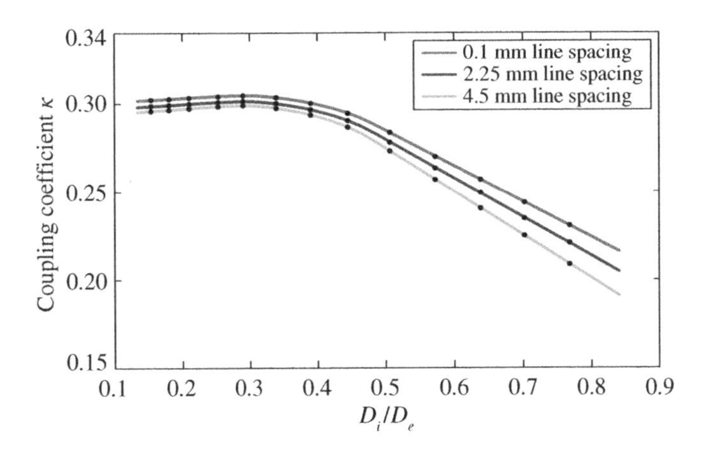

Figure 3.8 Coupling coefficient with respect to different line spacings and D_i/D_e (ferrite substrate: 2.5 mm; air gap: 100 mm).

the affection of the number of turns. Regarding the same coil topology, namely the same ratio D_i/D_e, a smaller line spacing means a greater number of turns, which leads to an increased self-inductance, which is also verified by the simulation results, as shown in Figure 3.7.

Figure 3.8 shows the change in the coupling coefficient with respect to different values of the line spacing and D_i/D_e. For the same D_i/D_e, the increasing of the line spacing causes a slight and even negligible decreasing of the coupling coefficient κ. Thus, it illustrates that the line spacing only affects the self-inductance while holds an ignorable impact on the coupling coefficient, which shows important reference meaning for the coil design.

3.2.2.2.3 Air Gap The air gap means the distance between the transmitting coil and the pickup coil, which determines the mutual inductance via the coupling coefficient and thus affects the transmission efficiency. Accordingly, this section gives the corresponding analysis as shown in Figure 3.9, especially for the impact of the air gap on the coupling coefficient. Figure 3.9 illustrates that the coupling coefficient declines notably with the increasing of the air gap, which quantitively verifies the importance of optimal design for the air gap in WPT systems. In order to ensure transmission efficiency, the air gap is usually required to be less than half of the external diameter of the circular planar spiral coil. Meanwhile, when the ratio D_i/D_e is in the range of 0.1–0.4, the coupling coefficient remains approximately constant regardless of the variation in the air gap. In practical, the external radius cannot be arbitrarily chosen but determined according to the requirement and the limitation of specific applications. The above results imply a design

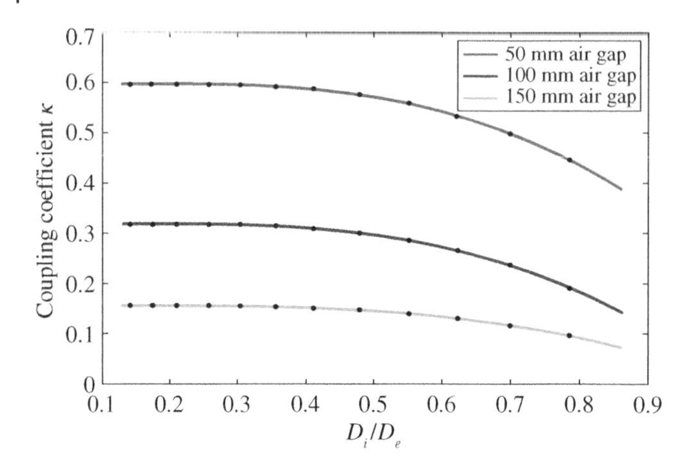

Figure 3.9 Coupling coefficient with respect to different air gaps and D_i/D_e (ferrite substrate: 2.5 mm; line spacing: 0.1 mm).

criterion to determine the internal radius after confirming the external radius of the coil.

Accordingly, the variation in the air gap results in the change in the coupling coefficient, namely the larger air gap, the smaller coupling coefficient. When the air gap is confirmed, however, the coupling coefficient is determined by the design of the coils. In order to ensure the coupling coefficient as high as possible, the ratio D_i/D_e should be maintained less than 0.4, which is the prerequisite to obtain the optimal transmission efficiency at a fixed air gap.

3.2.3 Shielding Methods

The WPT system can realize the transmission of electrical energy from the primary side to the secondary side over the air. Undoubtedly, the coupled magnetic field is the vital component of WPT systems. Being different from the traditional transformer, the WPT system falls into the loosely coupled system due to no core connecting the transmitting coils and the pickup coils. Hence, the magnetic flux leakage of WPT systems is far greater than that of the transformer, which inevitably results in concerns about the electromagnetic radiation as well as the external loss caused by the eddy current. Consequently, magnetic shielding is an important technical issue for WPT systems.

Figure 3.10 shows a typical primary pad, which mainly consists of the transmitting coil, the ferrite, and the shielding plate. As aforementioned, the density of the induced magnetic field can be enhanced using the ferrite, whereas the magnetic flux leakage is increased, accordingly. The aluminum shielding plate is commonly

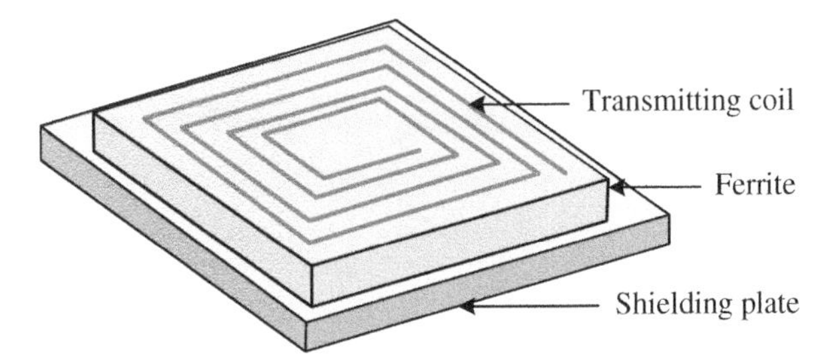

Figure 3.10 A typical coil with shielding plate utilized in WPT systems.

utilized to reduce the magnetic flux leakage for WPT systems. In general, the area of the aluminum shielding plates should be greater than that of the ferrite to achieve the expected shielding performance. However, the additional aluminum shielding plate inevitably reduces the transmission efficiency due to the eddy current in the aluminum [20]. The suppression of magnetic leakage is at the cost of the efficiency. Then, how to optimize the shield plate to reduce the magnetic flux leakage while maintaining the transmission efficiency? This is an important issue to which we should pay particularly attention.

The optimization of the shielding draws attention on the thickness of the shielding plate and the distance between the ferrite and the shielding plate on the transmission efficiency, respectively. As shown in Figure 3.11, it illustrates that the transmission efficiency decreases with the increasing of the thickness for the shielding plate, which is caused by the increased power loss of the eddy current. Thus, the thickness of the aluminum shielding plate should be maintained within an appropriate range. Additionally, Figure 3.12 depicts that the transmission efficiency is proportional to the distance between the aluminum shielding plate and the ferrite. Hence, the shielding plate should not be mounted clinging to

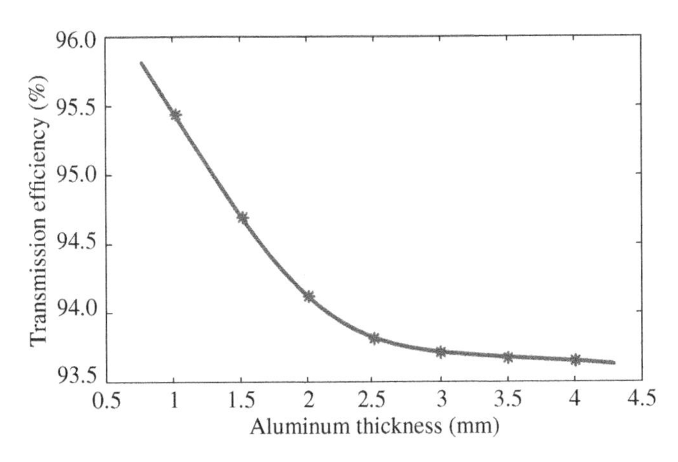

Figure 3.11 Transmission efficiency with respect to the thickness of aluminum plate.

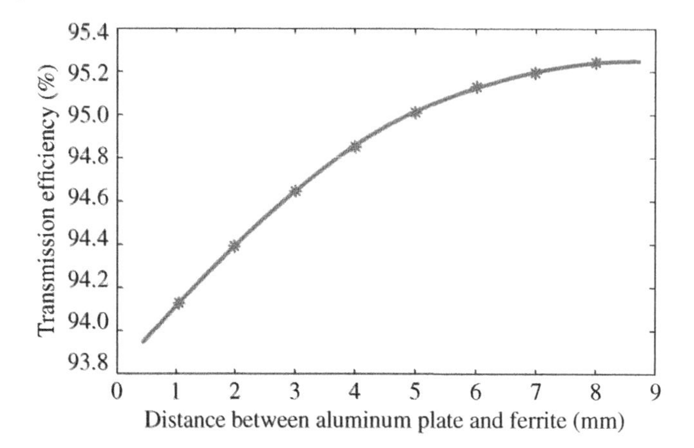

Figure 3.12 Transmission efficiency with respect to the distance between aluminum plate and ferrite.

the ferrite, which means an appropriate gap is necessary to prevent from the deterioration of the transmission efficiency.

3.3 Optimal Design

3.3.1 Quality Factor

After specifying the influence of the coil parameters on the transmission performance for WPT systems, this section elaborates on the optimal design of coils to achieve high transmission efficiency. According to Eq. (3.5), the coil design aims to increase the quality factor as high as possible. A high-quality factor means a low power loss. In other words, the key to coil optimization is how to obtain a great inductance with small resistance. In previous studies, a number of efforts have been made to improve the quality factor, which will be elaborated in this section, with emphasis on the coil windings.

3.3.1.1 Hollow Winding with Track-Width Ratio

In high-frequency applications, the AC resistance inevitably increases due to the skin effect and the proximity effect [21]. It indicates that the higher frequency and intensity of the magnetic field, the greater the eddy current loss. For planar spiral coils that are completely wound around the center, the distribution of the magnetic field is convex, which means that the peak value usually appears in the center region of the spiral winding. Figure 3.13 depicts the magnetic field intensity of the circular planar spiral coil. The intensity of the magnetic field around the central

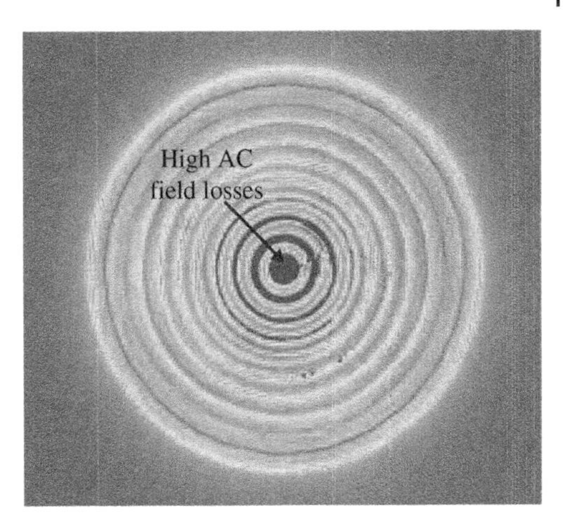

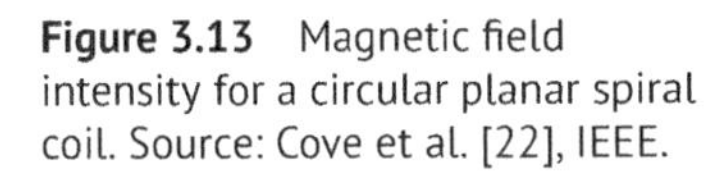
Figure 3.13 Magnetic field intensity for a circular planar spiral coil. Source: Cove et al. [22], IEEE.

region is the superposition of each turn of the magnetic field, which results in an enhanced magnetic field around the inner turns and thus increases the power loss. Besides, due to the limited area in the inner region, the inductance of the coil is mainly contributed by the outer turns rather than the inner turns. The above are the reasons why the quality factor of planar spiral coils decreases dramatically. In order to address this issue, the simplest way is to remove the central region. A coil winding comes along, namely the hollow winding. However, in order to achieve the maximum quality factor, the inductance of the hollow winding suffers from a reduction by 10–15%, which is not expected under this circumstance [7].

In addition, the track width, which represents the width of each turn, is another adjustment parameter to improve the quality factor of coils. The disadvantage of this pattern of winding is that the turns in the central area are retained so that the corresponding eddy current loss is increased due to the high density of the central magnetic field. Considering the disadvantage of these two windings, a new method combining the hollow winding and the varying track width is proposed to improve the quality factor of planar spiral coils, as shown in Figure 3.14.

Instead of removing the inner turns simply, the key of the hollow winding design is to enlarge the inner radius while keeping the turns of the spiral coil unchanged, which is able to reduce the high-frequency loss and meanwhile avoid the reducing of the inductance. What's more, the constant track-width ratio means that the track width reduces when the turns wind from the external side to the center. Such a way can effectively improve the quality factor of coils compared with the hollow winding and the varying track-width schemes.

Figure 3.15 shows the hollow winding with the constant track-width ratio, where the thickness of the spiral coil is constant, x_i and x_o represent the inside and the outside radii of the coil, C_l indicates the line spacing between the traces, N

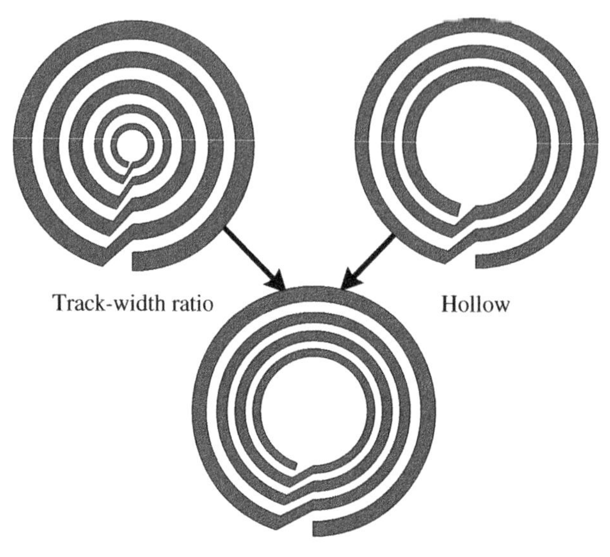

Figure 3.14 Hollow winding with track-width ratio.

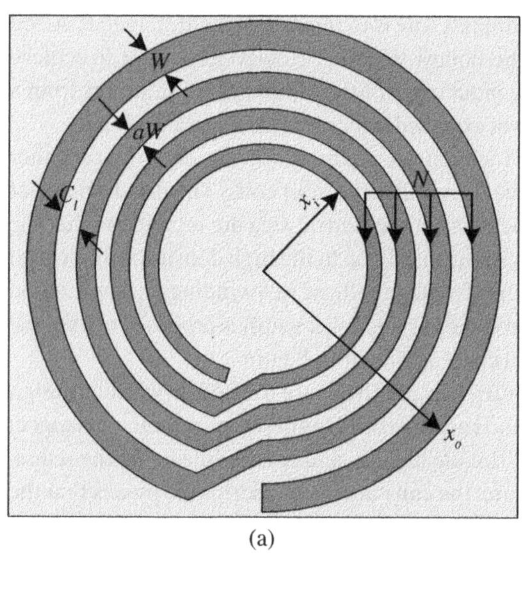

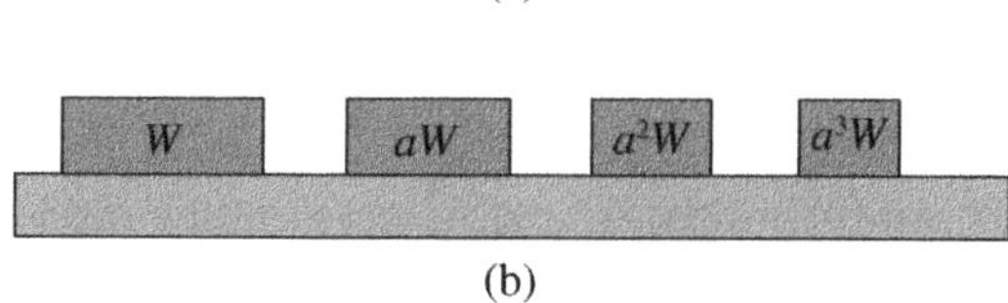

Figure 3.15 Diagram of hollow winding with track-width ratio: (a) isometric view and (b) vertical cross-section view.

is the number of turns, and W is the width of the outermost turn. The parameter a is defined as the width ratio between the adjacent turns. Then, the width of any turn can be expressed by

$$\omega_n = a^{n-1}W \tag{3.12}$$

Denote n as the sequence number of turns counting from the outer turns. Regarding the hollow winding with the constant track-width ratio, the case where $a < 1$ is only required to be investigated, namely the width of the turn decreases as it moves toward the center [22]. The highly intensive magnetic field in the center penetrates less, reducing high-frequency loss.

When the circular planar spiral coil is designed with a constant track-width ratio, each turn consists of a section of arc 330° and a section of the straight conductor. The constant track-width ratio means that the width of the turns varies along with the diameter of the coil. Hence, the straight conductor is used as a transition connecting the adjacent winding sections with different widths. Compared with ordinary planar spiral coils, the total length of the winding scheme with a constant track-width ratio inevitably produces a deviation to deteriorate the transmission performance, which is required to be as small as possible. As for hollow winding with track-width ratio, there is another advantage, that is due to the center of the spiral coil being hollow, it is convenient to place the ferrite in the center to enhance the coupling effect between coils.

According to the definition of the quality factor, which is determined by both the inductance and the resistance, the hollow winding with track-width ratio can reduce the AC resistance and thus significantly improve the quality factor by adjusting the inside radius x_i and the width ratio between adjacent turns a. Figure 3.16 depicts a three-dimensional plot reflecting the relationship among all the above parameters. It shows that the quality factor increases till reaching a

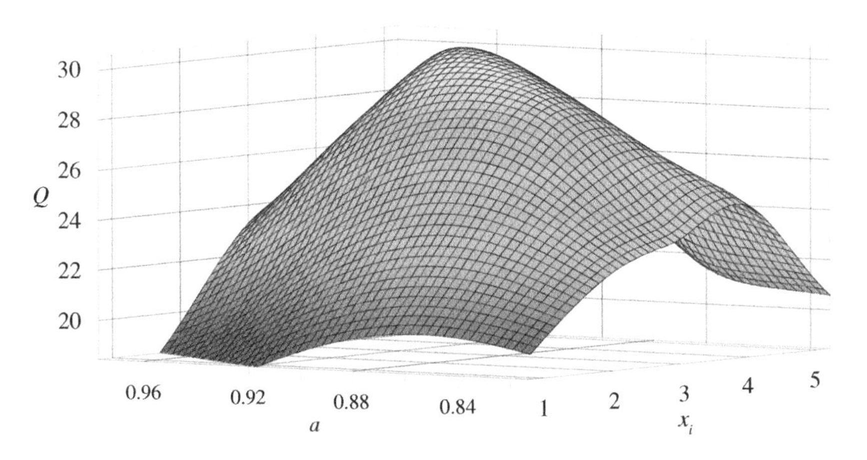

Figure 3.16 Quality factor with respect to inside radius and width ratio.

peak with the increasing of x_i and a. If the values of a and x_i increase continuously, the quality factor decreases after the peak. It indicates that there is an optimal parameter design to maximize the quality factor. Thus, how to design the inside radius and the width ratio is of vital importance for the optimal design of hollow winding coils.

3.3.1.2 Double-Layer Printed Spiral Coil

The commonly used coil of WPT systems adopts multiple individually insulated filaments called Litz wire, which is used to reduce the impact of the skin effect in high-frequency applications. However, this kind of coil usually requires a precise winding machinery; otherwise, each winding cannot guarantee the same characteristics. Besides, the size of the coils wound by Litz wire needs to be relatively large to ensure the transmission performance. It can be seen that the practical application of Litz wires is still limited though it addresses the issue about the skin effect for high-frequency applications. As a result, the printed spiral coil (PSC) has appeared as an alternative scheme. Compared with the traditional Litz wire winding coils, the PSC possesses advantages as follows. For one aspect, it is suitable for integration with other circuits within a limited space since the thickness of PSCs is extremely thin and even ignorable. For the other aspect, the PSC is more convenient for the design and the optimization of parameters compared with the Litz wire coils. It has been increasingly adopted for many applications because of its high precision, small footprint, and easy-to-fabricate characteristics although its transmission power capacity is limited [23]. Accordingly, how to optimize the performance of PSC, especially the quality factor, is an inevitable research topic. This section introduces a double-layer PSC for WPT systems and then focuses on the analysis of losses to specifically analyze the influence on the quality factor.

Figure 3.17 shows the structure of the double-layer PSCs. Compared with the commonly used single-layer coils, the upper and the lower layers of the printed circuit board (PCB) are both covered by the PSC. Meanwhile, the coils on both layers are connected in series by through-hole. Such a structure can offer a great inductance since it makes full use of the board space.

In order to further increase the quality factor, the power loss of the coils should be taken into account. In high-frequency applications, the loss of PSC mainly consists of three types, including the radiation loss, the resistance loss, and the dielectric loss.

For the strongly coupled magnetic resonant WPT system, the radiation loss is usually negligible. Instead, the main power loss comes from the resistance loss, which is mainly caused by the skin effect and the proximity effect. In order to facilitate the circuit analysis, the corresponding power loss can be simply explained using a resistor series connected with the inductor. In addition, the substrate also produces significant dielectric loss, especially at high frequency. Thus, the power

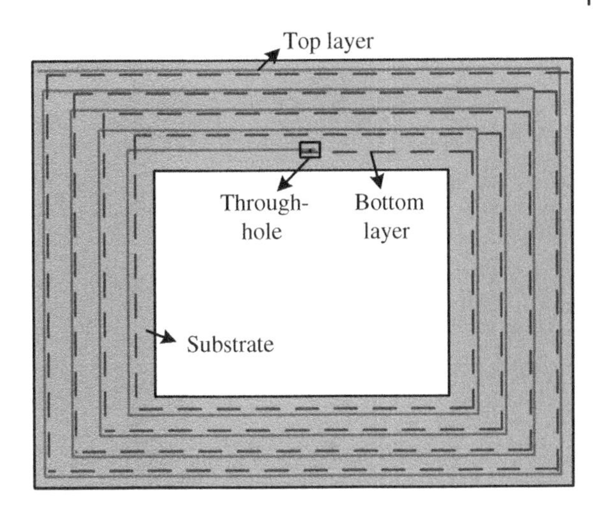

Figure 3.17 Structure of double-layer PSC.

loss analysis will be discussed from the perspective of the above three factors, that is the skin effect, the proximity effect, and the dielectric loss.

3.3.1.2.1 Skin Effect Being different from the Litz wire, the printed wire of PCBs is single stranded and more flattening. It leads to the high-frequency current passing through the skin of the wire instead of the middle, which is called the skin effect. By adopting the finite element method (FEM), it verifies that the current intensity on the surface of the wire is obviously higher than that in the interior. Then, the effective cross-sectional area of the conductor is reduced by the skin effect, which results in the increasing of the resistance as well as the extra power loss. Generally, the increasing of the cross-section area including the width and the thickness helps to reduce the resistance. The concept of wire width and wire thickness is introduced in Figure 3.18. On the one hand, the resistance decreases

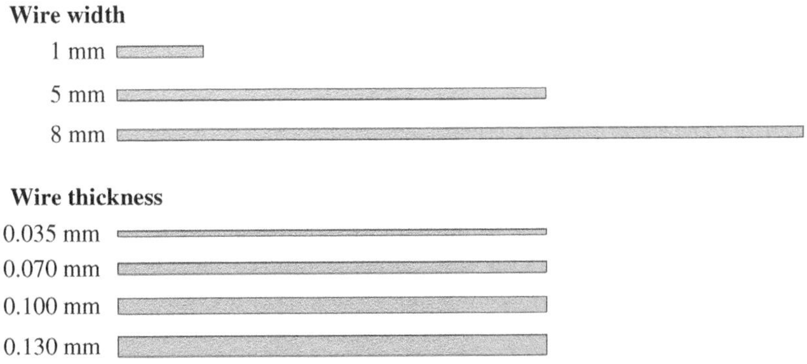

Wire width

1 mm

5 mm

8 mm

Wire thickness

0.035 mm

0.070 mm

0.100 mm

0.130 mm

Figure 3.18 The introduction of wire width and wire thickness.

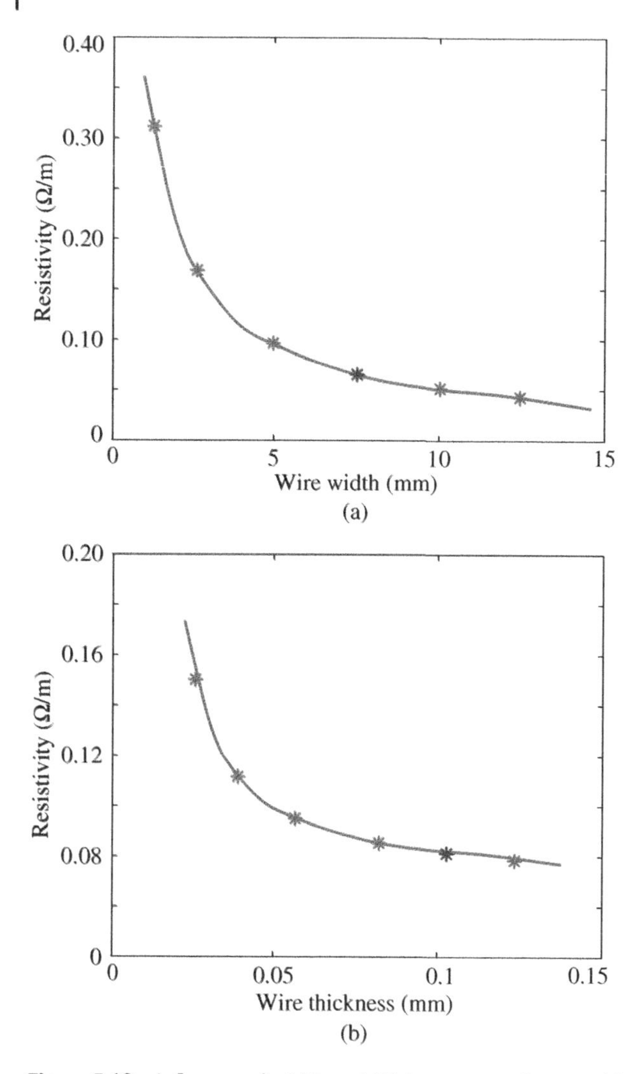

Figure 3.19 Influence of width and thickness on resistance: (a) with respect to wire width and (b) with respect to wire thickness.

dramatically with the increasing of the wire width at the initial stage and then enters into the saturation stage if further increased, just as shown in Figure 3.19a. The reason is that the middle of the wire cannot be utilized effectively if the wire width increases excessively. In addition, increasing the thickness of the wire is also an effective way to reduce the resistance, which can be verified according to Figure 3.19b. However, it should be noted that the thickness of the wire is generally limited to less than 0.5 mm on the PCB [5].

Based on the above analysis, the resistance can be effectively reduced by increasing the width and the thickness of the wire, but it does not mean the greater the better. In order to optimize the design of the wire on PCBs, the point of inflection on these curves as shown in Figure 3.19 is the optimal choice for the width and the thickness.

3.3.1.2.2 Proximity Effect The flowing current is no longer distributed evenly across the cross section of the wire but concentrated to one side due to the magnetic field induced by the adjacent conductor, which is called the proximity effect. The skin effect and the proximity effect are always concomitant.

The skin effect leads to the current concentrates on the surface, but the current density at both ends of the conductor is even. Meanwhile, the proximity effect redistributes the current in the conductor. For two adjacent conductors, the current densities are low on the adjacent side and high on the other side. With the increasing of the current frequency, the proximity effect is more obvious.

According to the principle of the proximity effect, it can be inferred that the resistance can be reduced by increasing the line space and the thickness of the substrate to enlarge the distance between the adjacent wires. Figure 3.20 confirms the influence of the line spacing and the substrate thickness, but the decreasing trend of the resistance is not as remarkable as expected. The reason is that the proximity effect of the conductors on different layers has little effect on the resistance due to the low current density in the middle of the conductor. Meanwhile, it should be noted that it is not advisable only increasing the line space excessively because it inevitably leads to the waste of space and thus reduces the inductance when the outer diameter of the coil is confirmed. In addition, there is no peak and the point of inflection on the curves as shown in Figure 3.20, which means no optimal values for the clearance and the thickness at all. Thus, users have to balance this contradiction according to the requirements and limitations of various applications.

3.3.1.2.3 Dielectric Loss For the double-layer PSC, the substrate of the PCB acts as the dielectric. Regarding high-frequency applications, the PCB has an inherent dissipation of electromagnetic energy, which cannot be ignored. In practical engineering applications, the dielectric loss is determined by the natural property of the dielectric itself and is usually represented by the tangent of dielectric loss. Figure 3.21 shows the variation in the PSC quality factor with respect to the tangent of dielectric loss. It illustrates that the quality factor increases dramatically within the range of the tangent from 10^{-2} to 10^{-4}. For the common materials of the PCB substrate, the tangent of the dielectric loss is in the scope of 10^{-2}–10^{-3}. Taking the most widely used material FR-4 as an exemplification, the tangent value of the dielectric loss is about 0.016. As aforementioned, the quality factor should

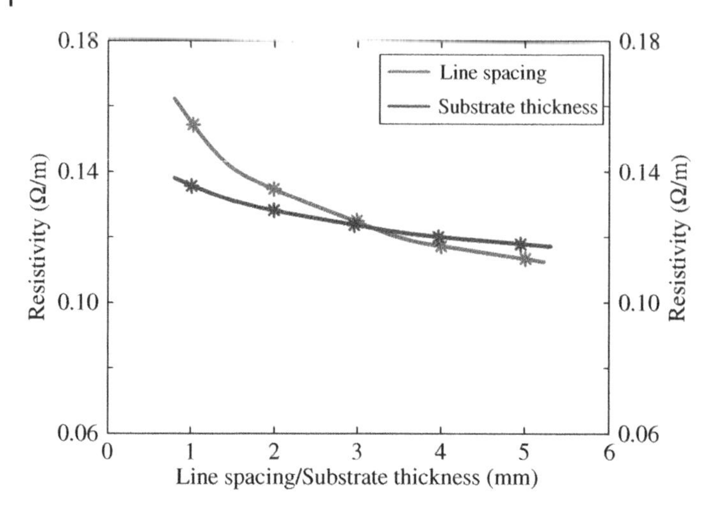

Figure 3.20 Influence of line spacing and substrate thickness on resistance.

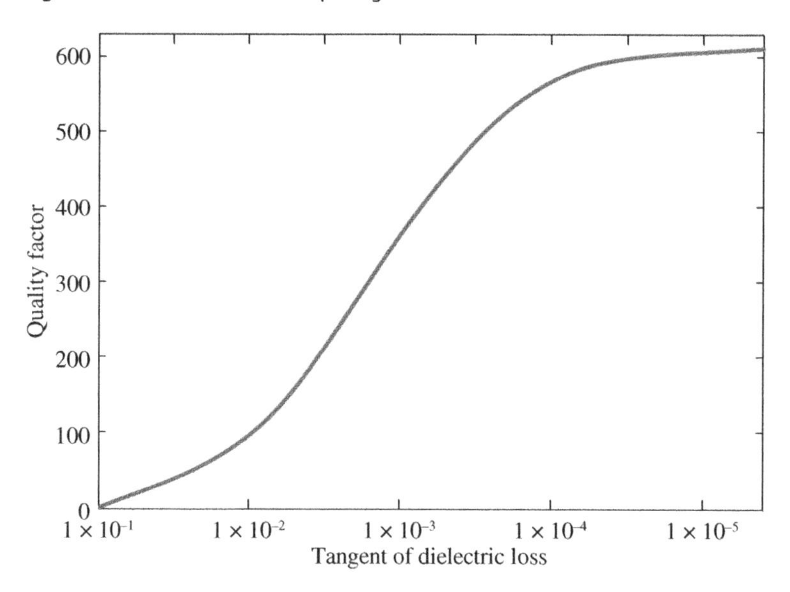

Figure 3.21 Quality factor versus tangent of dielectric loss.

be maintained as high as possible, aiming to ensure the transmission efficiency of WPT systems. Accordingly, from the perspective of the dielectric loss, the selection of materials is extremely essential to obtain an expected quality factor. The material with a low dielectric loss tangent is more suitable for the substrate utilized in high-frequency applications; for example, the dielectric loss tangent of F4BM is less than 7×10^{-4} to be commonly chosen for WPT systems.

3.3.2 Coupling Effect

According to Eq. (3.5), it is seen that apart from improving the quality factor Q of the coils, the other important aspect to improve the transmission efficiency is to increase the coupling coefficient κ between the transmitting and the pickup coils. Generally, the WPT falls into the loosely coupled system, which means a low value of the coupling coefficient (normally $\kappa < 0.2$). In order to maintain considerable transmission efficiency, the disadvantage of the low coupling coefficient can be compensated by increasing the quality factor of the coils. However, it is effective only for normal working conditions, which means that it cannot deal with the situation caused by the relatively long transmission or the misalignment. Accordingly, while the quality factor is one of the ways to enhance the transmission efficiency by reducing the loss, we have to find another way to fundamentally increase the capability of power transmission between the primary and the secondary sides, that is the coupling coefficient.

According to what has been discussed in Section 3.2, the coupling coefficient can be increased by optimizing the parameters of the coils such as the diameter and the number of turns. The optimization of the coil parameters is an effective way to enhance the coupling effect, especially for low-power, small-air gap, and little-misalignment applications such as the wireless charging for portable consumer electronics. For high-power or misaligned transmission applications such as wireless charging for EVs, however, it is not enough by only optimizing the diameter and the number of turns of the coils. Then, how to enhance the coupling effect for such a kind of application apart from simply optimizing coil parameters? Hence, in this section, the focus on enhancing the coupling effect will move toward the coil topology as well as the pad structure for EV applications.

3.3.2.1.1 Circular Pad Regarding the EV wireless charging system, various researchers prefer to use pads instead of traditional coils in such high-power WPT applications [24–26]. Then, what exactly is a coil pad? Taking the circular coil pad as an example, as depicted in Figure 3.22, the coil pad typically consists of six main components, including the plastic cover, the coil, the coil former, the ferrite, the aluminum ring, and the aluminum backing plate, which is commonly utilized for EV wireless charging systems. Simply speaking, the pad refers to not only a winding coil but also the magnetic and shielding materials [27, 28].

When the pad is adopted as the pickup mounted in EVs, the size and the weight of the pads should be particularly optimized to prevent from overoccupying the space inside the vehicle, respectively. In addition, the misalignment tolerance is another concern for the design of pads used in EV wireless charging systems.

According to the discussion in Section 3.2.2, the coupling effect can be enhanced by optimizing the design of various parameters, including the coil diameter, the

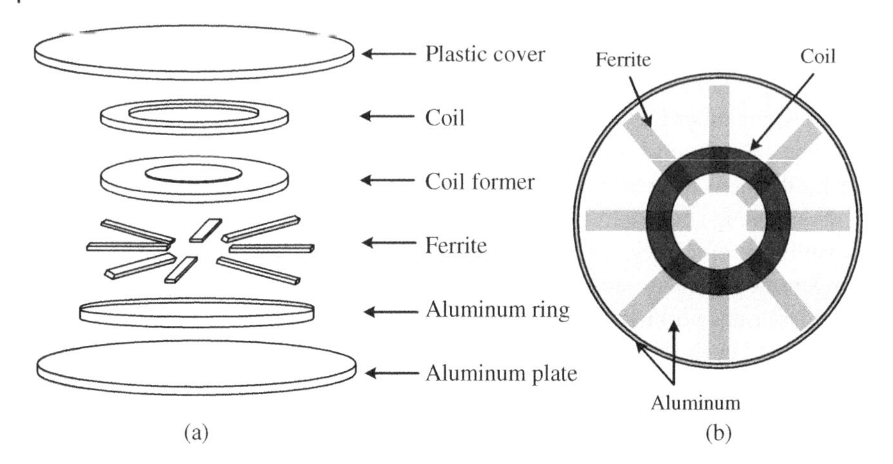

Figure 3.22 Schematics of circular pad: (a) exploded view and (b) top view.

ferrite, the line spacing, the number of turns, and the shielding methods. The ferrite is commonly utilized to reshape the distribution of the magnetic flux. For example, Figure 3.23a,b shows the distribution and the density of the magnetic flux by adopting the ferrite substrate or not, respectively. It illustrates that the ferrite can effectively guide the flux coming back from the secondary side to directly return to the primary side, which means that the leakage flux is significantly reduced, especially for the backside of the transmitting coil. Such a type of magnetic field is called as a single-sided magnetic field. It can increase the flux density around the coil and thus enhance the coupling effect between the transmitting and the pickup coils. However, it should be noted that the leakage flux still exists even though the flux is guided back to the coil via the ferrite substrate, which inevitably increases the extra power loss and even raises the concern about the human body due to the increased electromagnetic radiation. Accordingly, it is necessary to find a way to further reduce the leakage flux to address the issue above, that is the aluminum shielding, which is commonly utilized for the pad by two types, such as the aluminum plate and the aluminum ring. Figure 3.23c shows the distribution and the density of the magnetic field for the circular pad with the ferrite substrate and the aluminum plate. It illustrates that the leakage flux on the backside of the coils is almost eliminated. Meanwhile, the single-sided magnetic field generated by the ferrite can significantly reduce the power loss in the aluminum plate. Consequently, the additional component has little effect on the transmission efficiency of the WPT systems [28].

However, one application limitation should be taken into account when adopting the circular pad, that is the height of the magnetic flux path is usually a quarter of the coil diameter. The flux density drops dramatically due to the air gap

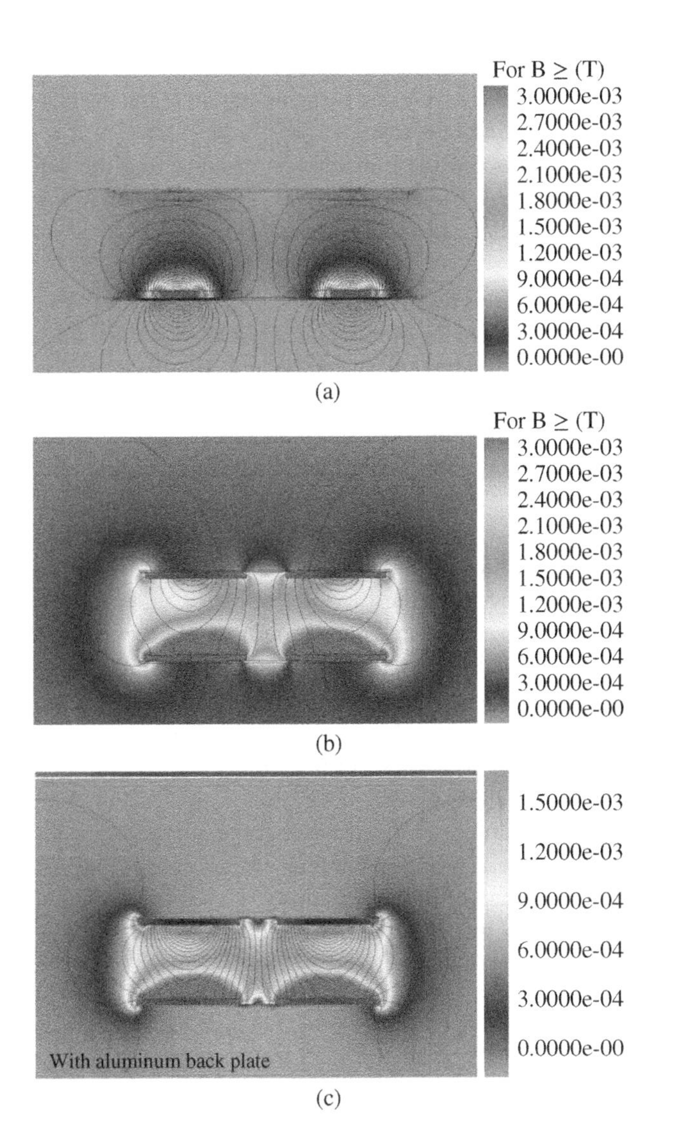

Figure 3.23 The schematic diagram of flux density: (a) traditional circular coil; (b) circular coil with ferrite; and (c) circular coil with ferrite and aluminum back plate (coil pad).

beyond the height limitation, which deteriorates the coupling coefficient between the transmitting and the pickup coils, accordingly. It means that the diameter of the coil should increase at least four times of the vertical height [12]. For example, a circular pad with a diameter of 700 mm can achieve a coupling coefficient of 0.17 over the 175-mm air gap. If the air gap increases to 200 mm, the diameter of the coil is required to be increased by almost 100 mm. In practical applications, it is impossible to extend the diameter excessively due to the limitation of the installation environment, the construction, and the cost.

Accordingly, how to balance between the transmission distance and the coil diameter or even breakthrough such a limitation and increase misalignment tolerance are key to the design of circular pads.

3.3.2.1.2 Flux Pipe In order to address the abovementioned issue, namely the air gap is limited by the diameter of coil and the misalignment tolerance is not high, a kind of bar topology was proposed by adopting a rectangular ferrite bar wounded with the coils in the midsection [12, 27, 29]. Although the height of the magnetic flux is increased, the flux density is not enhanced simultaneously. The traditional methods to solve this problem are simply increasing the coil coverage and adopting capacitance compensation [30]. However, these two methods inevitably increase the loss and cost significantly. This contradictory status leads to the development of the flux pipe which will be introduced as follows.

Instead of increasing the coil coverage, the flux pipe adopts a different coil arrangement method in which two subcoils are wound around the two ends of the ferrite bar rather than the midsection. Figure 3.24 depicts the schematics of the flux pipe structure, where the subcoils are magnetically connected in series to ensure the flux passing from one coil to the other as well as electrically connected in parallel to reduce the inductance. Figure 3.25 shows the 2-D magnetic flux distribution of the circular pad and the flux pipe.

Using the flux pipe structure, Figure 3.25 illustrates that the adopted subcoils can increase the length of the magnetic flux distribution in the ferrite cores, which can afford the air gap up to about half of the ferrite length. Accordingly, the flux pipe can effectively increase the power transmission range when adopting the same size as the circular pad.

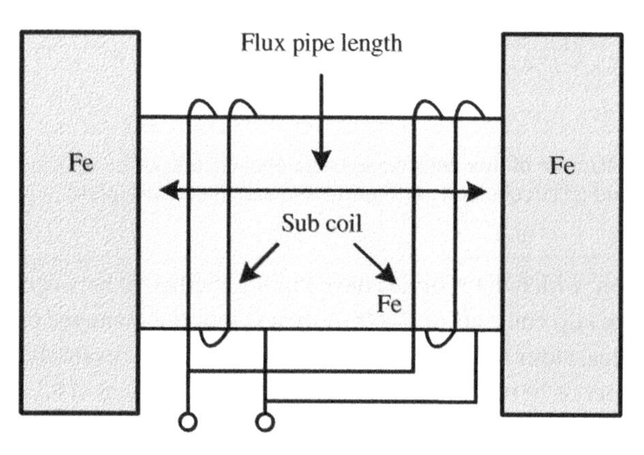

Figure 3.24 Schematics of flux pipe with two subcoils.

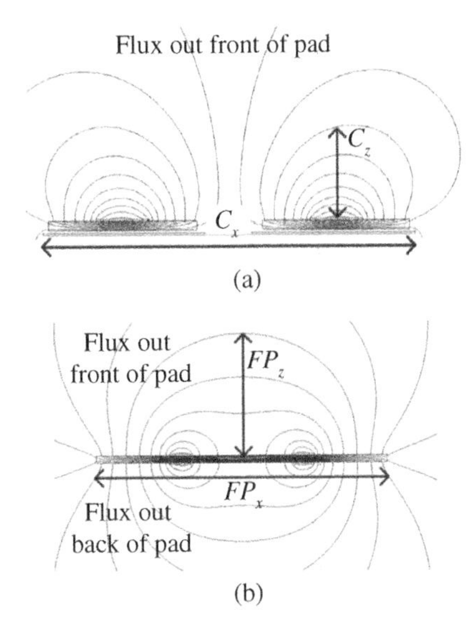

Figure 3.25 2-D flux path of the circular pad and the flux pipe: (a) circular and (b) flux pipe. Source: Budhia et al. [12].

Being different from the circular pad, the design of the flux pipe mainly concerns about the parameters of the subcoils, such as the number of turns, the total number, and the clearance. In Ref. [31], it implies that the coupling effect can be improved by simply increasing the number of subcoils on the same ferrite. Besides, the coupling strength and the misalignment tolerance are both determined by the clearance between the adjacent subcoils.

The flux pipe effectively increases the height of the induced magnetic field while avoiding increasing the diameter of the coil excessively as well as enhancing the misalignment tolerance compared with the circular pad. In Ref. [29], the comparative analysis between a 700-mm-diameter circular pad and a 550-mm-length flux pipe was carried out, with emphasis on the coupling performance. The results show that the coupling coefficient of the flux pipe is 0.2 at the 200-mm air gap, which is greater than that of the circular pad. In addition, the coupling coefficient of the flux pipe only drops to 0.17 under the 200-mm horizontal offset vertical to the direction of the ferrite bars. For the case with the 200-mm horizontal offset parallel to the direction of ferrite bars, the coupling coefficient of the flux pipe drops to 0.1, while the circular pad decreases to 0.05. Besides, the total coil length of the pipe is far less than the requirement of the circular pad, resulting in a lower loss and a lower cost for EV wireless charging systems.

As depicted in Figure 3.25, however, the flux pipe cannot produce the single-sided magnetic field as aforementioned. The amount of magnetic flux is identical between the front and the back of the coil, which means that if the aluminum plate is added to the backside of the flux pipe, the magnetic flux density will be weakened significantly. As a result, the quality factor of the flux pipe declines; for example, the quality factor decreases from 260 to 86 when adding the aluminum to the flux pipe [12]. Consequently, the shielding method adopted for circular pads cannot be directly utilized for the flux pipe as the aluminum inevitably deteriorates the power transmission performance of the EV wireless charging system.

Is it possible to design a new pad structure possessing both the single-sided shape and the increased height of the induced magnetic field? Section 3.3.2.1.3 will introduce a DD pad, responding to the above question.

3.3.2.1.3 Double-D Pad To combine the advantage of both the circular pad and the flux pipe, a new coil pad structure named DD pad was proposed as depicted in Figure 3.26, where two spiral coils lie on the ferrite bars and align to the vertical

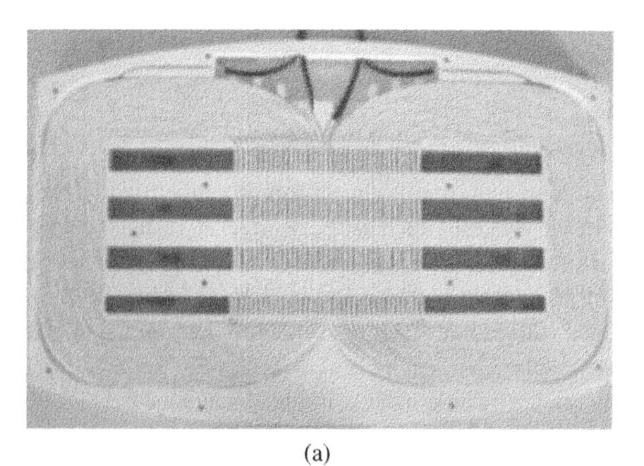

(a)

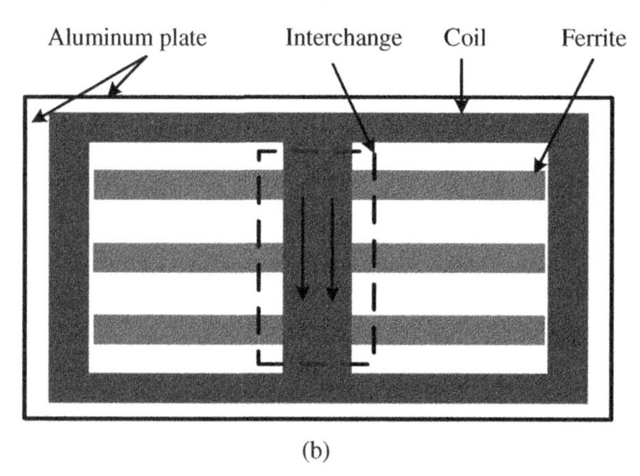

(b)

Figure 3.26 DD pad: (a) physical structure (Source: Covic and Boys [32], IEEE) and (b) schematic diagram.

direction of the ferrite bar. Each spiral coil is shaped like the letter "D," so this coil pad structure is called double-D pad or DD pad. How can DD pad produce a single-sided magnetic field as the same as the circular pad and increase the height of the induced magnetic field as the same as the flux pipe?

Figure 3.26b shows the schematic diagram of a DD pad, where the direction of the currents at the intersection of two coils is the same. It results in two coils connected serially in magnetic and parallelly in electric as same as the flux pipe. As a result, the part of the intersection can be considered as a flux pipe. Additionally, as shown in Figure 3.27, the magnetic field of the DD pad is single sided because both the coils lie on the ferrite like the circular pad, which allows the aluminum plate to be placed on the backside of the pad, which prevents the deterioration from the magnetic flux. Accordingly, as far as the design ideas are concerned, the DD pad successfully combines the advantages of the circular pad and the flux pipe.

The optimal design of the DD pads needs to adjust various parameters, such as the length or the width of the pad, the number of turns, and the ferrite. Being different from other types of the pad, the DD pad also needs to take into account the gap between the ferrite strips and the distance between the coil and the ferrite. The research team at The University of Auckland illustrates the effect of the above design parameters on the coupling coefficient of the DD pads [33–35]. For example, the coupling coefficient can be increased by enlarging the clearance of the ferrite strips as well as the width and the length of the pads. In addition, the pros and cons as well as the performance are also elaborated for DD pads using experimental verifications [12, 17]. For example, a 740-mm-length and 430-mm-width DD pad can achieve a coupling coefficient of almost 0.3 at the 200-mm air gap and higher misalignment tolerance than the circular pad.

Compared to the circular pads, the DD pad can effectively increase the coupling coefficient and enhance the misalignment tolerance for its increased height of the

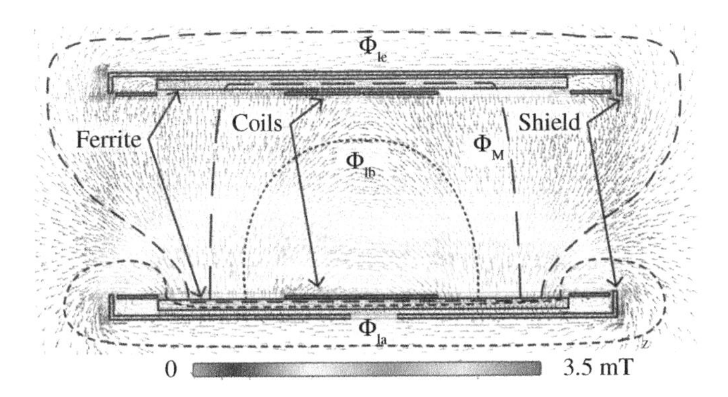

Figure 3.27 Magnetic flux. Source: Bhudia et al. [12].

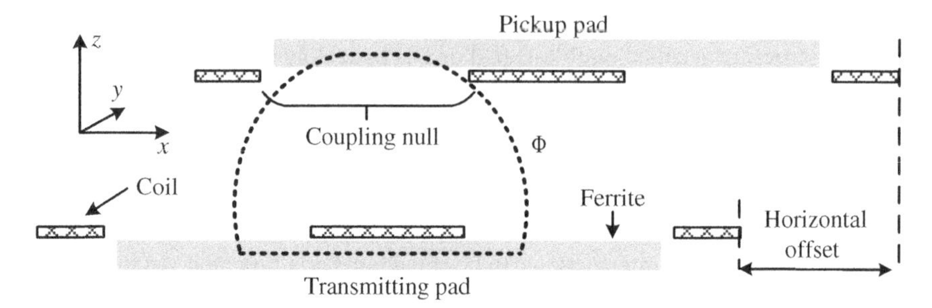

Figure 3.28 DD pad in the null position.

magnetic flux path. However, the DD pad cannot ensure the misalignment tolerance in the *x*-axis as high as that in the *y*-axis. It is caused by the occurrence of coupling null point when the horizontal offset of *x*-axis is greater than about 34% length of the pad, as shown in Figure 3.28. At this position, the flux goes in and out of the same coil, which results in no induced voltage. In other words, although the height of the magnetic flux path is increased, the width of the induced magnetic field in the *x*-axis direction is not increased. Accordingly, the DD pad is incapable of enhancing the misalignment tolerance in the *x*-axis. In order to address the problems of null magnetic position caused by the misalignment of the DD pads, some optimal structures are proposed based on the DD pad, named multicoil pad. Hence, Section 3.3.2.1.4 will introduce the multi-coil polarized pads.

3.3.2.1.4 Multicoil Pad Based on the DD pad, a number of derivative structures have been proposed for either the transmitter or the pickup to enhance the transmission performance of WPT systems, wherein the multicoil pad is one of the typical structures. Its design idea is to realize the magnetically independent structure by utilizing multiple decoupled coils. In such a way, the multicoil pad can acquire power from the magnetic flux parallel and vertical to the direction of the ferrite bars [36]. Especially in the secondary side, the multicoil pickup pad can acquire power under both the alignment and misalignment conditions. Thus, it is widely used for the static and dynamic wireless charging of EVs. The multicoil pad includes various types including the DDQ pad, the bipolar pad, and the tripolar pad [10, 36–38]. Figure 3.29 shows the schematic diagrams of the DDQ pad and the bipolar pad, which are two of the most commonly used types and elaborated in the following content.

DDQ Pad As shown in Figure 3.29a, the DDQ pad adopts a quadrature coil added to the backside of the DD coil. The quadrature arrangement of the additional coil and the DD coils means that the corresponding analysis can be carried out individually. In order to reveal the working mechanism of the quadrature coil, Figure 3.30

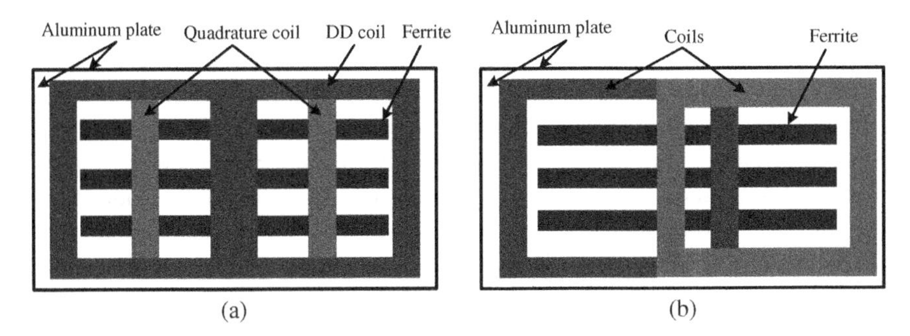

Figure 3.29 Simplified structure of DDQ and bipolar pad: (a) DDQ pad and (b) bipolar pad.

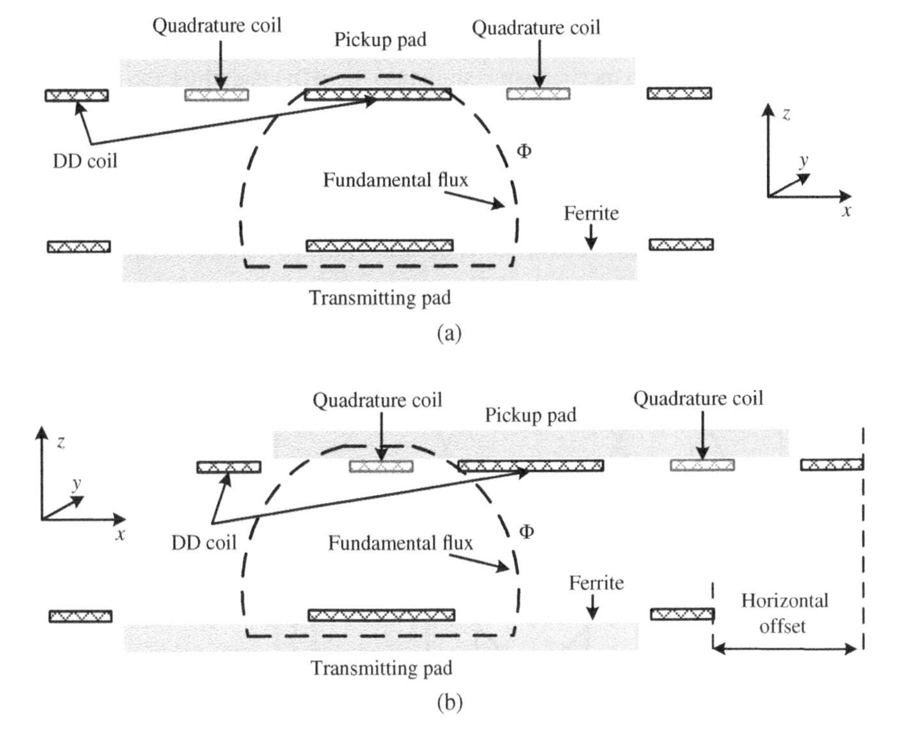

Figure 3.30 DDQ pickup pad: (a) horizontally aligned and (b) horizontally misaligned.

shows the DDQ pad at horizontally aligned and misaligned positions, respectively. It implies that the additional quadrature coil has little influence on the power acquired by the DD coil at the horizontally aligned position. When the pad exists in the position with a significant horizontal offset, the quadrature coil can pick up the flux. As far as the secondary side is concerned, the DDQ pad can significantly enhance the coupling coefficient, especially at horizontally misaligned positions. For the primary side, the power zone can be increased when the DDQ pad is used as the transmitting pad. Accordingly, the DDQ pad can effectively increase the misalignment tolerance for WPT systems, which shows significant meaning for EV wireless charging systems. Figure 3.31 illustrates that when the offset happens near the center, the main coil that receives power from the primary side is the DD coil. The Q coil centered on (230, 0) and (−230, 0) is used for the quadrature charge zones when the offset is beyond the boundary of the charging zone of the DD coil. Consequently, the whole charge zone of the DDQ pad can achieve a length of almost 800 mm and width of 460 mm, which is much greater than that of the circular pad.

In addition, a DD2Q pad is developed by adding two quadrature coils instead of one coil [39]. These two coils are placed in parallel under the double-D coil to further increase the coupling coefficient at misalignment positions, as shown in Figure 3.32. Its working mechanism is similar to the DDQ pad; thus, the previous paragraph can be referred for detailed information.

Bipolar Pad The other common type of multicoil pads is the bipolar pad, which is also called BP pad [40, 41]. As depicted in Figure 3.29b, the BP pad is composed of two independent coils of the same size. These two coils overlap partially

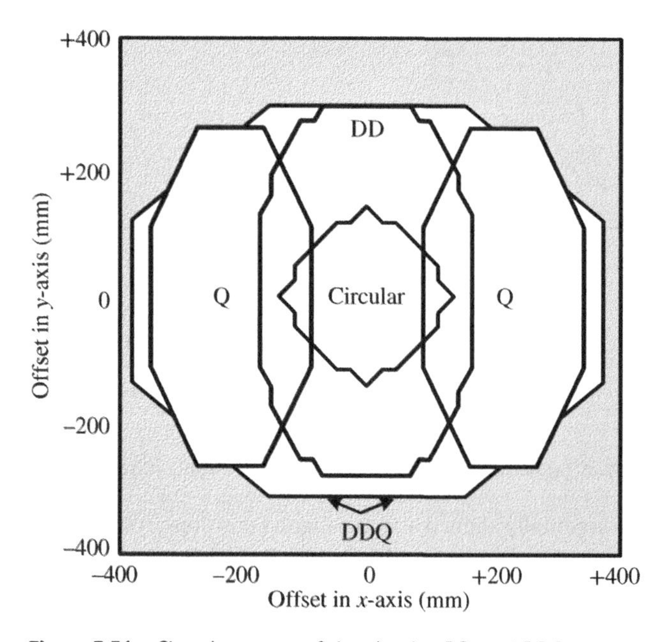

Figure 3.31 Charging zones of the circular, DD, and DDQ pads.

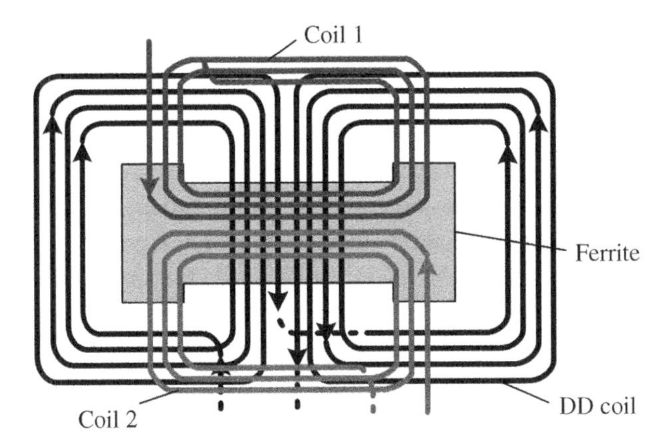

Figure 3.32 Simplified structure of DD2Q.

and lie on the ferrite bar. As a transmitting coil pad, the BP pad possesses a wide range of power transfer. When it is adopted as the pickup pad, it can effectively increase the coupling effect with the circular pad or the DD pad. Different from the DDQ pad that achieves sufficient tolerance to horizontal misalignment by an external coil called quadrature coil, the BP pad achieves this tolerance according to the design of the overlapping coils. As the inductive power decreases in one coil, the power in another coil is increased to provide adequate power to the load so that the influence of misalignment is effectively suppressed. What's more, the mutual coupling between the two overlapped coils is designed to ensure zero. By this way, two decoupled BP coils could be independently tuned and controlled. Although the performance of the BP pads is similar to the DDQ pad, the BP pad reduces the usage of copper by 30% compared with the DDQ pad, which means lower construction cost in the WPT systems.

The aforementioned multicoil polarized couplers are derived from the DD coil, and all of them have the advantage of capturing quadrature components of flux. Accordingly, a satisfied tolerance to misalignment is acquired for both the DDQ and the BP pad. In addition, the multicoil pad presents a tiny change in air gap spacing. Hence, the multicoil structure is normally suitable for primary and secondary pads, which has a great applicative prospect in EVs.

3.4 Summary

This chapter discusses the principle, influence of parameters and structures, the optimal design of coupled coils, and the charging pads for EV wireless charging systems. During the past years, several achievements have been made by various universities and research institutes all around the world, which significantly promote the development of WPT technologies. With the increasing number of wireless charging applications, the design and optimization of coupled coils are facing challenges and opportunities in future studies. Specifically speaking, the optimal design of coupled coils will extend from the simple studies on the coil parameter and structure to the comprehensive consideration on WPT technical level (such as the power transmission efficiency, the output power, the transmission distance, and the misalignment tolerance) and even the application level (such as the size, weight, and cost).

Apart from synthetically considering various factors mentioned above, the application of metamaterials gives a brand-new idea for the coil design and optimization [42]. Unlike the traditional Litz and PCB coils, the most obvious feature of metamaterials is the left-handed characteristic, which means that it can provide negative permeability and negative conductivity. This characteristic can improve the transmission efficiency and misalignment tolerance significantly. However, the metamaterials always operate at the MHz range, which makes it difficult to be adopted in high-power wireless charging systems. Thus, how to reduce the resonant frequency of the metamaterials and adapt it for the coil optimization of WPT systems is a very promising research direction in the future.

In addition, the innovation of the resonance method also provides new ideas for the optimal design of coils. The traditional *LC* resonant coil is widely used in the WPT systems, whereas additional lumped capacitance of the *LC* resonant coils increases the loss of WPT systems, thus resulting in lower efficiency. In response to this problem, the self-resonance coil has been obtaining more and more attention in recent years. The self-resonance coil works at the resonate state based on the parasitic capacitance instead of additional lumped capacitance. Thus, it has a higher quality factor and less power loss. However, the extremely low parasitic capacitance requires a high operating frequency, which is not suitable for high-power applications as similar to that of metamaterials, which should be particularly paid attention to in further studies.

Another troublesome issue is the coil design and optimization of EV dynamic wireless charging systems. Nowadays, the optimal design of coils mostly focuses on static wireless charging for EVs. However, with the increasing popularity of EVs, how to improve the coupling performance of coils for EV dynamic wireless charging systems is becoming more and more essential. In response to this problem, several research institutions such as ORNL and KAIST have launched corresponding studies and proposed novel solutions, wherein the power track is the most representative structure, which will be discussed in Chapter 9. However, there are still many challenges in the coil design for dynamic wireless charging such as the electromagnetic shielding performance, the continuous coils misalignment, and the cost of implementation. Consequently, although a great number of achievements have been made in previous studies, there are still a lot of work awaiting us to accomplish in future studies.

References

1 Sampath, J.P.K., Alphones, A., Vilathgamuwa, D.M. et al. (2014). Coil enhancements for high efficiency wireless power transfer applications. In: *IECON 2014 – 40th Annual Conference of the IEEE Industrial Electronics Society*, Dallas, TX, 2978–2983. IEEE.

2 Furusato, K., Imura, T., and Hori, Y. (2016). Design of multi-frequency coil for capacitor-less wireless power transfer using high order self-resonance of open end coil. In: *2016 IEEE Wireless Power Transfer Conference (WPTC)*, Aveiro, 1–4. IEEE.

3 Sample, P., Meyer, D.T., and Smith, J.R. (2011). Analysis, experimental results, and range adaptation of magnetically coupled resonators for wireless power transfer. *IEEE Transactions on Industrial Electronics* 58 (2): 544–554.

4 Zhang, Z., Pang, H., Georgiadis, A., and Cecati, C. (2019). Wireless power transfer—an overview. *IEEE Transactions on Industrial Electronics* 66 (2): 1044–1058.

5 Chen, K. and Zhao, Z. (2013). Analysis of the double-layer printed spiral coil for wireless power transfer. *IEEE Journal of Emerging and Selected Topics in Power Electronics* 1 (2): 114–121.

6 Craninckx, J. and Steyaert, M.S.J. (1997). A 1.8 GHz low-phase-noise CMOS VCO using optimized hollow spiral inductors. *IEEE Journal of Solid-State Circuits* 32 (5): 736–744.

7 Su, Y., Liu, X., Lee, C.K., and Hui, S.Y.R. (2012). On the relationship of quality factor and hollow winding structure of coreless printed spiral winding (CPSW) inductor. *IEEE Transactions on Power Electronics* 27 (6): 3050–3056.

8 Hsu, H.M. (2004). Analytical formula for inductance of metal of various widths in spiral inductors. *IEEE Transactions on Electron Devices* 51 (8): 1343–1346.

9 López-Villegas, J.M., Samitier, J., Cané, C. et al. (2000). Improvement of the quality factor of RF integrated inductors by layout optimization. *IEEE Transactions on Microwave Theory and Techniques* 48 (1): 76–83.

10 Moon, S., Kim, B.-C., Cho, S.-Y. et al. (2014). Analysis and design of a wireless power transfer system with an intermediate coil for high efficiency. *IEEE Transactions on Power Electronics* 61 (11): 5861–5870.

11 Jonah, O. and Georgakopoulos, S.V. (2013). Optimized helix with ferrite core for wireless power transfer via resonance magnetic. In: *IEEE Antennas and Propagation Society International Symposium*, 1040–1041. IEEE.

12 Budhia, M., Boys, J.T., Covic, G.A., and Huang, C. (2013). Development of a single-sided flux magnetic coupler for electric vehicle IPT charging systems. *IEEE Transactions on Industrial Electronics* 60 (1): 318–328.

13 Jinliang, L., Qijun, D., Wenshan, H., and Hong, Z. (2017). Research on quality factor of the coils in wireless power transfer system based on magnetic coupling resonance. In: *2017 IEEE PELS Workshop on Emerging Technologies: Wireless Power Transfer (WoW)*, Chongqing, 123–127. IEEE.

14 Li, S. and Mi, C.C. (2015). Wireless power transfer for electric vehicle applications. *IEEE Journal of Emerging and Selected Topics in Power Electronics* 3 (1): 4–17.

15 Hui, S.Y.R., Zhong, W., and Lee, C.K. (2014). A critical review of recent progress in mid-range wireless power transfer. *IEEE Transactions on Power Electronics* 29 (9): 4500–4511.

16 Hagiwara, N. (2013). Study on the principle of contactless electric power transfer via electromagnetic coupling. *Electrical Engineering in Japan* 182: 53–60.

17 Zhang, W., White, J.C., Abraham, A.M., and Mi, C.C. (2015). Loosely coupled transformer structure and interoperability study for EV wireless charging systems. *IEEE Transactions on Power Electronics* 30 (11): 6356–6367.

18 Wang, Y., Liu, W., and Xie, Y. (2019). Design and optimization for circular planar spiral coils in wireless power transfer system. In: *2019 22nd International Conference on Electrical Machines and Systems (ICEMS)*, Harbin, China, 1–4. IEEE.

19 Peng, W. and Chen, Z. (2018). Enhanced planar wireless power transfer systems with ferrite material. In: *2018 IEEE Wireless Power Transfer Conference (WPTC)*, Montreal, QC, Canada, 1–4. IEEE.

20 Kavitha, M., Bobba, P.B., and Prasad, D. (2016). Effect of coil geometry and shielding on wireless power transfer system. In: *2016 IEEE 7th Power India International Conference (PIICON)*, Bikaner, 1–6. IEEE.

21 Wang, S. (2003). Modeling and design of planar integrated magnetic components. Master thesis. Virginia Polytechnic Institute and State University.

22 Cove, S.R., Ordonez, M., Shafiei, N., and Zhu, J. (2016). Improving wireless power transfer efficiency using hollow windings with track-width-ratio. *IEEE Transactions on Power Electronics* 31 (9): 6524–6533.

23 Jolani, F., Yu, Y., and Chen, Z. (2014). A planar magnetically coupled resonant wireless power transfer system using printed spiral coils. *IEEE Antennas and Wireless Propagation Letters* 13: 1648–1651.

24 Miller, J.M. and Daga, A. (2015). Elements of wireless power transfer essential to high power charging of heavy duty vehicles. *IEEE Transactions on Transportation Electrification* 1 (1): 26–39.

25 Bosshard, R., Kolar, J.W., Mühlethaler, J. et al. (2015). Modeling and η-α-Pareto optimization of inductive power transfer coils for electric vehicles. *IEEE Journal of Emerging and Selected Topics in Power Electronics* 3 (1): 50–64.

26 Bosshard, R. and Kolar, J.W. (2016). Inductive power transfer for electric vehicle charging: technical challenges and tradeoffs. *IEEE Power Electronics Magazine* 3 (3): 22–30.

27 Budhia, M., Covic, G.A., and Boys, J.T. (2011). Design and optimization of circular magnetic structures for lumped inductive power transfer systems. *IEEE Transactions on Power Electronics* 26 (11): 3096–3108.

28 Patil, D., McDonough, M.K., Miller, J.M. et al. (2018). Wireless power transfer for vehicular applications: overview and challenges. *IEEE Transactions on Transportation Electrification* 4 (1): 3–37.

29 Budhia, M., Covic, G., and Boys, J. (2010). A new IPT magnetic coupler for electric vehicle charging systems. In: *IECON 2010 – 36th Annual Conference on IEEE Industrial Electronics Society*, Glendale, AZ, 2487–2492. IEEE.

30 Budhia, M., Covic, G.A., Boys, J.T., and Huang, C. (2011). Development and evaluation of single sided flux couplers for contactless electric vehicle charging. In: *2011 IEEE Energy Conversion Congress and Exposition*, Phoenix, AZ, 614–621. IEEE.

31 Sis, S.A. and Orta, E. (2019). Flux pipe pads with improved coupling characteristics. In: *2019 16th International Multi-conference on Systems, Signals & Devices (SSD)*, Istanbul, Turkey, 250–253. IEEE.

32 Covic, G.A. and Boys, J.T. (2013). Modern trends in inductive power transfer for transportation applications. *IEEE Journal of Emerging and Selected Topics in Power Electronics* 1 (1): 28–41.

33 Pearce, M.G.S., Covic, G.A., and Boys, J.T. (2016). Leakage and coupling of square and double D magnetic couplers. In: *Proceedings of IEEE 2nd Annual Southern Power Electronics Conference*, 1–6. IEEE.

34 Pearce, M.G.S., Covic, G.A., and Boys, J.T. (2019). Robust ferrite-less double D topology for roadway IPT applications. *IEEE Transactions on Power Electronics* 34 (7): 6062–6075.

35 Lin, F.Y., Covic, G.A., and Boys, J.T. (2015). Evaluation of magnetic pad sizes and topologies for electric vehicle charging. *IEEE Transactions on Power Electronics* 30 (11): 6391–6407.

36 Zaheer, A., Hao, H., Covic, G.A., and Kacprzak, D. (2015). Investigation of multiple decoupled coil primary pad topologies in lumped IPT systems for interoperable electric vehicle charging. *IEEE Transactions on Power Electronics* 30 (4): 1937–1955.

37 Zaheer, G., Covic, A., and Kacprzak, D. (2014). A bipolar pad in a 10-kHz 300-W distributed IPT system for AGV applications. *IEEE Transactions on Industrial Electronics* 61 (7): 3288–3301.

38 Kim, S., Covic, G.A., and Boys, J.T. (2017). Tripolar pad for inductive power transfer systems for EV charging. *IEEE Transactions on Power Electronics* 32 (7): 5045–5057.

39 Ke, G., Chen, Q., Gao, W. et al. (2016). Power converter with novel transformer structure for wireless power transfer using a DD2Q power receiver coil set. In: *2016 IEEE Energy Conversion Congress and Exposition (ECCE)*, Milwaukee, WI, 1–6. IEEE.

40 Zaheer, D.K. and Covic, G.A. (2012). A bipolar receiver pad in a lumped IPT system for electric vehicle charging applications. In: *IEEE Energy Conversion Congress and Exposition (ECCE)*, Raleigh, NC, 283–290. IEEE.

41 Covic, G.A., Kissin, M.L.G., Kacprzak, D. et al. (2011). A bipolar primary pad topology for EV stationary charging and highway power by inductive coupling. In: *2011 IEEE Energy Conversion Congress and Exposition*, Phoenix, AZ, 1832–1838. IEEE.

42 Zhang, Z., Zhang, B., Deng, B. et al. (2018). Opportunities and challenges of metamaterial-based wireless power transfer for electric vehicles. *Wireless Power Transfer* 5 (1): 9–19.

4

Design and Optimization for Power Circuits

In this chapter, the design and optimization of static compensation networks for wireless power transfer (WPT) systems based on the concept of impedance matching are comprehensively described in Section 4.1 at first. Then, in order to ensure the system performances, the tunable adjustment schemes for dynamic WPT systems are introduced. In addition, the design and control schemes of power inverters, which are also an important part for the design of power circuits in WPT systems, are analyzed in Section 4.2.

4.1 Impedance Matching

Impedance matching is defined as the process of making one impedance look like another using circuit-based design methods, aiming to achieve optimal system performance [1]. The impedance matching technology is commonly used in transmission lines and widely applied in various fields such as electrical engineering systems, acoustic systems, optical systems, and mechanical systems, which all involve the energy transmission from source to load. Generally, the way to implement a static impedance matching is designing a passive network between the source and the load, which is commonly called *compensation network* by some authors to refer to the impedance matching. Practically, various components and circuits can be used to realize the impedance matching.

For WPT systems, two of the most prominent sources causing the deterioration of system performances are frequency splitting and impedance mismatching. Frequency splitting is caused by the fluctuation in the coupling effect between the transmitting and the pickup coils, which will be elaborated in detail in Chapter 7. Impedance mismatching can be quantified as differences in the input impedances of two circuits, which may result in reflected signals from the receiver back into the transmitter in some high-frequency transmission applications [1]. In WPT systems, the requirement for impedance matching can be appreciated using the

Wireless Power Transfer: Principles and Applications, First Edition. Zhen Zhang and Hongliang Pang.

peer to peer system. In this section, the theoretical analysis is carried out with emphasis on the design of compensation networks from two aspects of the static impedance matching and the adaptive impedance matching.

4.1.1 Compensation Networks

4.1.1.1 Basic Topologies: SS/SP/PS/PP

The reason why the compensation capacitance is required for WPT systems is to match with the primary and the secondary inductances to work at the resonant state, which aims to ensure the resulted reactive power for the inductances to generate an adequate magnetic field [2]. The compensation can not only effectively reduce the volt–ampere (VA) rating of the power supply in the primary side [3, 4] but also maximize the transmission power capability by canceling the secondary inductance in the secondary side [5]. Apart from the inductance coils, the power source, and the load, a compensation network is necessary to ensure the transmission performance of WPT systems. Commonly, the combination of compensation capacitors and inductors is used to form compensation networks. There are four fundamental compensation topologies that are widely used for WPT systems by means of their advantages of simplicity and practicality. Based on the characteristics of the primary output, the four fundamental compensation topologies can be classified into two types, that is one is the primary voltage-source type including series–series (SS) and series–parallel (SP), while another is the primary current-source type including parallel–series (PS) and parallel–parallel (PP) [6]. The former letter represents the type of compensation in the primary side, and the latter denotes the type of compensation in the secondary side.

When the WPT system operates at the resonant frequency, the self-inductance of the pickup coil can be matched with the compensation capacitance in the secondary side. Then, the resonance capacitance can be applied for all four compensation topologies in the secondary side, which is given as

$$\begin{cases} j\omega L_s + \dfrac{1}{j\omega C_s} = 0 \\ C_s = \dfrac{1}{\omega^2 L_s} \end{cases} \tag{4.1}$$

4.1.1.1.1 *Voltage-Source Type: SS/SP Topologies* Based on Kirchhoff's law, as shown in Figure 4.1a, the circuit equations of the SS compensation topology can be obtained as

$$\begin{cases} \left[R_p + j\left(\omega L_p - \dfrac{1}{\omega C_p}\right)\right] I_p + j\omega M I_s = U_{in} \\ j\omega M I_p + \left[R_s + R_L + j\left(\omega L_s - \dfrac{1}{\omega C_s}\right)\right] I_s = 0 \end{cases} \tag{4.2}$$

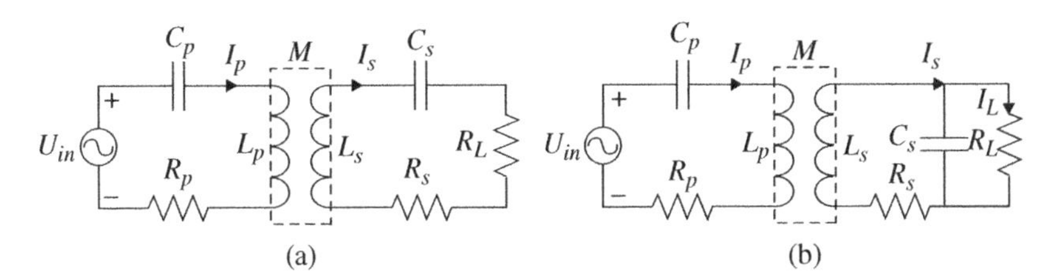

Figure 4.1 Compensation topologies: (a) SS; (b) SP.

Denote Z_p and Z_s as the primary and secondary equivalent impedances, respectively, which are given by

$$\begin{cases} Z_p = R_p + j\left(\omega L_p - \dfrac{1}{\omega C_p}\right) \\ Z_s = R_s + R_L + j\left(\omega L_s - \dfrac{1}{\omega C_s}\right) \end{cases} \tag{4.3}$$

By substituting (4.3) into (4.2), the primary current I_p and secondary current I_s can be calculated as

$$\begin{cases} I_p = \dfrac{U_{in}}{Z_p + \dfrac{\omega^2 M^2}{Z_s}} \\ I_s = -\dfrac{j\omega M U_{in}}{Z_p Z_s + \omega^2 M^2} \end{cases} \tag{4.4}$$

Define $Z_{r\text{-}ss}$ as the reflected impedance from the secondary to the primary side in the SS topology, which is given by

$$Z_{r-ss} = \frac{\omega^2 M^2}{Z_s} \tag{4.5}$$

Based on the equivalent M-model circuit of the SS topology as depicted in Figure 4.2a, $Z_{p\text{-}ss}$ is defined as the total equivalent impedance as

$$Z_{p-ss} = Z_p + Z_{r-ss} \tag{4.6}$$

In order to compensate the inductance, the imaginary part of the impedance $Im(Z_{p\text{-}ss})$ should be specified first, which can be calculated as

$$Im(Z_{p-ss}) = \frac{\omega^2 M^2 \left(\omega L_s - \dfrac{1}{\omega C_s}\right)}{\left(\omega L_s - \dfrac{1}{\omega C_s}\right)^2 + R_s + R_L} + \omega L_p - \frac{1}{\omega C_p} = \omega L_p - \frac{1}{\omega C_p} \tag{4.7}$$

Then, the imaginary part of the total equivalent impedance $Im(Z_{p\text{-}ss})$ should be eliminated completely, so that the transmission performance of WPT systems can

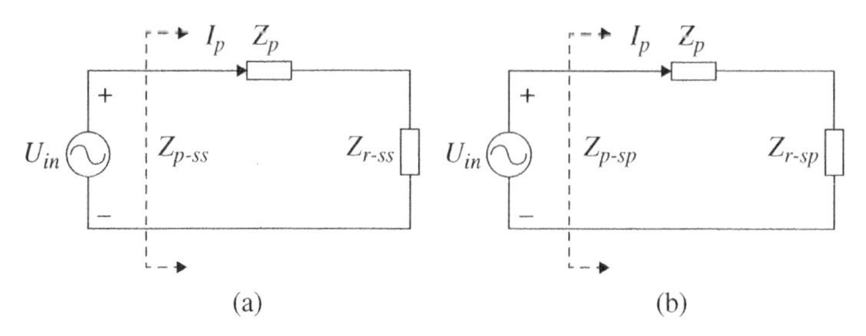

Figure 4.2 Equivalent M-model circuit: (a) SS; (b) SP.

be ensured. For facilitating the following theoretical analysis, the line resistance in the secondary side is ignored. Accordingly, the primary capacitance C_p can be deduced as

$$\begin{cases} C_p = \dfrac{1}{\omega^2 L_p} = \dfrac{C_s L_s}{L_p} \\ C_s = \dfrac{1}{\omega^2 L_s} \end{cases} \tag{4.8}$$

Considering the neglectable impact of the secondary line resistance, the resistance R_s is ignored in this chapter. Based on (4.2), the input power P_{in} and the output power P_{out} of SS topology can be expressed as

$$\begin{cases} P_{in-ss} = \dfrac{R_L U_{in}{}^2}{R_p R_L + \omega^2 M^2} \\ P_{out-ss} = \dfrac{\omega^2 M^2 R_L U_{in}{}^2}{(R_p R_L + \omega^2 M^2)^2} \end{cases} \tag{4.9}$$

Therefore, the transmission efficiency $\eta_{\_ss}$ can be obtained as

$$\eta_{\_ss} = \frac{P_{out-ss}}{P_{in-ss}} = \frac{\omega^2 M^2}{R_p R_L + \omega^2 M^2} \tag{4.10}$$

Similarly, the circuit equations of the SP compensation topology shown in Figure 4.1b can be obtained as

$$\begin{cases} \left[R_p + j\left(\omega L_p - \frac{1}{\omega C_p}\right)\right] I_p + j\omega M I_s = U_{in} \\ j\omega M I_p + R_L I_L + (R_s + j\omega L_s) I_s = 0 \\ R_L I_s + \frac{(I_L - I_s)}{j\omega C_s} = 0 \end{cases} \tag{4.11}$$

Then, Z_p and Z_s can be expressed as

$$\begin{cases} Z_p = R_p + j\left(\omega L_p - \dfrac{1}{\omega C_p}\right) \\ Z_s = R_s + j\omega L_s + \dfrac{R_L}{1 + j\omega C_s} \end{cases} \tag{4.12}$$

Additionally, the primary current I_p, the secondary current I_s, and the load current I_L can be deduced from (4.11) and (4.12) as

$$\begin{cases} I_p = \dfrac{U_{in}}{Z_p + \dfrac{\omega^2 M^2}{Z_s}} \\ I_s = -\dfrac{j\omega M U_{in}}{\dfrac{Z_s}{Z_p} + \left(\dfrac{\omega M}{Z_p}\right)^2} \\ I_L = \dfrac{I_s}{j\omega C_s R_L + 1} \end{cases} \tag{4.13}$$

As shown in Figure 4.2b, $Z_{r\text{-}sp}$ and $Z_{p\text{-}sp}$ are defined as the reflected impedance and the total equivalent impedance of the SP topology, respectively, which can be expressed as

$$Z_{r-sp} = \frac{\omega^2 M^2}{Z_s} \tag{4.14}$$

$$Z_{p-sp} = Z_p + Z_{r-sp} \tag{4.15}$$

Likewise, the imaginary part of the impedance $Im(Z_{p\text{-}sp})$ can be deduced as

$$Im(Z_{p-sp}) = -\frac{\omega^3 M^2 \left(C_s R_L{}^2(\omega^2 L_s C_s - 1) + L_s\right)}{(R_s + R_L - \omega^2 L_s C_s R_L)^2 + \omega^2 (L_s + C_s R_s R_L)^2} + \omega L_p - \frac{1}{\omega C_p} \tag{4.16}$$

In the analysis of the SP topology, the line resistances are also ignored. Let $Im(Z_{p\text{-}sp}) = 0$, the primary capacitance C_p can be obtained as

$$\begin{cases} C_p = \dfrac{L_s{}^2 C_s}{L_p L_s - M^2} \\ C_s = \dfrac{1}{\omega^2 L_s} \end{cases} \tag{4.17}$$

By ignoring the resistance R_s in (4.11), the input power P_{in} and the output power P_{out} of the SP topology can be deduced as

$$\begin{cases} P_{in-sp} = \dfrac{L_s{}^2 U_{in}{}^2}{L_s{}^2 R_p + M^2 R_L} \\ P_{out-sp} = \dfrac{L_s{}^2 M^2 R_L U_{in}{}^2}{\left(M^2 R_L + L_s{}^2 R_p\right)^2} \end{cases} \tag{4.18}$$

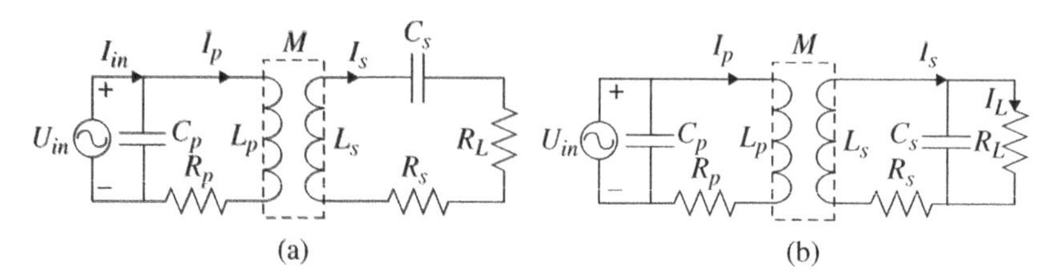

Figure 4.3 Compensation networks: (a) PS; (b) PP.

Therefore, the corresponding transmission efficiency $\eta_{\_sp}$ can be obtained as

$$\eta_{\_sp} = \frac{P_{out-sp}}{P_{in-sp}} = \frac{R_L M^2}{R_L M^2 + L_s{}^2 R_p} \tag{4.19}$$

4.1.1.1.2 *Current-Source Type: PS/PP Topologies* As shown in Figure 4.3a, the circuit equations of the PS compensation topology can be obtained as

$$\begin{cases} (R_p + j\omega L_p)I_p + j\omega M I_s = U_{in} \\ j\omega M I_p + \left[R_s + R_L + j\left(\omega L_s - \frac{1}{\omega C_s}\right)\right] I_s = 0 \\ (I_{in} - I_p) = j\omega C_p U_{in} \end{cases} \tag{4.20}$$

Denote Z_p and Z_s as the primary and the secondary equivalent impedances, respectively, which are given by

$$\begin{cases} Z_p = R_p + j\omega L_p \\ Z_s = R_s + R_L + j\left(\omega L_s - \frac{1}{\omega C_s}\right) \end{cases} \tag{4.21}$$

By substituting (4.21) into (4.20), the primary current I_p, the secondary current I_s, and the power supply current I_{in} can be expressed as

$$\begin{cases} I_p = \dfrac{U_{in}}{Z_p + \dfrac{\omega^2 M^2}{Z_s}} \\ I_s = -\dfrac{j\omega M U_{in}}{\dfrac{Z_s}{Z_p} + \left(\dfrac{\omega M}{Z_p}\right)^2} \\ I_{in} = \left(\dfrac{Z_s}{Z_p Z_s + \omega^2 M^2} + j\omega C_p\right) U_{in} \end{cases} \tag{4.22}$$

Denote $Z_{r\text{-}ps}$ as the reflected impedance of the PS topology from the secondary to the primary side, which is given by

$$Z_{r-ps} = \frac{\omega^2 M^2}{Z_s} \tag{4.23}$$

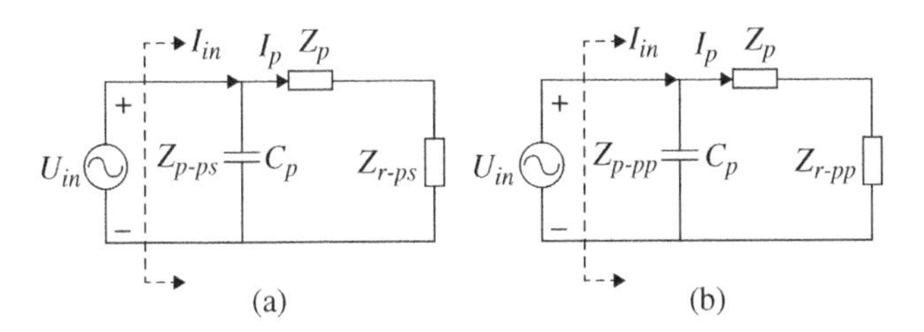

Figure 4.4 Equivalent M-model circuits: (a) PS; (b) PP.

Figure 4.4a depicts the equivalent M-model circuit of the PS topology, where $Z_{p\text{-}ps}$ is the total equivalent impedance,

$$Z_{p-ps} = Z_p + Z_{r-ps} \tag{4.24}$$

In order to compensate the inductance, the imaginary part of the impedance $Im(Z_{p\text{-}ps})$ should be first calculated as

$$Im(Z_{p-ps}) = \frac{\omega\left[L_p{R_L}^2(1-\omega^2 L_p C_p) - C_p(R_p R_L + \omega^2 M^2)^2\right]}{{R_L}^2(1-\omega^2 L_p C_p)^2 + \omega^2 {C_p}^2 (R_p R_L + \omega^2 M^2)^2} \tag{4.25}$$

For achieving excellent transfer performances, such as high efficiency and transmission power, the imaginary part of the total equivalent impedance should be eliminated.

By ignoring the line resistance and eliminating the imaginary part of the impedance, the primary capacitance C_p and the secondary capacitance C_s can be obtained as

$$\begin{cases} C_p = \dfrac{L_s C_s}{\dfrac{M^4}{L_p L_s C_s {R_L}^2} + L_p} \\ C_s = \dfrac{1}{\omega^2 L_s} \end{cases} \tag{4.26}$$

Then, the input power P_{in} and the output power P_{out} of the PS topology can be deduced as

$$\begin{cases} P_{in-ps} = \dfrac{R_L(\omega^2 M^2 + R_L R_p){U_{in}}^2}{(\omega^2 M^2 + R_L R_p)^2 + \omega^2 {L_p}^2 {R_L}^2} \\ P_{out-ps} = \dfrac{\omega^2 M^2 R_L {U_{in}}^2}{(R_p R_L + \omega^2 M^2)^2 + \omega^2 {L_p}^2 {R_L}^2} \end{cases} \tag{4.27}$$

Therefore, the corresponding transmission efficiency $\eta_{\_ps}$ can be obtained as

$$\eta_{\_ps} = \frac{P_{out-ps}}{P_{in-ps}} = \frac{\omega^2 M^2}{\omega^2 M^2 + R_L R_p} \tag{4.28}$$

Similarly, the circuit equations of the PP compensation topology as shown in Figure 4.3b can be obtained as

$$\begin{cases} (R_p + j\omega L_p)I_p + j\omega M I_s = U_{in} \\ (I_{in} - I_p)\dfrac{1}{j\omega C_p} = U_{in} \\ j\omega M I_p + R_L I_L + (R_s + j\omega L_s)I_s = 0 \\ I_L R_L = -\dfrac{(I_L - I_s)}{j\omega C_s} \end{cases} \tag{4.29}$$

Then, Z_p and Z_s can be expressed as

$$\begin{cases} Z_p = R_p + j\omega L_p \\ Z_s = R_s + j\omega L_s + \dfrac{R_L}{1 + j\omega C_s R_L} \end{cases} \tag{4.30}$$

Additionally, the primary current I_p, the secondary current I_s, and the load current I_L can be deduced from (4.29) and (4.30) as

$$\begin{cases} I_p = \dfrac{U_{in}}{Z_p + \dfrac{\omega^2 M^2}{Z_s}} \\ I_s = -\dfrac{j\omega M U_{in}}{\dfrac{Z_s}{Z_p} + \left(\dfrac{\omega M}{Z_p}\right)^2} \\ I_L = \dfrac{I_s}{j\omega C_s R_L + 1} \end{cases} \tag{4.31}$$

As depicted in Figure.4.4b, $Z_{r\text{-}pp}$ and $Z_{p\text{-}pp}$ are the reflected impedance and the total equivalent impedance of the PP topology, respectively,

$$Z_{r-pp} = \frac{\omega^2 M^2}{Z_s} \tag{4.32}$$

$$Z_{p-pp} = Z_p + Z_{r-pp} \tag{4.33}$$

Likewise, the imaginary part of the impedance $Im(Z_{p\text{-}pp})$ can be deduced as

$$Im(Z_{p-ps}) = \frac{\begin{matrix} \omega L_s^2(L_pL_s - M^2)(\omega^2 M^2 C_p - \omega^2 L_p L_s C_p + L_s) \\ -\omega C_p\left(M^2 R_L + R_p L_s^2\right)^2 \end{matrix}}{\left[L_s^2 - \omega^2 C_p L_s (L_p L_s - M^2)\right]^2 + \omega^2 C_p^2\left(L_s^2 R_p + M^2 R_L\right)^2} \tag{4.34}$$

By ignoring the line resistances, the primary capacitance C_p and the secondary capacitance C_s can be obtained based on $Im(Z_{p\text{-}pp}) = 0$ as

$$\begin{cases} C_p = \dfrac{L_s^3 C_s (L_p L_s - M^2)}{M^4 C_s R^2 + L_s (L_p L_s - M^2)^2} \\ C_s = \dfrac{1}{\omega^2 L_s} \end{cases} \tag{4.35}$$

The input power P_{in} and the output power P_{out} of the PP topology can be deduced as

$$\begin{cases} P_{in-pp} = \dfrac{\left(R_L M^2 + L_s^2 R_p\right) U_{in}^2 L_s^2}{\omega^2 L_s^2 (L_p L_s - M^2)^2 + \left(L_s^2 R_p + M^2 R_L^2\right)} \\ P_{out-pp} = \dfrac{L_s^2 M^2 R_L U_{in}^2}{\left(R_p L_s^2 + R_L M^2\right)^2 + \omega^2 L_s^2 (L_p L_s - M^2)^2} \end{cases} \tag{4.36}$$

Therefore, the transmission efficiency $\eta_{\_pp}$ can be obtained as

$$\eta_{\_pp} = \frac{P_{out-pp}}{P_{in-pp}} = \frac{M^2 R_L}{M^2 R_L + L_s^2 R_p} \tag{4.37}$$

Table 4.1 shows the comparison results of the four basic topologies. It can be obviously seen that the reflected resistance and system efficiency show the same expression in SS, PS topologies and SP, PP topologies, respectively. Additionally, the input, output, and other characteristics of the four basic topologies are listed in Table 4.2. The three main differences of the four basic topologies are given:

(1) The independence on coupling coefficient only exists in the SS topology for the four basic topologies.
(2) The SS and SP topologies can realize both the load-independent constant voltage and the constant-current output, while the PS and PP topologies can only

Table 4.1 Four fundamental compensation networks.

Topology	Primary capacitance	Secondary capacitance	Reflected resistance	System efficiency
SS	$\dfrac{C_s L_s}{L_p}$	$\dfrac{1}{\omega^2 L_s}$	$\dfrac{\omega^2 M^2}{R_L}$	$\dfrac{\omega^2 M^2}{R_p R_L + \omega^2 M^2}$
SP	$\dfrac{L_s^2 C_s}{L_p L_s - M^2}$	$\dfrac{1}{\omega^2 L_s}$	$\dfrac{M^2 R_L}{L_s^2}$	$\dfrac{R_L M^2}{R_L M^2 + L_s^2 R_p}$
PS	$\dfrac{L_s C_s}{\dfrac{M^4}{L_p L_s C_s R_L^2} + L_p}$	$\dfrac{1}{\omega^2 L_s}$	$\dfrac{\omega^2 M^2}{R_L}$	$\dfrac{\omega^2 M^2}{R_p R_L + \omega^2 M^2}$
PP	$\dfrac{L_s^3 C_s (L_p L_s - M^2)}{M^4 C_s R^2 + L_s (L_p L_s - M^2)^2}$	$\dfrac{1}{\omega^2 L_s}$	$\dfrac{M^2 R_L}{L_s^2}$	$\dfrac{R_L M^2}{R_L M^2 + L_s^2 R_p}$

Table 4.2 Comparison of compensation characteristics.

Topology	SS	SP	PS	PP
Power source type	Voltage	Voltage	Current	Current
Load-independent output type	Voltage and current	Voltage and current	Only voltage	Only current
Independence on coupling coefficient	Yes	No	No	No
Efficiency under maximum output power	50%	50%	—	—
Forbidden case	Secondary open circuit	Secondary short circuit	Secondary short circuit	Secondary open circuit
Total copper mass at 200 kW [7]	Less copper mass	4.6% more than SS	30% more than SS	24% more than SS

realize the load-independent constant voltage and the constant-current output, respectively.

(3) On the condition of maximum output power, the transmission efficiency of the SS and SP topologies is only 50%, while the PS and PP topologies can reach over 90%.

4.1.1.2 Hybrid Topologies

The four fundamental compensation topologies show salient advantages of simplicity, flexibility, and accuracy. As the saying goes, every coin has two sides, the single-compensator-based topologies are no exception. The main drawback is the limited number of adjustable resonant parameters, which significantly limits the application of these fundamental compensation topologies. To address this issue, the hybrid topologies are proposed to increase the controlling degrees of freedom and satisfy the critical design criteria. Especially for the dynamic charging of electric vehicles (EVs) [8, 9] and the driving system of light-emitting diodes [10], various hybrid compensation topologies have been proposed and implemented to realize advanced functionalities of the constant output and the dynamic charging [11, 12].

The hybrid topologies will be discussed and classified according to two aspects of input-source types and output characteristics in this section.

4.1.1.2.1 L-Networks Figure 4.5 depicts the basic configurations of L-networks, which are classified into four types based on how capacitors and inductors are connected in the circuit.

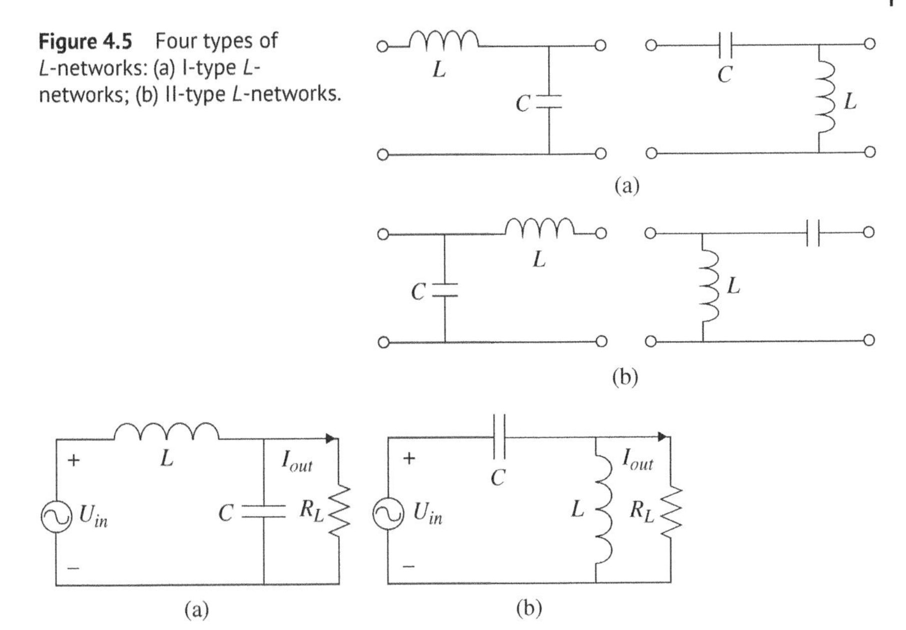

Figure 4.5 Four types of *L*-networks: (a) I-type *L*-networks; (b) Π-type *L*-networks.

Figure 4.6 I-type *L*-networks using input voltage source: (a) *LC* tank; (b) *CL* tank.

When adopting the voltage source as the power supply, the I-type *L*-networks as shown in Figure 4.6 are utilized to obtain a constant-current output. The corresponding output current I_{out} and the input voltage U_{in} shown in Figure 4.6a can be represented as

$$I_{out} = U_L\left(j\omega C + \frac{1}{j\omega L}\right) - U_{in}j\omega C \tag{4.38}$$

When the L and C resonate at the operating frequency ω, the I_{out} can be rewritten as

$$\begin{cases} \omega = \sqrt{\dfrac{1}{LC}} \\ I_{out} = -U_{in}j\omega C \end{cases} \tag{4.39}$$

Similarly, the corresponding output current as shown in Figure 4.6b can be expressed as

$$\begin{cases} \omega = \sqrt{\dfrac{1}{LC}} \\ I_{out} = -\dfrac{1}{j\omega C}U_{in} \end{cases} \tag{4.40}$$

From (4.39) and (4.40), it can be found that the output current is determined by the input voltage, the coupling coefficient, the compensation capacitance, and

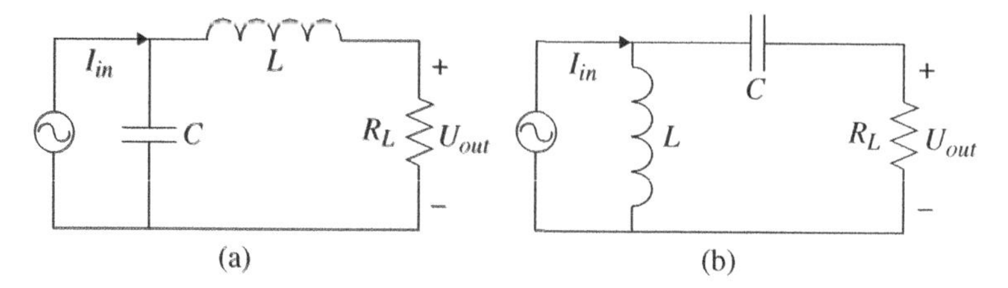

Figure 4.7 Π-type L-networks using input current source: a) CL tank; (b) LC tank.

the compensation inductance. It means that the I-type L-network can realize the load-independent constant-current output when adopting the voltage source excitation.

When adopting the current source as the power supply, the Π-type L-networks as shown in Figure 4.7 are utilized to realize the constant voltage output. The output voltage U_{out} and the input current I_{in} of Π-type L-network shown in Figure 4.7a can be represented as

$$U_{out} = \frac{1}{j\omega C}(I_{in} - I_{out}) - j\omega L I_{out} = \frac{1}{j\omega C} I_{in} - \left(\frac{1}{j\omega C} + j\omega L\right) I_{out} \tag{4.41}$$

When the L and C resonate at the operating frequency ω, the U_{out} can be rewritten as

$$\begin{cases} \omega = \sqrt{\dfrac{1}{LC}} \\ U_{out} = \dfrac{1}{j\omega C} I_{in} \end{cases} \tag{4.42}$$

Similarly, the corresponding output voltage as shown in Figure 4.6b can be expressed as

$$\begin{cases} \omega = \sqrt{\dfrac{1}{LC}} \\ U_{out} = j\omega L I_{in} \end{cases} \tag{4.43}$$

From (4.42) and (4.43), it is illustrated that the output voltage is independent of the load R_L, which is determined by the input voltage, the coupling coefficient, the compensation capacitance, and the compensation inductance. Thus, the Π-type L-network can realize the load-independent constant voltage output when adopting the current source as the power supply.

4.1.1.2.2 T- and Π-networks

Apart from L-networks, the T- and Π-networks are commonly used for WPT systems. When the voltage source is used as the power supply, the T-network as shown in Figure 4.8a is adopted to obtain a constant

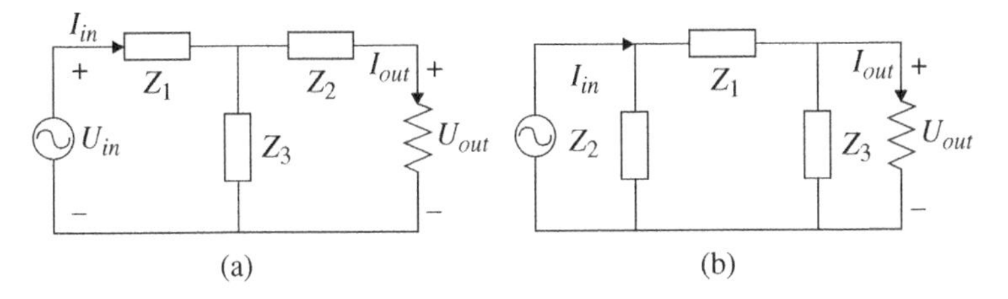

Figure 4.8 *T* and *Π* networks: (a) *T*-network; (b) *Π*-network.

voltage output. The voltage equation can be obtained as

$$\begin{cases} U_{out} = I_{out}R_L \\ U_{in} = I_{in}Z_1 + (I_{in} - I_{out})Z_3 \\ U_{in} = I_{in}Z_1 + I_{out}Z_2 + U_{out} \end{cases} \tag{4.44}$$

Then, the relationship between the output voltage U_{out} and the input voltage U_{in} can be deduced as

$$U_{in} = U_{out}\left(1 + \frac{Z_1}{Z_3} + \frac{Z_1Z_2 + Z_2Z_3 + Z_1Z_3}{Z_3R_L}\right) \tag{4.45}$$

From (4.45), the output voltage is independent of the load R_L only if the condition $Z_1Z_2 + Z_2Z_3 + Z_1Z_3 = 0$ makes sense. As shown in Figure 4.9, this chapter summarizes the typical topologies of the *T*-network which are used to realize the constant voltage output with respect to the voltage power source. When using the *T-LCL* topology shown in Figure 4.9a, the condition can be expressed as

$$L_1 + L_2 - \omega^2 L_1 L_2 C = 0 \tag{4.46}$$

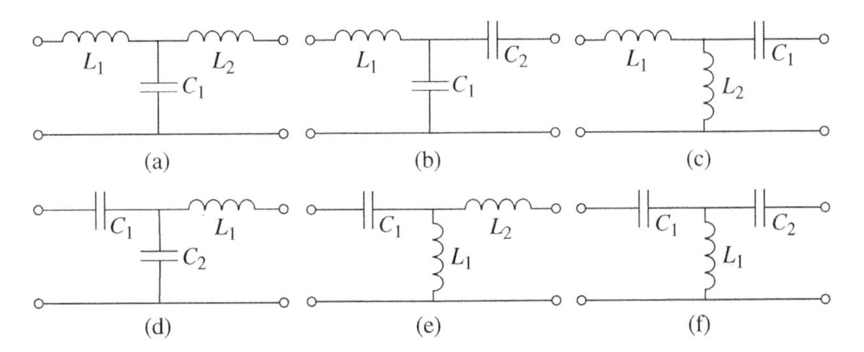

Figure 4.9 *T*-type topologies for constant voltage output using voltage power source: (a) *T-LCL* topology; (b) *T-LCC* topology; (c) *T-LLC* topology; (d) *T-CCL* topology; (e) *T-CLL* topology; (f) *T-CLC* topology.

Table 4.3 T and Π networks.

Topology T-V-V	Coupling coefficient	Topology Π-C-C	System efficiency
T-LCL	$\omega = \sqrt{\frac{1}{L_1C} + \frac{1}{L_2C}}$	Π-LCL	$\omega = \frac{1}{\sqrt{CL_1 + CL_2}}$
T-LCC	$\omega = \frac{1}{\sqrt{LC_1 + LC_2}}$	Π-LCC	$\omega = \sqrt{\frac{1}{LC_1} + \frac{1}{LC_2}}$
T-LLC	$\omega = \sqrt{\frac{1}{L_1C} + \frac{1}{L_2C}}$	Π-LLC	$\omega = \frac{1}{\sqrt{CL_1 + CL_2}}$
T-CCL	$\omega = \frac{1}{\sqrt{LC_1 + LC_2}}$	Π-CCL	$\omega = \sqrt{\frac{1}{LC_1} + \frac{1}{LC_2}}$
T-CLL	$\omega = \sqrt{\frac{1}{L_1C} + \frac{1}{L_2C}}$	Π-CLL	$\omega = \frac{1}{\sqrt{CL_1 + CL_2}}$
T-CLC	$\omega = \frac{1}{\sqrt{LC_1 + LC_2}}$	Π-CLC	$\omega = \sqrt{\frac{1}{LC_1} + \frac{1}{LC_2}}$

Accordingly, the required operating frequency that can ensure the constant output voltage can be deduced as

$$\begin{cases} \omega = \sqrt{\frac{1}{L_1C} + \frac{1}{L_2C}} \\ U_{out} = \dfrac{U_{in}L_1}{L_1 + L_2} \end{cases} \tag{4.47}$$

In addition, the related constraints of other typical T topologies as depicted in Figure 4.9b–f are listed in Table 4.3.

When the current source is used as the power supply, the Π-network as shown in Figure 4.8b is adopted for the constant-current output. The current equation can be obtained as

$$\begin{cases} U_{out} = I_{out}R_L \\ I_{in} = \dfrac{U_{in}}{Z_2} + \dfrac{U_{in} - U_{out}}{Z_1} \\ I_{in} = \dfrac{U_{in}}{Z_2} + I_{out} + \dfrac{U_{out}}{Z_3} \end{cases} \tag{4.48}$$

Then, the relationship between the output voltage I_{out} and the input voltage I_{in} can be deduced as

$$I_{in} = I_{out}\frac{Z_1 + Z_2}{Z_2} + I_{out}\frac{R_L(Z_1 + Z_2 + Z_3)}{Z_2Z_3} \tag{4.49}$$

From (4.49), it is shown that the condition to ensure the output current independent of the load R_L is $Z_1 + Z_2 + Z_3 = 0$. The corresponding deduction of the operating frequency is similar to that of T topologies as aforementioned. In order

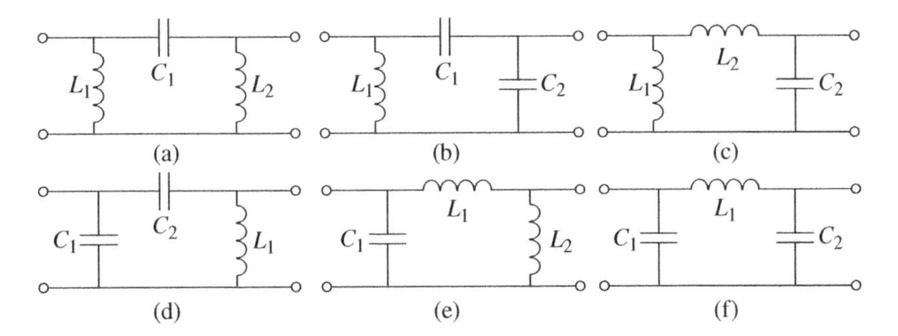

Figure 4.10 *Π*-type topologies for constant-current output using current power source: (a) *Π-LCL* topology; (b) *Π-LCC* topology; (c) *Π-LLC* topology; (d) *Π-CCL* topology; (e) *Π-CLL* topology; (f) *Π-CLC* topology.

to facilitate readers' references, all related constraints of *Π* topologies as shown in Figure 4.10a–f are listed in Table 4.3.

4.1.1.2.3 LCC-S/P/LCC Topologies *L*-, *T*-, and *Π*-networks can be used to simplify the analysis of hybrid topologies. Considering that the voltage source is commonly used as the power supply for WPT systems, the primary- and double-sided *LCC* topologies are elaborated with emphasis on the circuit, the resonant frequency, and the summary in this section.

The *LCC* topology in the primary side with the secondary series or parallel compensations is commonly used to realize load-independent constant voltage or constant-current output. Figure 4.11 depicts the equivalent *T* model of the *LCC-S* compensation network. For facilitating the circuit analysis, it is divided into three parts. According to the analysis in Part 1, Module 1 belongs to the I-type *L*-network which can ensure constant-current output under the voltage-source supply. Module 2 is a series of capacitors and inductors, where the primary capacitance C_p can be in resonance with the leakage inductance $L_p - L_m$ at the operating frequency. Hence, the primary output performs the constant-current characteristics. Similarly, on the basis of the analysis in Part 1, Module 3 belongs to II-type *L*-network which can realize the constant voltage output under the current power supply. In this module, the secondary capacitance C_s and the secondary inductance L_s are designed to be resonant at the same operating

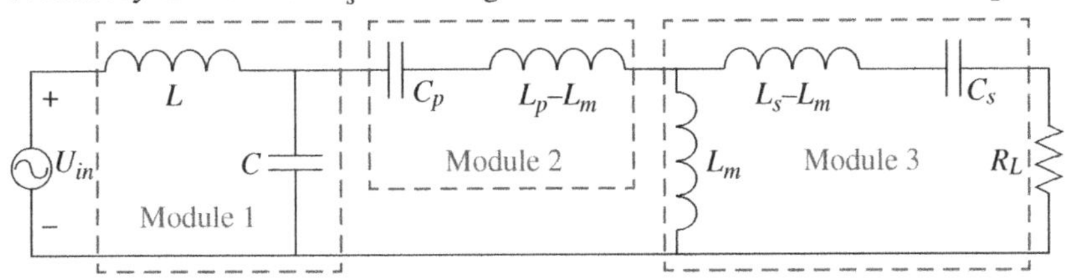

Figure 4.11 Equivalent T model of *LCC-S* compensation network.

frequency with the primary side. Accordingly, in order to realize the constant voltage output, the operating frequency of the *LLC-S* topology can be expressed as

$$\omega = \frac{1}{\sqrt{LC}} = \frac{1}{\sqrt{(L_p - L_m)C_p}} = \frac{1}{\sqrt{L_s C_s}} \tag{4.50}$$

Figure 4.12 depicts the equivalent *T* model of the *LCC-P* compensation network, which has the same Module 1 and Module 2 as the *LCC-S* network. Then, the primary output can be still regarded as a constant-current source. Module 3 belongs to the *Π-LLC* topology, which means that it can realize constant-current output under the current-source power supply. Accordingly, in order to realize the constant-current output, the operating frequency of the *LLC-P* topology can be obtained as

$$\omega = \frac{1}{\sqrt{LC}} = \frac{1}{\sqrt{(L_p - L_m)C_p}} = \frac{1}{\sqrt{L_s C_s}} \tag{4.51}$$

Figure 4.13 depicts the equivalent *T* model of the double-sided *LCC* compensation network. In the primary side, Module 1 and Module 2 are both the same as that of the *LCC-S* network. Then, the primary output can be regarded as a constant-current source. Module 3 belongs to the *Π*-type topology, which means that it can realize the constant-current output under the current-source power supply. The inductance L_l of Module 4 has no impact on the constant-current output. Accordingly, in order to realize the constant-current output, the operating frequency of the double-sided *LCC* topology can be obtained as

$$\omega = \frac{1}{\sqrt{LC}} = \frac{1}{\sqrt{(L_p - L_m)C_p}} = \sqrt{\frac{1}{L_s C_{s1}} + \frac{1}{L_s C_{s2}}} \tag{4.52}$$

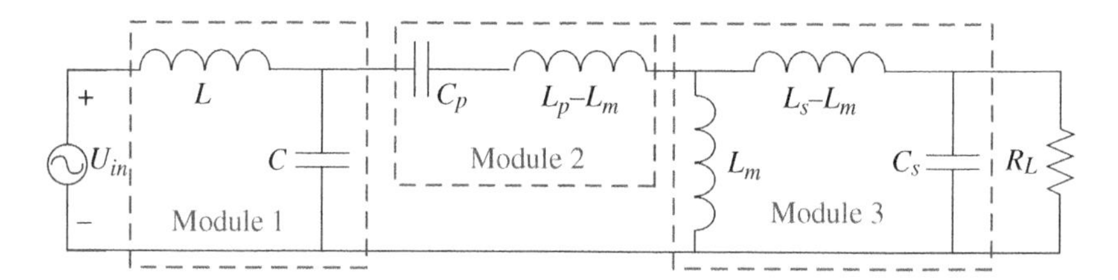

Figure 4.12 Equivalent T model of *LCC-P* compensation network.

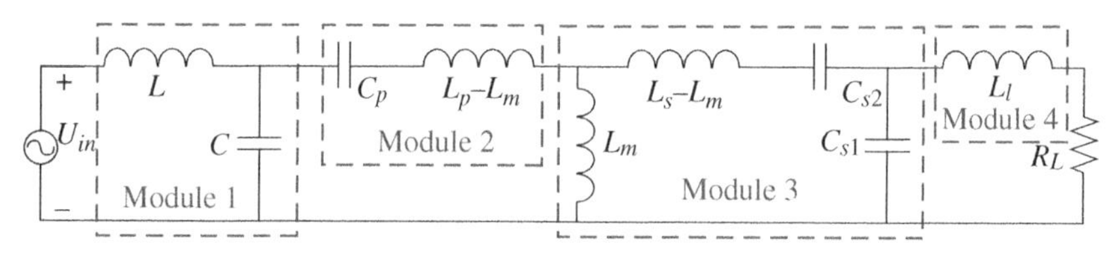

Figure 4.13 Equivalent T model of double-sided *LCC* compensation network.

4.1.2 Tunable Impedance Matching Networks

Static impedance matching networks have gained fruitful achievements through researchers' tireless efforts over the years. However, for practical WPT applications, the transmission performances such as the power and transmission efficiency will be seriously affected by the disturbance of the external environment and the drift of the internal parameters. For example, in 1-to-n WPT systems [13], the impedance matching network needs to be adjusted along with the shifting of the system operating frequency caused by the random access of loads, uncertain quantity of load, and the cross-coupling effects. Accordingly, a more flexible and feasible impedance matching strategy comes into our mind. Aiming at the solution of adaptive impedance matching under the shifting of system operating frequency, this section elaborates on three aspects, which are discrete wide-range adjustment scheme, continuous adjustment scheme, and hybrid adjustment scheme.

4.1.2.1 Discontinuous Adjustment-Capacitor Array

4.1.2.1.1 Capacitor Array In recent years, a great number of efforts have been made to solve the problem of impedance mismatch of WPT systems. In Ref. [14], an automated impedance matching which can effectively enhance the transmission efficiency was proposed. In 2014, Heebl et al. designed a varactor-based impedance adjustment system which can achieve a simultaneous conjugate impedance matching to cope with variations of the mutual inductance [15]. Considering that the fundamental of compensation networks in the WPT system is the compensated function of the compensation capacitor, the most direct way is adding adjustable capacitor which can be realized by capacitor array. In Ref. [16], an adaptive impedance matching network based on a novel capacitor matrix was proposed. Additionally, Zhang et al. proposed a topology-reconfigurable capacitive compensation network which can realize the energy encryption for multiple-pickup WPT systems in Ref. [17]. The proposed capacitor matrix, which will be introduced in this section, can significantly extend the capacitive compensating range to cover a wide-range varying frequency while using the compensation capacitors as little as possible.

Figure 4.14 depicts an equivalent circuit model of the WPT system, which mainly includes the power supply, the matching networks, the coils (L_p, L_s), and the load. It can be observed that the matching networks of the double side are modeled as a combination of capacitors arranged in L-type, which are named as C_{p1}, C_{p2} and C_{p1}, C_{p2}, respectively.

The structure of the proposed capacitor matrix is shown in Figure 4.15, which consists of $M \times N$ (M rows and N columns) capacitors connected to each other in series or parallel. Different modes can be switched by the selection of the switch array. For $M \times N$ capacitances with the conterminous switches, there are $2^{(M \times N)}$

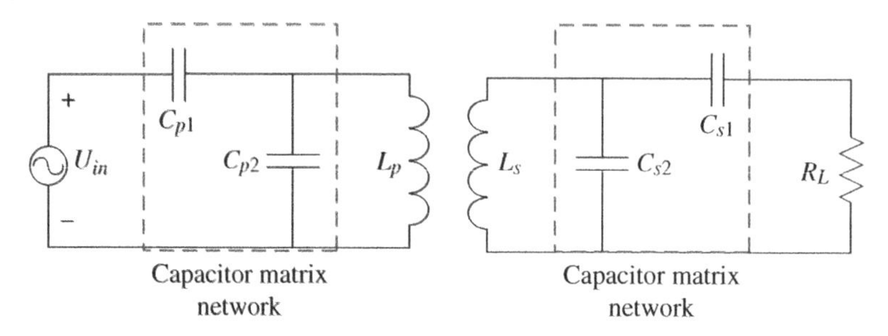

Figure 4.14 Equivalent WPT system with proposed capacitor matrix.

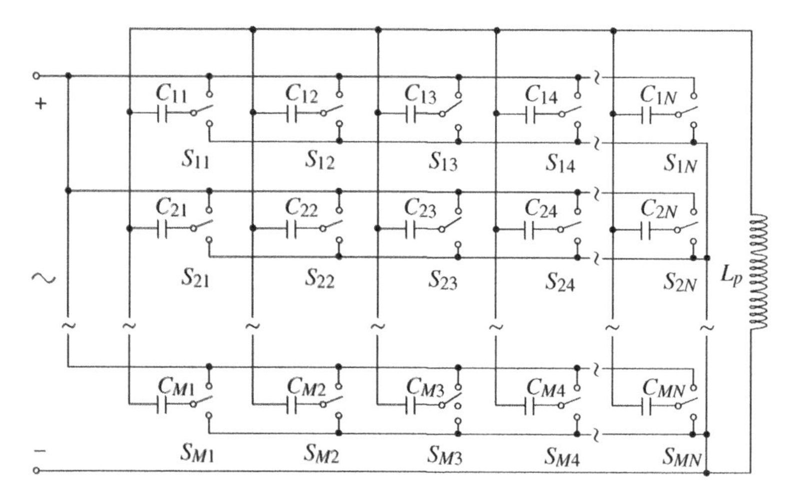

Figure 4.15 $M \times N$ capacitor matrix.

different combinations with $2^{(M \times N)}$ different values of capacitance available. As an example, Figure 4.16a depicts a 2×3 capacitor matrix with the series capacitors C_1, C_4 and the parallel capacitors C_2, C_3, C_5, and C_6. The equivalent circuit is presented in Figure 4.16b. It can be seen that the capacitor matrix can be equivalent to the series–parallel topology where the value of C_{p1} is equal to $C_1 + C_4$, and the value of C_{p2} is equal to $C_2 + C_3 + C_5 + C_6$. Additionally, there are 64 combinations available based on the 2×3 capacitor matrix shown in Figure 4.16a.

4.1.2.1.2 Topology-Reconfigurable Capacitor Matrix A topology-reconfigurable capacitive compensation network is proposed based on the capacitor matrix as mentioned above. As shown in Figure 4.17, an additional connection network, which is highlighted in red, is added on the original capacitor matrix to enrich the topology. Figure 4.18a depicts the exemplified 2×4 reconfigurable capacitive topology. In comparison with the original capacitor

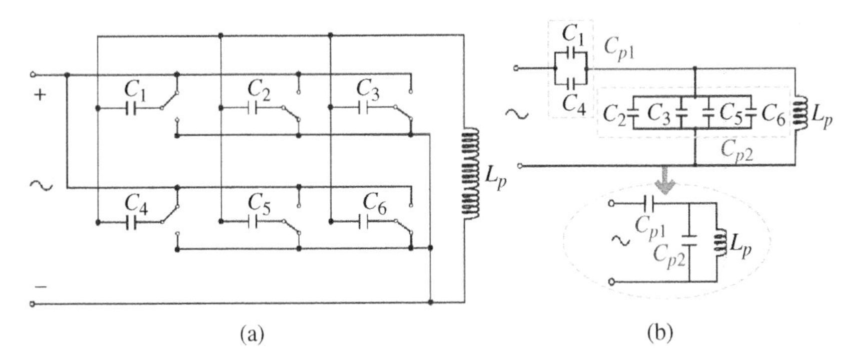

Figure 4.16 Exemplified 2×3 capacitor matrix: (a) combination scheme; (b) equivalent circuit.

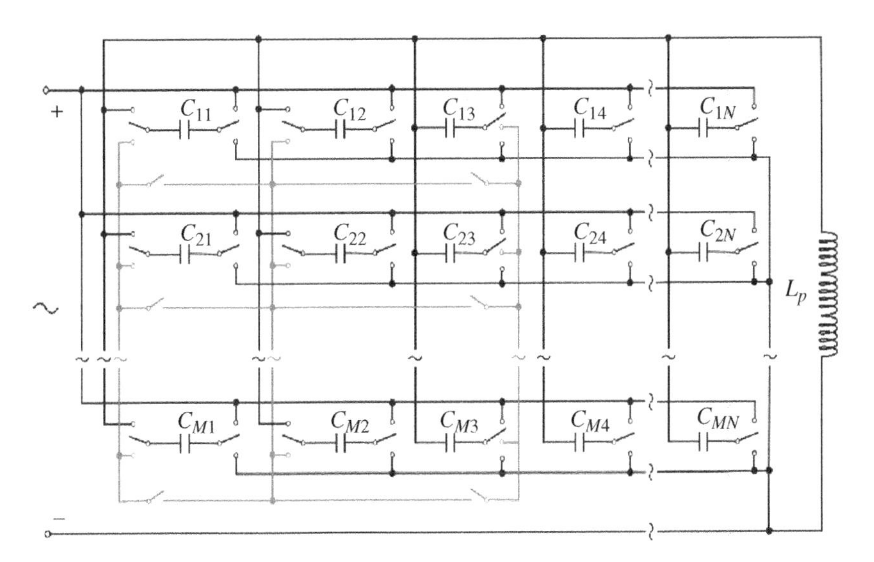

Figure 4.17 Topology-reconfigurable capacitor matrix.

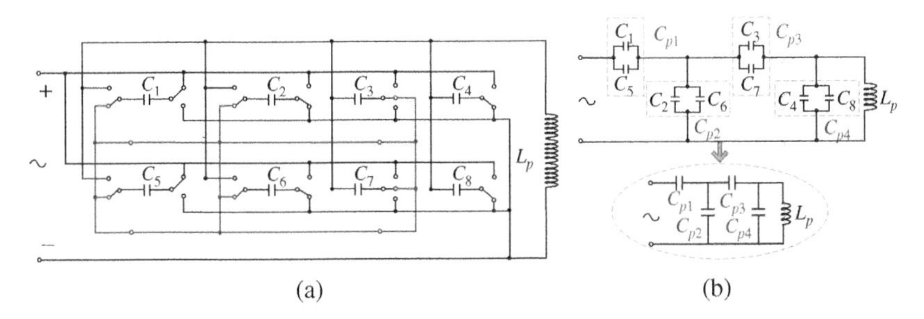

Figure 4.18 Exemplified 2×4 reconfigurable topology capacitor matrix: (a) combination scheme; (b) equivalent circuit.

matrix, the topology-reconfigurable capacitor matrix can be equivalent to the series–parallel–series–parallel topology as depicted in Figure 4.18b. In other words, the topology-reconfigurable capacitor matrix can realize more combinations, which means that it can cover a wider capacitance range. Denote C_{ps} and $C_{ps}{}^{2}$ as the total capacitance value of the series–parallel topology and the series–parallel–series–parallel topology, respectively. Then, the calculation formulas are given by

$$C_{sp} = C_{MN}^{1} \sum_{i=1}^{MN-1} C_{MN-1}^{i} + C_{MN}^{2} \sum_{i=1}^{MN-1} C_{MN-2}^{i} + \cdots + C_{MN}^{j} \sum_{i=1}^{MN-j} C_{MN-j}^{i} + C_{MN}^{MN-1} C_{1}^{1} \tag{4.53}$$

$$C_{sp^2} = \left(\sum_{i=1}^{M} C_{M}^{i} \right)^{3} \cdot \sum_{i=1}^{MN-3M} C_{MN-3M}^{i} \tag{4.54}$$

In summary, the topology-reconfigurable capacitor matrix can effectively extend the capacitive compensating range; especially, when the series–parallel topology cannot maintain the resonant working state, the series–parallel–series–parallel topology will be used. However, the proposed method based on the capacitor matrix is only able to realize discrete adjustment within the covered range.

4.1.2.2 Continuous Adjustment-Virtual Impedance

4.1.2.2.1 Question As previously mentioned, the existing impedance adjustment method based on the capacitor array is only able to realize discrete adjustment. In addition, due to the limited capacitance values of the available capacitors, it requires a great number of physical capacitors to cover a wide-range fluctuation of the system which not only results in the degradation of the system performances but also brings increased cost. Accordingly, a continuously adjustable impedance matching method, which can be independent of physical capacitors for multiple-pickup WPT system, is urgently in need. Now, the question comes: is it possible to imitate the characteristics of the capacitors by circuit design? A concept of "virtual capacitor" was firstly proposed in Ref. [18], which realizes a capacitive output impedance at power frequencies (50 Hz) by adding an integrator block. However, due to the limitation of the high operating frequency in the WPT system, the use of proposed virtual capacitors as a potential solution may pose new challenges of the ultrahigh-frequency switching and high-speed processing. Then, here comes another question: How to realize "virtual capacitor" to deal with the high operating frequency (20–300 kHz) in practical WPT system? In order to imitate the phase characteristics of the real capacitor, a self-balancing virtual impedance network based on current compensation was proposed in Ref. [19]. According to the reasonable circuit design of the power-electronic-based strategy, the proposed scheme can maintain the resonate state for the expected operating frequency. The detailed introduction follows.

4.1.2.2.2 Solution Figure 4.19 depicts the equivalent circuit of the multiple-pickup WPT system with the proposed self-balancing network. The self-balancing network, which consists of the inductor L_v, the capacitor C_v, and the line resistance R_v, is connected to the primary-side compensation capacitor L_p. By means of injecting current into the primary-side compensation capacitor to eliminate the reactive impedance, the primary-side compensation circuit can be readjusted to the resonant state. For the convenience of theoretical derivation, the equivalent circuit of the proposed self-balancing network is shown in Figure 4.20, where U_p and I_p are the primary exciting voltage and current, and U_v and I_v are the output voltage and current of the full-bridge inverter of the self-balancing network. Denote Z_p, Z_v, and Z_{Cp} as the equivalent impedances of the primary circuit, self-balancing network, and the compensation capacitor C_p, respectively, which can be presented as

$$\begin{cases} Z_p = j\omega L_p + \dfrac{1}{j\omega C_p} + R_p \\ Z_v = j\omega L_v + \dfrac{1}{j\omega C_v} + R_v \\ Z_{Cp} = \dfrac{1}{j\omega C_p} \end{cases} \tag{4.55}$$

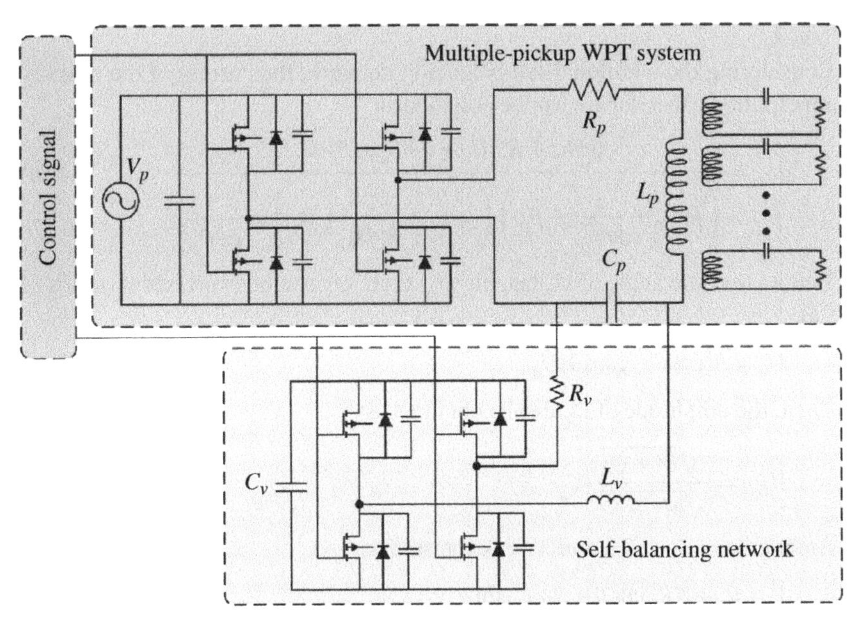

Figure 4.19 Equivalent circuit of multiple-pickup WPT system with self-balancing network.

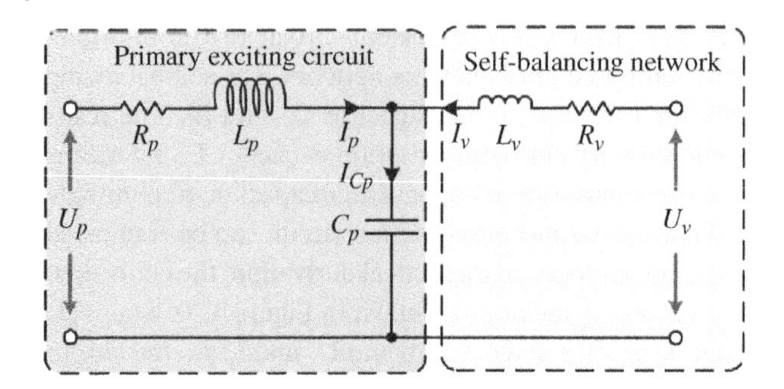

Figure 4.20 Equivalent circuit of the proposed self-balancing network.

According to Kirchhoff laws, the circuit equations can be obtained as

$$\begin{cases} U_p = Z_p I_p + Z_{Cp} I_p \\ U_v = Z_{Cp} I_v + Z_v I_v \end{cases} \tag{4.56}$$

Then, the currents of I_p and I_v can be deduced as

$$\begin{cases} I_p = \dfrac{1}{Z_p Z_v - {Z_{Cp}}^2}(Z_v U_p - Z_{Cp} U_v) \\ I_v = \dfrac{1}{Z_p Z_v - {Z_{Cp}}^2}(Z_p U_v - Z_{Cp} U_p) \end{cases} \tag{4.57}$$

Considering the additional self-balancing network, the current of the primary compensation capacitor I_{Cp} can be obtained as

$$I_{Cp} = \frac{(j\omega L_p + R_p)U_v + (j\omega L_v + R_v)U_p}{\left(j\omega L_p + \dfrac{1}{j\omega C_p} + R_p\right)\left(j\omega L_v + \dfrac{1}{j\omega C_p} + R_v\right) + \dfrac{1}{\omega^2 {C_p}^2}} \tag{4.58}$$

Denote θ as the angle of U_v lagging U_p; then, U_v can be expressed in terms of U_p as

$$U_v = A(\cos\theta - j\sin\theta)U_p \tag{4.59}$$

Then, the amplitude of I_{Cp} can be obtained as

$$|I_{Cp}| = \frac{\sqrt{P^2 + Q^2}}{\sqrt{M^2 + N^2}}|U_p| \tag{4.60}$$

Among them, P, Q, M, and N are expressed as

$$P = R_v + AR_p\cos\theta + A\omega L_p\sin\theta \tag{4.61}$$

$$Q = A\omega L_p\cos\theta - AR_p\sin\theta + \omega L_v \tag{4.62}$$

$$M = \frac{L_v + L_p}{C_p} 2 + R_v R_p - \omega^2 L_p L_v - \frac{1}{\omega^2 {C_p}^2} \tag{4.63}$$

$$N = \omega L_v R_p + \omega L_p R_v - \frac{R_v + R_p}{\omega C_p} \tag{4.64}$$

The initial value of the current I_{Cp} named as I_{Cp0} can be expressed as

$$I_{Cp0} = \frac{U_p}{j\omega L_p + R_p + \frac{1}{j\omega C_p}} \tag{4.65}$$

Denote κ as the current gain after compensation, which can be calculated as

$$\kappa = \frac{\sqrt{P^2 + Q^2}\sqrt{{R_p}^2 + \left(\omega L_p - \frac{1}{\omega C_p}\right)^2}}{\sqrt{M^2 + N^2}} \tag{4.66}$$

In order to adjust the equivalent impedance of the virtual impedance network by injecting compensating current, it should be ensured that κ can be altered according to the requirement. From Eqs. (4.61), (4.64), and (4.66), it is shown that L_v and R_v are key to ensure $\kappa > 1$ or $\kappa < 1$ with respect to the arbitrary amplitude A and the phase angle θ under the circumstance of inductive load and capacitive load, respectively.

Considering the reflected impedance of multiple pickups, denote U_M as the reflected voltage of multiple pickups, and suppose there exist N pickups, then U_M can be expressed as

$$U_M = \sum_{i=1}^{N} \frac{dI_{si}}{dt} \tag{4.67}$$

As the impedance characteristics of multiple loads are unknown, U_M can be considered as an external random input variable with an upper bound. Then, the dynamic equation of the system can be obtained as

$$\begin{cases} U_p = L_p \frac{dI_p}{dt} + R_p I_p + U_M \\ U_v = L_v \frac{dI_v}{dt} + R_v I_v + U_{Cp} \\ \frac{dU_{Cp}}{dt} = \frac{1}{C_p}(I_p + I_v) \end{cases} \tag{4.68}$$

Consider I_p, I_v, and U_{Cp} as the state variables; thus, the state matrix can be expressed as

$$\begin{bmatrix} I_p \\ I_v \\ U_{Cp} \end{bmatrix} = \begin{bmatrix} -\frac{R_p}{L_p} & 0 & -\frac{1}{L_p} \\ 0 & -\frac{R_v}{L_v} & -\frac{1}{L_v} \\ \frac{1}{C_p} & \frac{1}{C_p} & 0 \end{bmatrix} \begin{bmatrix} I_p \\ I_v \\ U_{Cp} \end{bmatrix} + \begin{bmatrix} \frac{1}{L_p} & 0 & -\frac{1}{L_p} \\ 0 & \frac{1}{L_v} & 0 \\ 0 & 0 & 0 \end{bmatrix} \begin{bmatrix} U_p \\ U_v \\ U_M \end{bmatrix} \tag{4.69}$$

In order to theoretically demonstrate the stability of the proposed virtual impedance, the positive definite Lyapunov function candidate is considered as

$$V = L_p I_p{}^2 + L_v I_v{}^2 + C_p U_{Cp}{}^2 \tag{4.70}$$

The stability analysis of the above equation can be carried out using the derivative of the Lyapunov function candidate, which is given as

$$\dot{V} = 2\left(L_p I_p \left(-\frac{R_p}{L_p} I_p - \frac{1}{L_p} U_{Cp} \right) + L_v I_v \left(-\frac{R_v}{L_v} I_v - \frac{1}{L_v} U_{Cp} \right) + C_p U_{Cp} \left(\frac{I_p}{C_p} + \frac{I_v}{C_p} \right) \right) \tag{4.71}$$

Then, the differential of the energy function can be simplified as

$$\dot{V} = -2\left(R_p I_p{}^2 + R_v I_v{}^2 \right) < 0 \tag{4.72}$$

According to Eqs. (4.70)–(4.72), the unique equilibrium point is globally asymptotically stable at the state of $I_p = 0$, $I_v = 0$, and $U_{Cp} = 0$ based on the Lyapunov stability theorem. In addition, the system has different stable equilibrium points with different inputs of U_p, U_v, and U_M. Therefore, the analysis of the transition state can be ignored from the initial state when the system is at the equilibrium point of nonresonant status to the steady state when the system is at the resonant status. Consequently, the proposed self-balancing network can ensure the stability and eventually converge to the stable equilibrium resonant point of the system under arbitrary input conditions of U_p, U_v, and U_M. Accordingly, the key point of the self-balancing control is to calculate the phase and amplitude of the compensated current which is injected by self-balancing network. Consider the following example aiming at the mismatch caused by inductive reflected impedance. The vector diagram of the primary-side mismatching caused by the inductive component in the reflected impedance is depicted in Figure 4.21a. Denote the phase angle of the primary exciting voltage U_p and the primary exciting current I_p as β, the primary-side equivalent inductance is L_{pr}, which can be represented as

$$jU_p \sin\beta = ji_p\omega(L_{pr} - L_p) \tag{4.73}$$

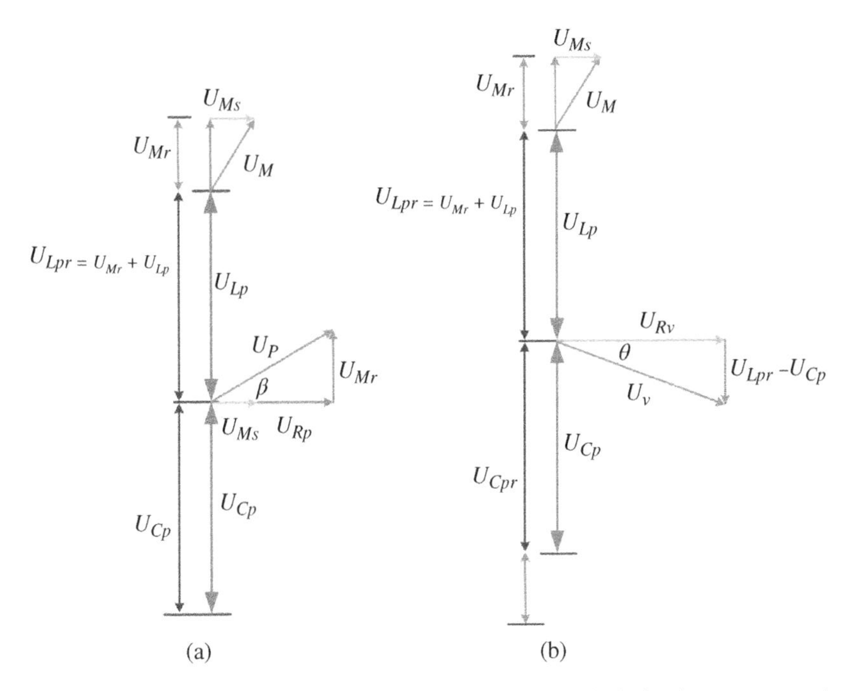

Figure 4.21 Vector schematic: (a) mismatching caused by the inductive component in reflected impedance; (b) reactive impedance matching with the proposed self-balancing network.

As shown in Figure 4.21b, the compensated U_{Cp} is equal in the opposite direction to U_{Lpr}.

$$j\omega I_p L_{pr} = I_{Cp} \frac{1}{j\omega C_p} \tag{4.74}$$

Additionally, I_p and I_v are in the same direction at the resonant state, and the corresponding relations can be obtained as

$$\begin{cases} I_{Cp} = I_p + I_v \\ U_{Rv} = I_v R_v \\ U_{Lv} = j\omega L_v I_v \\ U_{Cp} = I_{Cp} \dfrac{1}{j\omega C_p} \end{cases} \tag{4.75}$$

According to Figure 4.21b, the lagging angle θ and the input voltage U_v of the self-balancing network can be represented as

$$\begin{cases} \tan\theta = \dfrac{U_{Lpr} - U_{Cp}}{U_{Rv}} \\ U_v = \sqrt{(U_{Lpr} - U_{Cp})^2 + {U_{Rv}}^2} \end{cases} \tag{4.76}$$

Accordingly, by measuring the primary exciting voltage u_p, the primary exciting current I_p, and the corresponding phase angle β between u_p and I_p, the equivalent inductance value L_{pr} can be calculated according to Eq. (4.73). Then, the angle θ can be obtained based on Eq. (4.76). Meanwhile, in terms of implementation, the gate signal of the inverter in virtual impedance network is controlled to realize the angle θ lagging behind the gate signal of the inverter in the primary exciting circuit.

4.1.2.2.3 Implementation The simulation verification of the self-balancing virtual impedance network is shown based on MATLAB®/SIMULINK in this section. In order to verify the feasibility of the proposed virtual impedance scheme in the condition of inductive and capacitive operating states, two main case studies of both inductive and capacitive operating states with the key parameters listed in Table 4.4 are carried out:

Case 1 – The WPT system with one inductive pickup under the load value of 5 Ω.

Case 2 – The WPT system with multiple resistive pickups (two or three) under the load value of 5 Ω.

***Case 1** – One inductive pickup under the load value of 5 Ω.*

Table 4.5 lists the simulation results with one inductive pickup, which includes the compared items of the primary exciting current, the total pickup power, and the transmission efficiency. It can be seen that the primary exciting current rises from 3.0 to 3.3 A (10%), and the pickup power boosts from 16.7 to 19.5 W (16.8%). It can be obviously seen from the results that the virtual impedance scheme can effectively boost the primary exciting current and the total pickup power, which are significantly reduced due to the inductive variation in the resonance state. Meanwhile, the system efficiency varies from 72.1% to 70.1%, which means that

Table 4.4 Key parameters of the transmitting and the pickup coils.

Item	Value
Inductance of the transmitting coil (L_p)	60 μH
Capacitance of the transmitting side (C_p)	42.2 nF
Inductance of the one-pickup coil (L_s)	60 μH
Capacitance of the inductive one-pickup side (C_s)	46.42 nF
Inductances of the two-pickup coils ($L_{s21} = L_{s22}$)	60 μH
Capacitances of the two-pickup sides ($C_{s21} = C_{s22}$)	42.2 nF
Inductances of the three-pickup coils ($L_{s31} = L_{s32} = L_{s33}$)	60 μH
Capacitances of the three-pickup sides ($C_{s31} = C_{s32} = C_{s33}$)	42.2 nF

Table 4.5 One pickup with inductive characteristics under the frequency of 100 kHz (coupling $\kappa = 0.1$, DC = 10 V).

Load value	Item	Without virtual impedance	With virtual impedance
5 Ω	Primary current	3.0 A	3.3 A
	Pickup power	16.7 W	19.5 W
	Efficiency	72.1%	70.1%

Table 4.6 Two pickups with resistive characteristics under the frequency of 100 kHz (coupling $\kappa = 0.1$, cross-coupling $\kappa = 0.02$, DC = 10 V).

Load value	Item	Without virtual impedance	With virtual impedance
2 Ω	Primary current	0.71 A	0.74 A
	Total pickup power	4.69 W	5.08 W
	Efficiency	77.6%	77.6%

the transmission efficiency can be successfully maintained with the additional virtual impedance network. Then, here comes the question: whether the virtual impedance scheme can realize the regulation under the capacitive fluctuation of the resonance state?

Case 2 *– Multiple resistive pickups (two and three) under the load value of 2 Ω.*

As aforementioned in Chapter 2, the system resonance state in the primary side presents weak capacitive characteristics due to the cross-coupling effect in the multiple-pickups WPT systems. Then, the simulations of multiple pickups (two and three) with virtual impedance scheme are carried out with the results including the compared items of the primary exciting current, the total pickup power, and the transmission efficiency, which are shown in Tables 4.6 and 4.7, respectively. It can be seen that the primary exciting current rises from 0.71 to 0.74 A (4.23%) and 0.51 to 0.56 A (9.8%), respectively. Moreover, the total pickup power boosts from 4.69 to 5.08 W (8.32%) and 3.34 to 3.99 W (19.5%), respectively. In conclusion, the virtual impedance scheme can also improve the transmission performances which are adversely affected due to the capacitive variation in the resonance state. The corresponding theoretical analysis of the regulation process with capacitive reflected impedance is encouraged for readers to explore.

Table 4.7 Three pickups with resistive characteristics under the frequency of 100 kHz (coupling $\kappa = 0.1$, cross-coupling $\kappa_{12} = 0.02$, $\kappa_{13} = 0.015$, $\kappa_{23} = 0.01$, DC = 10 V).

Load value	Item	Without virtual impedance	With virtual impedance
2 Ω	Primary current	0.51 A	0.56 A
	Total pickup power	3.34 W	3.99 W
	Efficiency	80.3%	79.8%

4.1.2.3 Hybrid Adjustment

Previous studies have introduced two methods to compensate the mismatch state of the dynamic WPT system. However, there exist both pluses and minuses in the two methods, where the scheme based on the topology-reconfigurable capacitor matrix can realize a wide-range but discontinuous adjustment, and the scheme based on the self-balancing network can realize a continuous but smaller range adjustment compared with the prior one. Accordingly, in order to realize a wide-range and continuous adjustment, a combination of the above methods is proposed, which can theoretically offer unlimited capacitance values. The proposed hybrid impedance-adjusting scheme adopts the topology-reconfigurable capacitor array for large-scale tuning, and the proposed continuously adjustable scheme for fine-tuning.

As shown in Figure 4.22, the proposed hybrid impedance-adjusting system consists of two modules of topology-reconfigurable capacitor matrix and

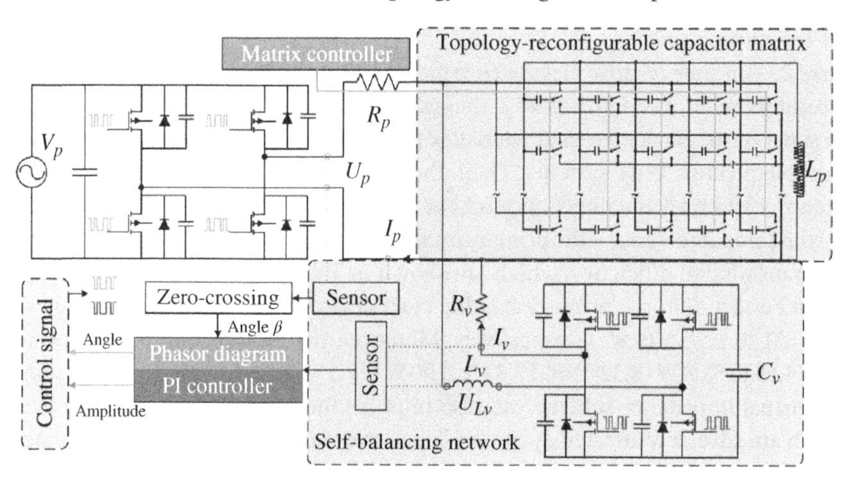

Figure 4.22 Block diagram of the proposed hybrid impedance matching method.

self-balancing network. The control signals of the H-bridge switches are generated by a phasor-diagram-based controller, where the amplitude and the phase angle of the compensating capacitance voltage U_{Lv} are calculated according to the measured primary voltage U_p, current I_p, and phase angle β. In addition, the topology-reconfigurable capacitor matrix, which is regulated by the matrix controller, is operated before the self-balancing network to realize the wide-range impedance adjustment and thus reduce DC power supply for the proposed continuously adjustable capacitor.

4.2 DC/AC Inverters

4.2.1 Introduction

The converters of DC to AC are known as inverters, which are used to change a DC input voltage to an AC output voltage with desired amplitude and frequency and widely used in industrial applications such as motor drives, renewable energy, and induction heating [20]. The amplitude and frequency of the output voltage can be controlled by the amplitude of the input DC voltage and controllable switching signal. Theoretically, the output voltage waveforms should be sine waves, but in fact, there exist harmonics in the waveforms of practical inverters, which will bring unfavorable influences in some high-frequency and -power applications. Accordingly, the output waveforms with lower harmonics are required in the high-frequency and -power applications. With the availability of high-speed power semiconductor devices, the harmonic contents of the output voltage can be minimized or reduced significantly by switching techniques.

Inverters can be broadly classified into two types as single- and three-phase inverters. In order to facilitate the analysis, this chapter only introduces single-phase inverters, which can be further classified into two categories according to the type of the input source. In other words, one is called the voltage-fed inverter when the input voltage is maintained constant, and the other is called current-fed inverter when the input current remains constant. Generally, the inversion is realized by the use of power semiconductor devices, which will be minutely introduced in Section 4.2.2. In addition, the AC output voltage is usually controlled by Pulse-Width Modulation (PWM) signals; the corresponding control schemes will be described in Section 4.2.5.

4.2.2 Wide-Bandgap Semiconductor Devices

With the increasing attention of WPT systems, the demand for high-frequency and -density power inverters is on the rise. According to the Qi standard of Wireless Power Consortium, the specified transmission frequency ranges from 87

Table 4.8 Main characteristics of the existing semiconductor materials.

Parameters	Si	GaAs	4H-SiC	6H-SiC	2H-GaN
Energy bandgap (eV)	1.12	1.42	3.26	3.0	3.4
Electric field breakdown (MV/cm)	0.25	0.6	2.2	3	3
Saturated electron drift (cm/s)	1×10^7	1.2×10^7	2×10^7	2×10^7	2.2×10^7
Thermal conductivity (W/(cm K))	1.5	0.5	4.9	4.9	1.3

to 205 kHz [21]. The increase in the switching frequency can reduce the energy storage in the passive components, which will contribute to the reduction in volume and cost. However, the high switching frequency not only requires faster switching devices but also brings higher switching losses, which will result in the decrease in the output power and system efficiency. In addition, for the practical charging applications such as high-power electric vehicles (EVs) and low-power small mobile devices, the requirements of high-voltage stresses and high power density for switching devices become more intense.

Power transistors as switching devices in power electronics systems determine the ceilings of switching frequency and output power. For example, the rated voltage level of silicon (Si)-based MOSFETs is about 600 V, which almost reaches its limit [22]. In order to pursue a breakthrough in the performances of power transistors, better materials or more optimal device structures are required. Accordingly, the demand for high-frequency and -density power systems has accelerated the exploration of new devices and new topologies [23]. The advent of wide-bandgap semiconductors provides a good solution to overcome the above difficulties, thanks to the intrinsic characteristics. Wide-bandgap semiconductors, such as silicon carbide (SiC) and gallium nitride (GaN) as the third-generation semiconductors, are defined as semiconductor materials with bandgap width greater than 3.0 eV [24]. In comparison with the previous two generations of semiconductor materials, the wide-bandgap semiconductor materials show better performances.

Table 4.8 lists the main characteristics of the existing semiconductor materials. The advantages of wide-bandgap semiconductor materials can be summarized as follows:

(1) ***Wider energy bandgap*** – Due to the wider energy bandgap than the Si-based devices, the wide-bandgap semiconductor materials have higher breakdown electric field, higher electron mobility, and higher thermal conductivity. Meanwhile, the wider energy bandgap also improves the operating temperature and radiation hardness.

(2) ***Higher breakdown electric field*** – Since more energy is required for electrons to transit from the valence band to the conduction band, the wide-bandgap semiconductor materials can endure stronger electric field under the same size. In addition, the thinner depletion layer brings lower on-resistance of wide-bandgap semiconductor materials, which means the conduction losses are lower than the Si-based devices.
(3) ***Higher electron mobility*** – The higher electron mobility means higher saturated electron drift, which will contribute to the increase in the operating frequency.
(4) ***Higher thermal conductivity*** – The higher thermal conductivity improves the performance of heat dissipation, which can ensure the operation at higher power density.

As can be seen from Table 4.9, SiC and GaN come out on top as the materials of choice for power devices according to the above advantages for high-frequency and -density power inverters. Still, the power devices based on the above two materials are suitable for different applications according to their detailed features. SiC-based power devices are suitable for high-voltage applications as the vertical structure leads to high voltage ratings (650–1700 V) [25]. However, the large gate charge requires high-power gate drivers, which are usually difficult to design. Therefore, SiC-based power devices are often used in low-frequency applications. Unlike SiC, GaN-based power devices are mostly used in relatively low-voltage and high-frequency applications. Since GaN layers are epitaxially grown on other

Table 4.9 Characteristics of different high-frequency inverters.

Topology	Voltage-fed full bridge	Current-fed full bridge	Voltage-fed half bridge	Current-fed half bridge	Class EF_2	Class E
Number of switches	4	4	2	2	1	1
Voltage stress	V_s	$\pi V_s/2$	V_s	πV_s	$2.316V_s$	$3.56V_s$
Current stress	$\pi I_0/2$	I_0	πI_0	I_0	$3.263I_0$	$2.86I_0$
Operating frequency	20–500 kHz	20–500 kHz	100 kHz–10 MHz	100 kHz–20 MHz	1–20 MHz	1–20 MHz
Output power	1–10 kW	1 kW–1 MW	<1 kW	<1 kW	1–100 W	1–100 W
Forbidden case	Short circuit	Open circuit	Short circuit	Open circuit	Open circuit	Open circuit

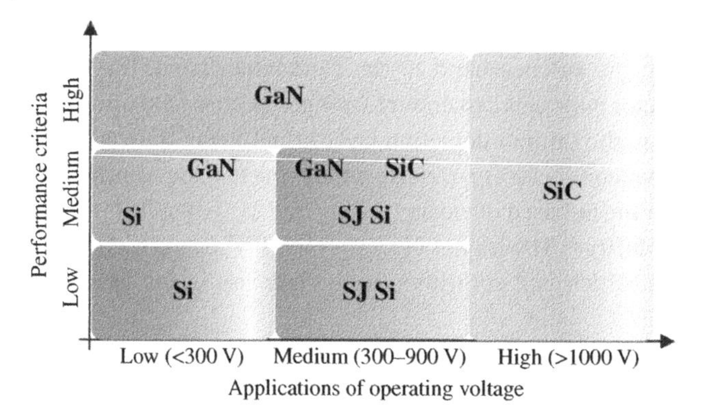

Figure 4.23 The respective applications of power semiconductor devices.

substrates (Si, SiC, and sapphire), the peak electric field occurs at the surface of lateral GaN-based power devices [26]. Accordingly, due to the lateral structure, the limitations in the maximum electric field lead to relatively low-voltage ratings (<650 V) in GaN-based power devices.

Figure 4.23 depicts the respective applications of power semiconductor devices in consideration of the performances of the operating voltage, operating frequency, and power density. Due to the advantages of the manufacturing processes and costs, the Si-based and super junction (SJ) Si-based devices are widely used in the applications of medium and low performances. For the higher performance applications, GaN-based power devices are preferred due to their lower switching losses. However, the SiC-based power devices present an overwhelming advantage on high-operating-voltage (>1000 V) applications on account of the vertical structure with high breakdown electric field.

4.2.3 Architectures

4.2.3.1 Single-Phase Bridge Inverters

Bridge inverters are one of the most commonly used inverters for WPT systems, which are classified into half-bridge inverters [27–29] and full-bridge inverters [30–32] according to the structure. Each structure can be divided into two types as voltage fed and current fed, based on the type of the equivalent input source. The input DC power supply of the voltage-fed inverter is in parallel with a large capacitor, which is equivalent to a voltage source. Commonly, voltage-fed inverters operate with an inductive load, and as a result, it requires an antiparallel diode to follow current, which is shown in Figures 4.24a and 4.26a. Additionally, a current-fed inverter consists of a DC power in series with a large inductor, which is equivalent to a current source. Current-fed inverters usually operate with a capacitive

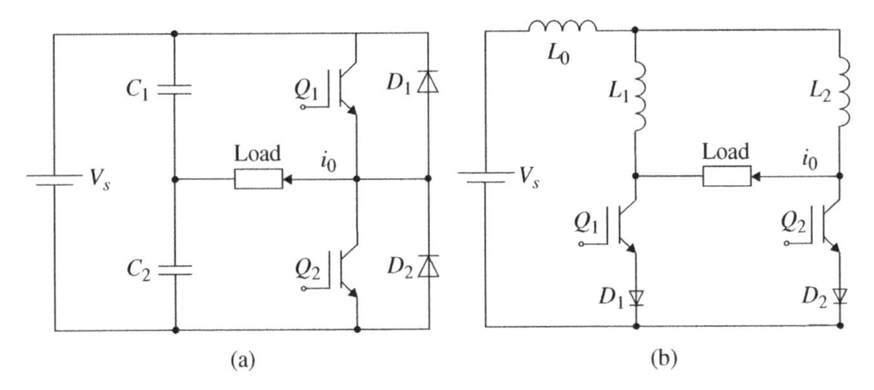

Figure 4.24 Schematic diagram of single-phase half-bridge inverters: (a) voltage-fed inverter; (b) current-fed inverter.

load, which will result in reverse voltage in each arm. Accordingly, the switches require a series connection to fast recovery diodes, as can be seen in Figure 4.24b and 4.26b.

As shown in Figure 4.24a, a single-phase voltage-fed half-bridge inverter is composed of two arms. Each arm consists of a controllable switch device and an antiparallel diode. The alternating current is generated by the alternating conduction of the two bridge arms. The two switches named as Q_1 and Q_2 are controlled by a set of complementary PWM signals. The load voltage operates at $V_s/2$, when the switch Q_1 turns on during a half cycle from 0 to $T/2$. Similarly, when the switch Q_2 turns on from $T/2$ to T, the load voltage operates at $-V_s/2$. Figure 4.25 depicts the waveforms of the voltage and fundamental current with the resistive load of a voltage-fed half-bridge inverter. In addition, the phase angle between the voltage and fundamental current is 0° only when the load is resistive. The root mean square (RMS) value of load voltage can be expressed as

$$V_{load} = \left(\frac{2}{T}\int_0^{T/2} \frac{V_s^2}{4} dt\right)^{1/2} = \frac{V_s}{2} \tag{4.77}$$

Then, the expression based on Fourier series can be obtained as

$$v_{load} = \frac{a_0}{2} + \sum_{n=1}^{+\infty}(a_n \cos(n\omega t) + b_n \sin(n\omega t)) \tag{4.78}$$

As the voltage waveform shows the characteristics of the odd function, the expression can be simplified as

$$v_{load} = \sum_{n=1}^{+\infty} b_n \sin(n\omega t) \tag{4.79}$$

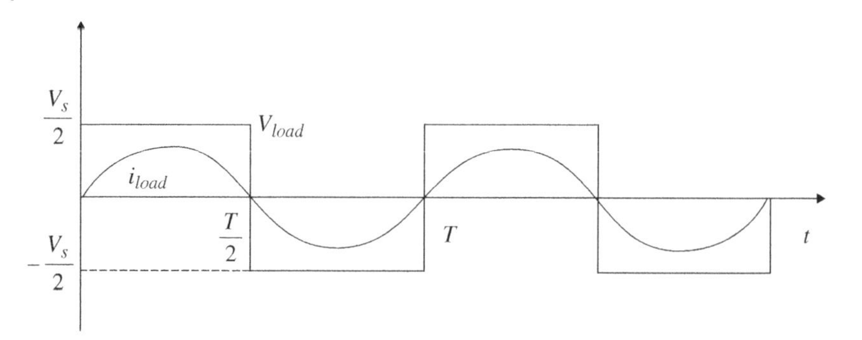

Figure 4.25 Waveforms of voltage and fundamental current with resistive load of voltage-fed half-bridge inverters.

where b_n can be obtained as

$$b_n = \frac{1}{\pi}\left[\int_{-\frac{\pi}{2}}^{0} -\frac{V_s}{2}\sin(n\omega t)d(\omega t) + \int_{0}^{\frac{\pi}{2}} \frac{V_s}{2}\sin(n\omega t)d(\omega t)\right] \tag{4.80}$$

Accordingly, the transient value of the load voltage can be expressed as

$$v_{load} = \sum_{n=1,3,5\ldots}^{+\infty} \frac{2V_s}{n\pi}\sin(n\omega t) \tag{4.81}$$

The fundamental RMS value of the load voltage can be obtained when $n = 1$ as

$$V_{load0} = \frac{2V_s}{\sqrt{2}\pi} = 0.45V_s \tag{4.82}$$

As can be seen in Figure 4.26a, a single-phase voltage-fed full-bridge inverter is composed of four arms with controllable switch devices and antiparallel diodes. The diagonal switches named as Q_1, Q_4 and Q_2, Q_3 constitute two sets of bridge arms, which are controlled by a set of complementary PWM signals, respectively. The load voltage operates at V_s, when the switches Q_1 and Q_4 turn on during a half cycle from 0 to $T/2$. Similarly, when the switches Q_2 and Q_3 turn on from $T/2$ to T, the load voltage operates at $-V_s$. Figure 4.27 depicts the waveforms of the voltage and fundamental current with resistive load of a voltage-fed full-bridge inverter. Additionally, the phase angle between the voltage and fundamental current is 0° only when the load is resistive. The RMS value of the load voltage can be expressed as

$$V_{load} = \left(\frac{2}{T}\int_{0}^{T/2} V_s^2 dt\right)^{1/2} = V_s \tag{4.83}$$

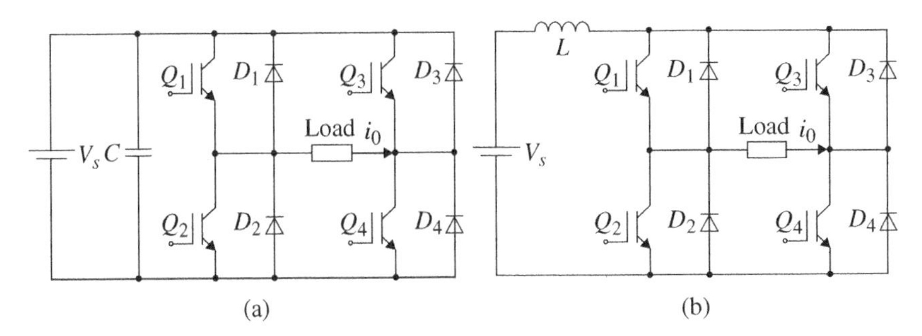

Figure 4.26 Schematic diagram of single-phase full-bridge inverters: (a) voltage-fed inverter; (b) current-fed inverter.

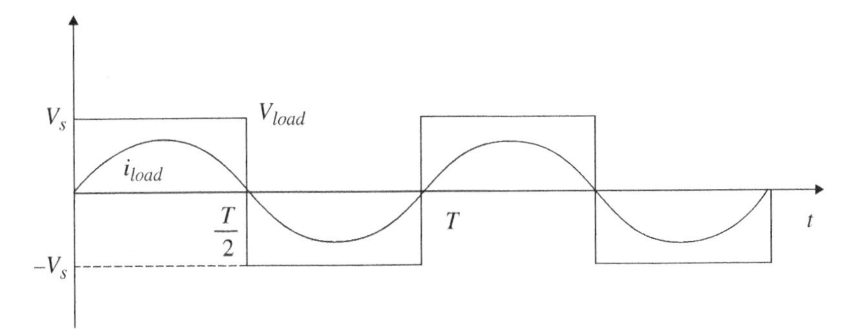

Figure 4.27 Waveforms of voltage and fundamental current with resistive load of voltage-fed full-bridge inverters.

Then, the expression based on Fourier series can be obtained as

$$v_{load} = \sum_{n=1,3,5\cdots}^{+\infty} \frac{4V_s}{n\pi} \sin(n\omega t) \tag{4.84}$$

The RMS value of the fundamental load voltage can be obtained when $n = 1$ as

$$V_{load0} = \frac{4V_s}{\sqrt{2}\pi} = 0.9V_s \tag{4.85}$$

In comparison with half-bridge inverters, the switches withstand half current (voltage) stress of voltage-fed (current-fed) full-bridge inverters. In order to ensure that the arms do not open at the same time, a dead zone should be set in PWM signals for voltage-fed bridge inverters. For current-fed bridge inverters, an overlapping zone should be set in PWM signals to prevent open circuit of bridge arms.

4.2.3.2 Class-E Inverters

Figure 4.28 depicts the circuit topology of class-E inverters, which consist of only one switch tube Q [33]. The DC power is in series with a large inductor, which is

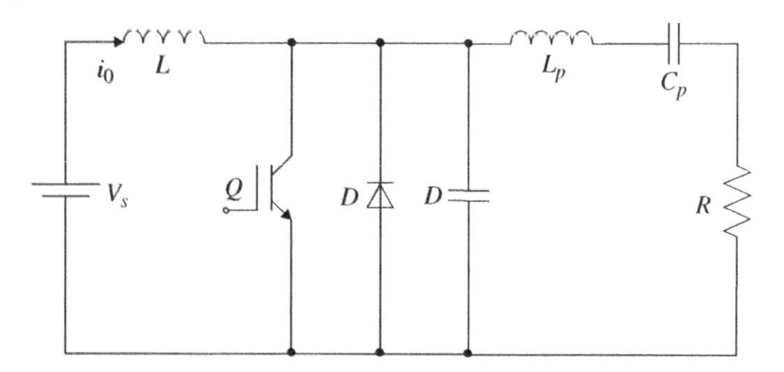

Figure 4.28 Schematic diagram of class-E inverters.

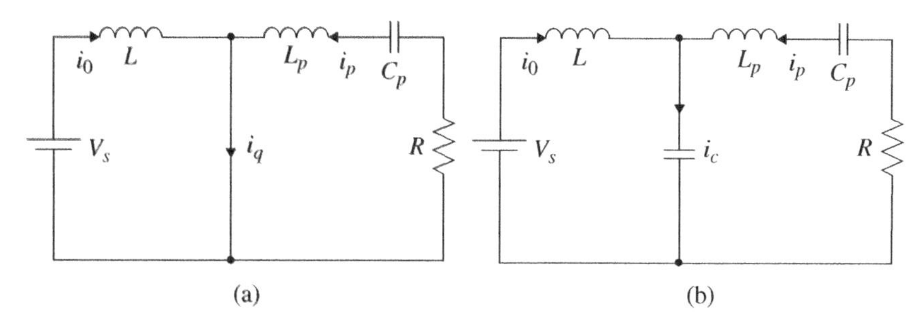

Figure 4.29 Equivalent circuit model of class-E inverters: (a) transistor Q turns on; (b) transistor Q turns off.

equivalent to a current source. The capacitor C is in parallel with the switch, which constitute the resonant network with primary inductor L_p and primary capacitor C_p. The operation model of class-E inverters can be divided into two stages. As shown in Figure 4.29a, transistor Q turns on when the current i_q consists of the source current i_0 and the load current i_p. Accordingly, the primary inductor L_p and the primary capacitor C_p constitute a series resonant loop, and the resonant frequency f_1 can be obtained as

$$f_1 = \frac{1}{2\pi\sqrt{L_p C_p}} \tag{4.86}$$

Then, the capacitor C is charged when the transistor Q turns off; the current i_c is composed of the source current i_0 and the load current i_p, which is shown in Figure 4.29b. When the capacitor C is added to the resonant loop, the resonant frequency f_2 can be obtained as

$$f_2 = \frac{1}{2\pi\sqrt{L_p \frac{CC_p}{(C + C_p)}}} \tag{4.87}$$

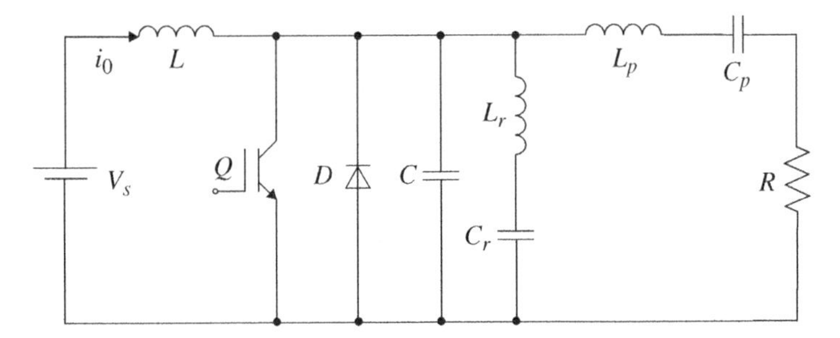

Figure 4.30 Schematic diagram of class-EF_2 inverters.

Since only one switch tube is used, the class-E inverter is easier to realize higher frequency output current with lower switching losses. Additionally, the soft switching condition can be realized when the inverter is working in an ideal state. However, the switch tube needs to withstand large voltage and current stress, which greatly limit the capacity of the output power. In order to improve the capacity of the output power, a new class-EF_2 inverter is proposed, which can be seen in Figure 4.30. On the basis of the class-E inverter, a parallel LC branch is added to the circuit in the class-EF_2 inverter [34]. The natural frequency of the newly added branch is twice as much as the operating frequency, which can provide the secondary harmonic component of the voltage superimposed on the switch tube. Thus, the voltage stress of the switch tube is reduced with the reduction in the voltage peak.

As can be seen from Table 4.9, different types of inverters are suitable for different applications in view of the operating frequency and power demand. Full-bridge inverters are generally applicable to high-power (higher than 1 kW) applications due to the low stress of the voltage and current. However, restricted by dead zone and diode reverse recovery, the operation frequency of full-bridge inverters is limited within 500 kHz. By contrast, due to the higher stress of voltage and current than full-bridge inverters, half-bridge inverters and class-E inverters are suitable for medium-power (lower than 1 kW) and low-power (lower than 100 W) applications, respectively, while the operating frequency can reach 20 MHz because of the lower switching losses and voltage variation rate of the switches.

4.2.4 Soft Switching

Switching losses are one of the major sources of losses in inverters, which are caused by the turn-on and turn-off transitions of power semiconductor devices. Even though the switching times only require times of tens of nanoseconds to microseconds, the resulting power losses can be significant due to the high operating frequency. To switch a semiconductor device between the on and off

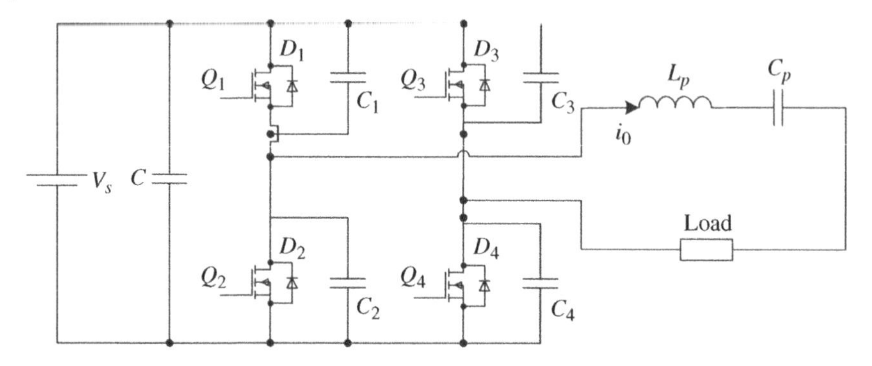

Figure 4.31 A full-bridge inverter with LC-series resonant circuit.

states, the insertion and removal of the controlling charge are required; hence, the amount of controlling charge influences both the switching time and the switching losses. Soft switching techniques known as zero-current switching (ZCS) [35] and zero-voltage switching (ZVS) [36–38] can effectively reduce the switching losses and improve the system efficiency. In the majority of applications, where the diode recovered charge and semiconductor output capacitances are the dominant sources of PWM switching loss, ZVS is preferred [39].

4.2.4.1 Zero-Current Switching (ZCS)

ZCS means the current of power transistors becoming zero before its turning off. The ZCS phenomenon occurs when the system is operated below the resonant frequency, which is also equal to a capacitive impedance characteristic of the equivalent circuit. The analysis is under the consideration for the operation of full-bridge inverter with LC-resonant circuit, which is shown in Figure 4.31. Figure 4.32a depicts the output voltage u_{out}, output current i_{out}, and the switching signals of the full-bridge inverter, where t_d is the interlock time. If the operating frequency is lower than the resonant frequency, the equivalent impedance of the resonant tank presents capacitive, which leads to a leading phase between the output current and the output voltage. The voltage and current waveforms of transistors and the conducting situation are shown in Figure 4.32b; the transistors ($Q_{1,4}$ or $Q_{2,3}$) are turned off while the corresponding body diodes ($D_{1,4}$ or $D_{2,3}$) conduct, which means the transistors can be turned off without incurring switching losses. However, due to the reverse recovery current of diodes, the switching losses will occur when the transistors turn on, which is the major disadvantage of the ZCS operation. Commonly, ZCS operation is used to mitigate the switching loss caused by stray inductances and current tailing in insulated-gate bipolar transistors (IGBTs), which can also be used for commutation of silicon-controlled rectifiers (SCRs).

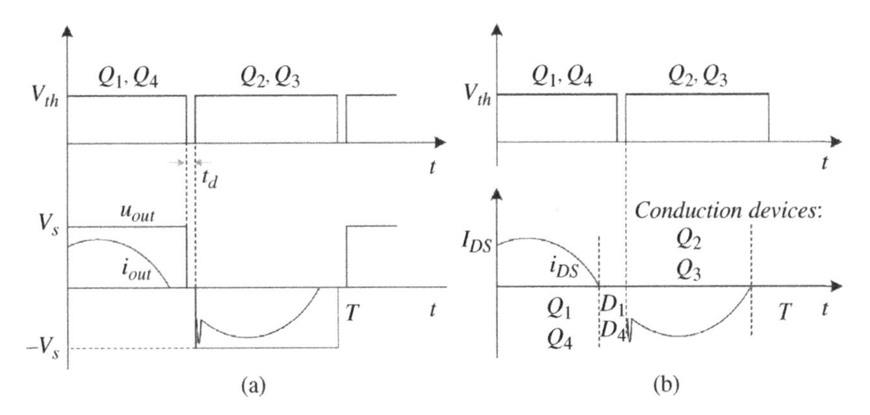

Figure 4.32 Schematic diagram of ZCS: (a) Output waveforms of full-bridge inverter; (b) voltage and current waveforms of transistor Q.

4.2.4.2 Zero-Voltage Switching (ZVS)

ZVS means the voltage of power transistors becoming zero before its turning on. The ZVS phenomenon occurs when the operating frequency is higher than the resonant frequency, which is also equal to an inductive impedance characteristic of the equivalent circuit. Figure 4.31 depicts a full-bridge inverter with LC-series resonant circuit, where C_1, C_2, C_3, and C_4 are the parallel combination of C_{gd} (gate-to-drain capacitance) and C_{ds} (drain-to-source capacitance), and D_1, D_2, D_3, and D_4 are body diodes (or using additional fast recovery diodes) of the respective MOSFETs Q_1, Q_2, Q_3, and Q_4. The ZVS operation requires that the parasitic capacitances ($C_{1,4}$ or $C_{2,3}$) are discharged, and the corresponding body diodes ($D_{1,4}$ or $D_{2,3}$) are on before turning the pair of MOSFETs ($Q_{1,4}$ or $Q_{2,3}$) on. Figure 4.33a–d depicts the whole commutation process of the ZVS operation. As can be seen in Figure 4.33b, the capacitors C_1 and C_4 are charged, and C_2 and C_3 are discharged, which leads to the conduction of the body diodes D_2 and D_3, as shown in Figure 4.33c. Accordingly, the MOSFET can turn on at approximately zero voltage, ignoring the diode drop voltage.

Figure 4.34a depicts the output voltage u_{out}, output current i_{out}, and the switching signals of the full-bridge inverter, where t_d and t_θ are the interlock time and lagging time. If the operating frequency is higher than the resonant frequency, the equivalent impedance of the resonant tank presents inductive, which leads to the lagging angle θ between the output current and the output voltage. The voltage and current waveforms of transistors and the conducting situation are shown in Figure 4.34b; in order to realize the commutation process shown in Figure 4.33, the discharge of the capacitance and the conduction of body diode must take place during the interlock time t_d which should be smaller than the lagging time t_θ.

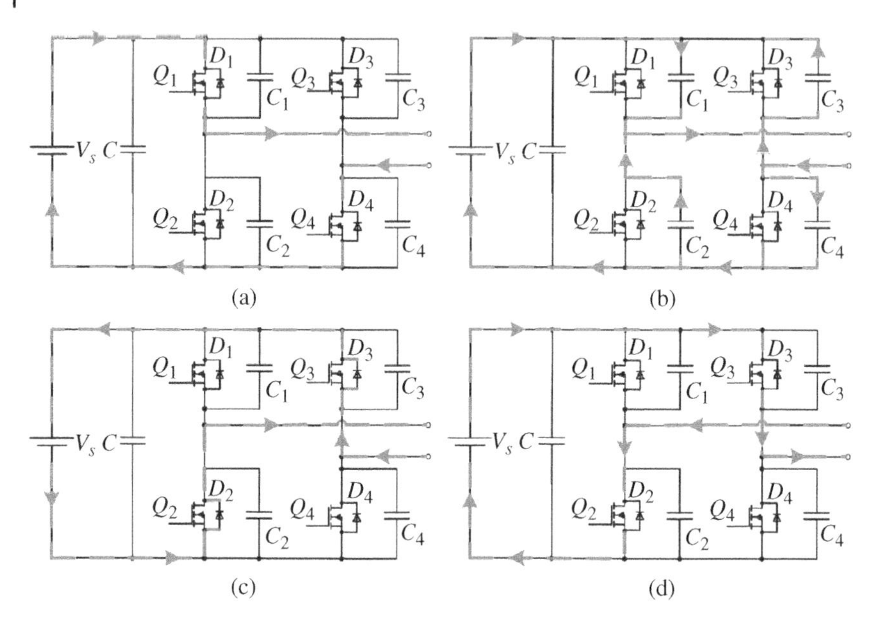

Figure 4.33 Commutation of ZVS operation: (a) $Q_{1,4}$ turns on, $Q_{2,3}$ turns off; (b) $Q_{1,2,3,4}$ turns off; (c) $Q_{2,3}$ turns on, $Q_{1,4}$ turns off; (d) $Q_{1,2,3,4}$ turns off.

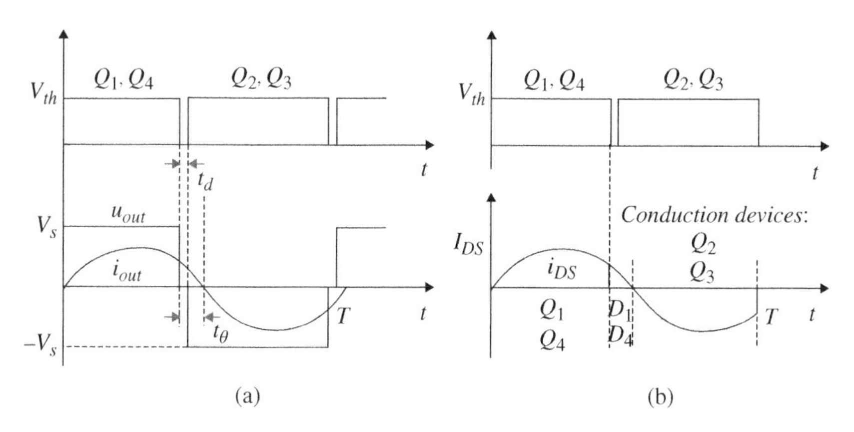

Figure 4.34 Schematic diagram of ZVS: (a) Output waveforms of full-bridge inverter; (b) voltage and current waveforms of transistor Q.

Therefore, sufficient inductive energy and interlock time must be ensured for ZVS operation.

For the switching devices containing MOSFETs and diodes, ZVS can effectively mitigate the switching losses caused by the diode recovered charge and parasitic capacitance. An additional advantage of ZVS is the reduction in Electromagnetic

Interference (EMI) associated with device capacitances, which is commonly generated by the rapid charging and discharging of the semiconductor device capacitances.

4.2.5 Control Schemes

In order to achieve the system requirements, such as the regulation of output voltage and the control of the operating frequency, the control schemes to adjust the output characteristics of inverters are extremely required in industrial applications. The commonly used method is to incorporate PWM control [40–42] and phase-shift control within the inverters [43–45].

4.2.5.1 Pulse-Width-Modulation Control

The commonly used PWM control techniques include single-pulse-width modulation, multiple-pulse-width modulation, and sinusoidal-pulse-width modulation. The PWM control signals are used as the gate driving signals, which are generated by comparing a reference signal with a triangular carrier wave. The output frequency of the controlled inverter is set by the frequency of the reference signal. In addition, the output voltage is determined by the peak amplitude of the reference signal and the modulation index. Accordingly, the output frequency and voltage of controlled inverter can be regulated by the frequency and the peak amplitude of the reference signal, respectively.

For WPT systems, due to the high-frequency sensitivity, the output frequency of inverters has great influence on the output performances. When the natural resonant frequency of the system is shifted, the output power will sharply decrease. Furthermore, the phenomenon of frequency splitting may occur as the variation of coupling state. Accordingly, the control of the system operating frequency is not only beneficial to ensure the recovery of the system resonant state but also conductive to ensure the optimal output power with the fluctuation in the coupling effect.

4.2.5.2 Phase-Shift Control

Phase-shift control method is mainly used to adjust the output voltage by changing its duty width for full-bridge inverters. The schematic diagram of the phase-shift control scheme is shown in Figure 4.35, where α is the phase-shift angle, Q_1, Q_2 are the leading switches, and Q_3, Q_4 are the lagging switches. As can be seen from Figure 4.35b, the phase-shift angle α is the lagging angle between the leading switches and the lagging switches with the value range of 0°–180°. Based on the Fourier series, the output voltage can be calculated by

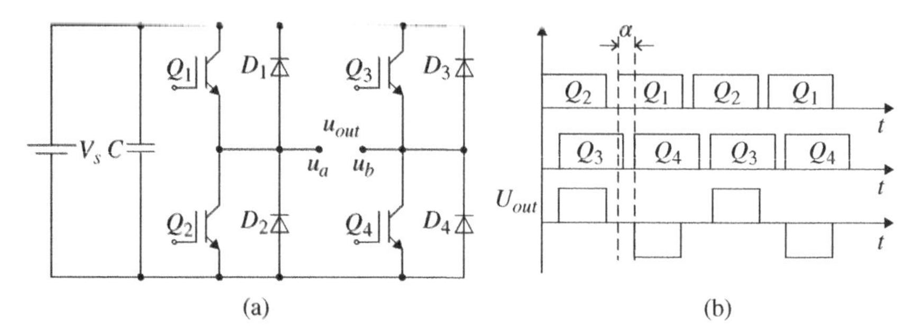

Figure 4.35 Schematic diagram of phase-shift control scheme for full-bridge inverter: (a) full-bridge inverters; (b) waveforms of control signals and output voltage.

$$\begin{cases} u_a = \sum\limits_{n=1,3,5\ldots}^{+\infty} \frac{2V_s}{n\pi} \sin(n\omega t) \\ u_b = \sum\limits_{n=1,3,5\ldots}^{+\infty} \frac{2V_s}{n\pi} \sin n(\omega t - \pi + \alpha) \end{cases} \tag{4.88}$$

Then, the output voltage u_{out} can be obtained as

$$u_{out} = u_a - u_b = \sum_{n=1,3,5,\ldots}^{+\infty} \frac{4V_s}{n\pi} \sin\left(\frac{n\pi - n\alpha}{2}\right) \cos\left[n\left(\omega t - \frac{\pi - \alpha}{2}\right)\right] \tag{4.89}$$

The RMS value of the fundamental output voltage can be obtained when $n = 1$ as

$$V_{out0} = \frac{2\sqrt{2}V_s}{\pi} \cos\frac{\alpha}{2} \tag{4.90}$$

Accordingly, the control of the output voltage can be realized by the regulation of the phase-shift angle α. However, it can be seen that the regulation range is limited within 0–V_s, which means there exist limitations for the scope of applications. In addition, the harmonic content of the output voltage will increase with the increase in the phase-shift angle. Moreover, the phase-shift control scheme only has regulation function in the output voltage and output power but no effect on the system efficiency.

References

1 Johnson, I.A. (2016). *Wireless Power Transfer*, 2e. Aalborg: River Publishers.
2 Chun, T.R. and Chris, M. (2017). *Wireless Power Transfer for Electric Vehicles and Mobile Devices*. Hoboken, NJ: Wiley.

3 Wang, C.S., Stielau, O.H., and Covic, G.A. (2005). Design considerations for a contactless electric vehicle battery charger. *IEEE Transactions on Industrial Electronics* 52 (5): 1308–1314.
4 Wang, C.S., Covic, G.A., and Stielau, O.H. (2004). Power transfer capability and bifurcation phenomena of loosely coupled inductive power transfer systems. *IEEE Transactions on Industrial Electronics* 51 (1): 148–157.
5 Keeling, N.A., Covic, G.A., and Boys, J.T. (2010). A unity-power-factor IPT pickup for high-power applications. *IEEE Transactions on Industrial Electronics* 57 (2): 744–751.
6 Zhang, Z., Pang, H., Georgiadis, A., and Cecati, C. (2019). Wireless power transfer – an overview. *IEEE Transactions on Industrial Electronics* 66 (2): 1044–1058.
7 Sallan, J., Villa, J.L., Llombart, A., and Sanz, J.F. (2009). Optimal design of ICPT systems applied to electric vehicle battery charge. *IEEE Transactions on Industrial Electronics* 56 (6): 2140–2149.
8 Al-Karakchi, A.A.A., Lacey, G., and Putrus, G. (2015). A method of electric vehicle charging to improve battery life. In: *2015 50th International Universities Power Engineering Conference (UPEC)*, Stoke on Trent, UK, 1–3.
9 Gu, W., Sun, Z., Wei, X., and Dai, H. (2014). A new method of accelerated life testing based on the grey system theory for a model-based lithium-ion battery life evaluation system. *Journal of Power Sources* 267 (3): 366–379.
10 Steigerwald, D.A., Bhat, J.C., Collins, D. et al. (2002). Illumination with solid state lighting technology. *IEEE Journal of Selected Topics in Quantum Electronics* 8 (2): 310–320.
11 Zhang, W. and Mi, C.C. (2016). Compensation topologies of high-power wireless power transfer systems. *IEEE Transactions on Vehicular Technology* 65 (6): 4768–4778.
12 Hou, J., Zhang, Z., Wong, S., and Tse, C.K. (2018). Analysis of output current characteristics for higher order primary compensation in inductive power transfer system. *IEEE Transactions on Power Electronics* 33 (8): 6807–6821.
13 Zhang, Y., Lu, T., Zhao, Z. et al. (2015). Selective wireless power transfer to multiple loads using receivers of different resonant frequencies. *IEEE Transactions on Power Electronics* 30 (11): 6001–6005.
14 Beh, T.C., Kato, M., Imura, T. et al. (2013). Automated impedance matching system for robust wireless power transfer via magnetic resonance coupling. *IEEE Transactions on Industrial Electronics* 60 (9): 3689–3698.
15 Heebl, J.D., Thomas, E.M., Penno, R.P., and Grbic, A. (2014). Comprehensive analysis and measurement of frequency-tuned and impedance-tuned wire-less non-radiative power-transfer systems. *IEEE Antennas and Propagation Magazine* 56 (5): 131–148.

16 Lim, Y., Tang, H., Lim, S., and Park, J. (2014). An adaptive impedance-matching network based on a novel capacitor matrix for wireless power transfer. *IEEE Transactions on Power Electronics* 29 (8): 4403–4413.

17 Zhang, Z., Ai, W., Liang, Z., and Wang, J. (2018). Topology-reconfigurable capacitor matrix for encrypted dynamic wireless charging of electric vehicles. *IEEE Transactions on Vehicular Technology* 67 (10): 9284–9293.

18 Zhong, Q.C. and Zeng, Y. (2014). Control of inverters via a virtual capacitor to achieve capacitive output impedance. *IEEE Transactions on Power Electronics* 29 (10): 5568–5578.

19 Zhang, Z. and Pang, H. (2020). Continuously-adjustable capacitor for multiple-pickup wireless power transfer under single-power-induced energy field. *IEEE Transactions on Industrial Electronics* 67 (8): 6418–6427.

20 Muhammad, H.R. (1993). *Power Electronics: Circuits, Devices, and Applications*. Englewood Cliffs, New Jersey: Prentice Hall.

21 Wireless Power Consortium. (2015). Qi Specification. The Wireless Power Consortium (WPC) website: http://www.wirelesspowerconsortium.com.

22 Lamichhane, R.R., Ericsson, N., Frank, S. et al. (2014). A wide bandgap silicon carbide (SiC) gate driver for high-temperature and high-voltage applications. In: *IEEE 26th International Symposium on Power Semiconductor Devices & IC's (ISPSD)*, 414–417.

23 Choi, J., Tsukiyama, D., Tsuruda, Y., and Davila, J.M.R. (2018). High-frequency, high-power resonant inverter with eGaN FET for wireless power transfer. *IEEE Transactions on Power Electronics* 33 (3): 1890–1896.

24 Kaminski, N. and Hilt, O. (2014). SiC and GaN devices-wide bandgap is not all the same. *IET Circuits, Devices and Systems* 8 (8): 227–236.

25 Mudholkar, M. and Mantooth, H.A. (2013). Characterization and modeling of 4H-SiC lateral MOSFETs for integrated circuit design. *IEEE Transactions on Electron Devices* 60 (6): 1923–1930.

26 Xu, J., Gu, L., Ye, Z. et al. (2020). Cascode GaN/SiC: a wide-bandgap heterogenous power device for high-frequency applications. *IEEE Transactions on Power Electronics* 35 (6): 6340–6349.

27 Yeo, T.D., Kwon, D., Khang, S.T. et al. (2017). Design of maximum efficiency tracking control scheme for closed-loop wireless power charging system employing series resonant tank. *IEEE Transactions on Power Electronics* 32 (1): 471–478.

28 Li, H., Wang, K., Huang, L. et al. (2015). Dynamic modeling based on coupled modes for wireless power transfer systems. *IEEE Transactions on Power Electronics* 30 (11): 6245–6253.

29 Ahn, D. and Mercier, P.P. (2016). Wireless power transfer with concurrent 200-kHz and 6.78-MHz operation in a single-transmitter device. *IEEE Transactions on Power Electronics* 31 (7): 5018–5029.

30 Fujita, T., Yasuda, T., and Akagi, H. (2017). A dynamic wireless power transfer system applicable to a stationary system. *IEEE Transactions on Industry Applications* 53 (4): 3748–3757.

31 Feng, H., Cai, T., Duan, S. et al. (2016). An LCC compensated resonant converter optimized for robust reaction to large coupling variation in dynamic wireless power transfer. *IEEE Transactions on Industrial Electronics* 63 (10): 6591–6601.

32 Samanta, S., Rathore, A.K., and Thrimawithana, D.J. (2017). Analysis and design of current-fed half-bridge (C)(LC)-(LC) resonant topology for inductive wireless power transfer application. *IEEE Transactions on Industry Applications* 53 (4): 3917–3926.

33 Aldhaher, S., Luk, C.K., and Whidborne, J.F. (2014). Tuning class E inverters applied in inductive links using saturable reactors. *IEEE Transactions on Power Electronics* 29 (6): 2969–2978.

34 Aldhaher, S., Yates, D.C., and Mitcheson, P.D. (2016). Design and development of a class EF_2 inverter and rectifier for multimegahertz wireless power transfer systems. *IEEE Transactions on Power Electronics* 31 (12): 8138–8150.

35 Sen Tang, C., Sun, Y., Su, Y.G. et al. (2009). Determining multiple steady-state ZCS operating points of a switch-mode contactless power transfer system. *IEEE Transactions on Power Electronics* 24 (2): 416–425.

36 Jiang, Y.B., Wang, L., Wang, Y. et al. (2019). Analysis, design, and implementation of accurate ZVS angle control for EV battery charging in wireless high-power transfer. *IEEE Transactions on Industrial Electronics* 66 (5): 4075–4085.

37 Li, H., Wang, K., Fang, J., and Tang, Y. (2019). Pulse density modulated ZVS full-bridge converters for wireless power transfer systems. *IEEE Transactions on Power Electronics* 34 (1): 369–377.

38 Tan, P., He, H., and Gao, X. (2016). Phase compensation, ZVS operation of wireless power transfer system based on SOGI-PLL. In: *2016 IEEE Applied Power Electronics Conference and Exposition (APEC)*, Long Beach, CA, USA, 3185–3188.

39 Robert, W.E. and Dragan, M. (2001). *Fundamentals of Power Electronics*. New York: Penguin.

40 Gati, E., Kampitsis, G., and Manias, S. (2017). Variable frequency controller for inductive power transfer in dynamic conditions. *IEEE Transactions on Power Electronics* 32 (2): 1684–1696.

41 Huang, Z., Wong, S., and Tse, C.K. (2018). Control design for optimizing efficiency in inductive power transfer systems. *IEEE Transactions on Power Electronics* 33 (5): 4523–4534.

42 Iguchi, S., Yeon, P., Fuketa, H. et al. (2015). Wireless power transfer with zero-phase-difference capacitance control. *IEEE Transactions on Circuits and Systems I: Regular Papers* 62 (4): 938–947.

43 Zhong, W. and Hui, S.Y.R. (2015). Maximum energy efficiency tracking for wireless power transfer systems. *IEEE Transactions on Power Electronics* 30 (7): 4025–4034.

44 Berger, A., Agostinelli, M., Vesti, S. et al. (2015). A wireless charging system applying phase-shift and amplitude control to maximize efficiency and extractable power. *IEEE Transactions on Power Electronics* 30 (11): 6338–6348.

45 Nguyen, B.X., Vilathgamuwa, D.M., Foo, G.H.B. et al. (2015). An efficiency optimization scheme for bidirectional inductive power transfer systems. *IEEE Transactions on Power Electronics* 30 (11): 6310–6319.

Part III

Control

5

Control for Single Pickup

In this chapter, the review of control schemes for wireless power transfer (WPT) systems is introduced in Section 5.1. Then, the closed-loop control schemes to obtain optimal transmission performances, especially maximizing efficiency control schemes, are described in Section 5.2. In addition, the design examples of the maximum efficiency tracking using perturbation and observation (P&O) method and mutual inductance estimation method are given comprehensively.

5.1 Review of Control Schemes

5.1.1 Factors Affecting Transmission Performances

In order to achieve the optimal output performances for WPT systems such as the maximum achievable efficiency, the stable power transfer, and the constant output current or voltage, the factors affecting system performances should be specified and considered in the design stage. Commonly, the system performances significantly depend on the magnetic resonant state, the mutual coupling state, and the equivalent load resistance, which will be analyzed on the basis of constant primary excitation in this section.

5.1.1.1 Effects of Magnetic Resonant State

For magnetic coupling resonant WPT systems, the fluctuation in the resonant operating state would adversely affect the capability of power transfer. Based on the analysis in Chapter 4, the secondary current I_s and the output power P_{out} of the series–series (SS) topology can be expressed as

$$\begin{cases} I_s = -\dfrac{j\omega M U_{in}}{Z_p Z_s + \omega^2 M^2} \\ P_{out} = \dfrac{\omega^2 M^2 {U_{in}}^2 R_L}{(Z_p Z_s + \omega^2 M^2)^2} \end{cases} \tag{5.1}$$

Wireless Power Transfer: Principles and Applications, First Edition. Zhen Zhang and Hongliang Pang.

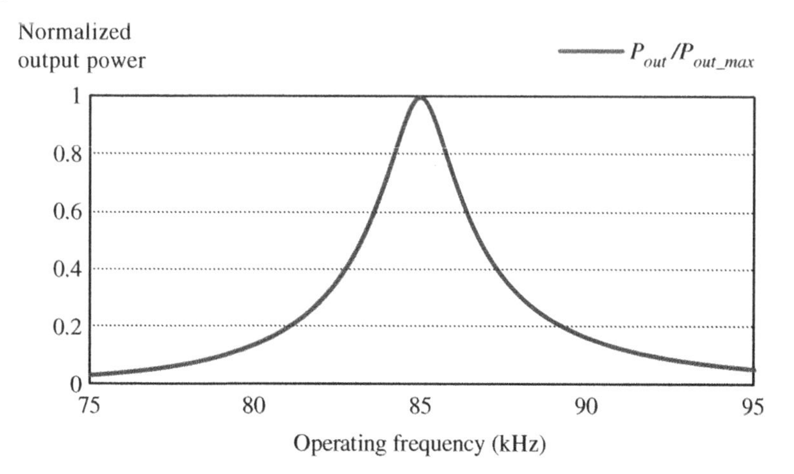

Figure 5.1 Normalized output power versus operating frequency.

Herein, Z_p and Z_s are the primary and secondary equivalent impedances, respectively.

$$\begin{cases} Z_p = R_p + j\left(\omega L_p - \dfrac{1}{\omega C_p}\right) \\ Z_s = R_s + R_L + j\left(\omega L_s - \dfrac{1}{\omega C_s}\right) \end{cases} \tag{5.2}$$

Figure 5.1 depicts the normalized output power varying with the fluctuation in the operating frequency. It can be observed that the power transfer capability drops dramatically when the operating frequency deviates from the resonant frequency. Due to the high sensitivity of the resonant state, it can be concluded that the resonant operating state is the prime condition for the normal operation of WPT systems.

5.1.1.2 Effects of Magnetic Coupling Coefficient and Load Resistance

5.1.1.2.1 Effects of Magnetic Coupling Coefficient The horizontal and vertical misalignments between the transmitting and the receiving coils result in the substantial change in the coupling coefficient. The following analysis is provided to illustrate the degradation of the output power and the transmission efficiency. Given that the WPT system operates at the resonant frequency, the output power P_{out} in Eq. (5.1) can be simplified as

$$P_{out} = \frac{\omega^2 \kappa^2 L_p L_s U_{in}{}^2 R_L}{(R_p R_s + R_p R_L + \omega^2 \kappa^2 L_p L_s)^2} \tag{5.3}$$

where κ is the coupling coefficient between the transmitting and the pickup coils. It can be expressed as

$$\kappa = \frac{M}{\sqrt{L_p L_s}} \tag{5.4}$$

Additionally, by solving the derivative equation

$$\frac{\partial P_{out}}{\partial \kappa} = 0 \tag{5.5}$$

the optimal value of the coupling coefficient κ for maximizing the output power can be derived as

$$\kappa = \sqrt{\frac{R_p R_s + R_p R_L}{\omega^2 L_p L_s}} \tag{5.6}$$

By substituting (5.6) into (5.3), the maximum output power can be calculated by

$$P_{out\_max} = \frac{{U_{in}}^2 R_L}{4(R_p R_s + R_p R_L)} \tag{5.7}$$

Regarding the transmission efficiency of WPT systems, η_p and η_s are denoted as the efficiency in the equivalent primary circuit and the efficiency in the secondary circuit of the SS topology. It can be separately calculated as

$$\begin{cases} \eta_p = \dfrac{Z_r}{R_p + Z_r} \\ \eta_s = \dfrac{R_L}{R_L + R_s} \end{cases} \tag{5.8}$$

Denote Z_r as the reflected impedance from the secondary to the primary side. Based on the SS topology, it can be given by

$$Z_r = \frac{\omega^2 M^2}{Z_s} \tag{5.9}$$

Accordingly, the transmission efficiency η can be obtained as

$$\eta = \eta_p \eta_s = \frac{\omega^2 \kappa^2 L_p L_s R_L}{(R_L + R_s)^2 R_p + \omega^2 \kappa^2 L_p L_s (R_L + R_s)} \tag{5.10}$$

It can be seen from the derivative equation $\frac{\partial \eta}{\partial \kappa} > 0$ that the transmission efficiency increases monotonically with the increase in the coupling coefficient κ. To intuitively show the relationship between the system performances and the coupling coefficient, the efficiency and the normalized output power versus the coupling coefficient are plotted in Figure 5.2. It shows that the normalized load power experiences the process of dramatically increasing and then relatively slowly decreasing, while the efficiency increases in a nonlinear trend with the increase in the coupling coefficient.

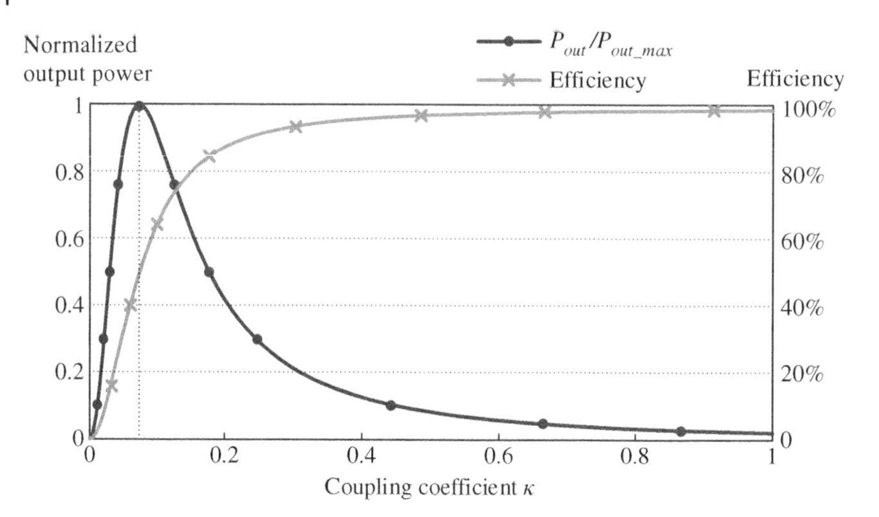

Figure 5.2 Efficiency and normalized output power versus coupling coefficient.

5.1.1.2.2 Effects of Load Resistance WPT technology has been widely used in the applications of battery loads [1], where the equivalent load resistance will vary with the charging stages. The changing equivalent load resistance results in the deviation from the optimal power transfer state. The following analysis is also provided according to two indexes of output power and transmission efficiency. By solving the following derivative equation:

$$\frac{\partial P_{out}}{\partial R_L} = 0 \tag{5.11}$$

the optimal value of load resistance for maximizing the output power can be derived as

$$R_{L\_P_{out}} = R_s + \frac{\omega^2 M^2}{R_p} \tag{5.12}$$

By substituting (5.12) into (5.3), the maximum output power with the variation in load resistance can be given by

$$P_{out\_max} = \frac{\omega^2 M^2 U_{in}{}^2}{4R_p(R_p R_s + \omega^2 M^2)} \tag{5.13}$$

To facilitate the analysis, the transmission efficiency η can be simplified using quality factors [2] as

$$\eta = \frac{1 - \dfrac{Q_R}{Q_s}}{1 + \dfrac{1}{\kappa^2 Q_R Q_p}} \tag{5.14}$$

Denote Q_p, Q_s, and Q_R as the quality factors of the primary coil, secondary coil, and secondary loop, respectively, which are given as follows:

$$\begin{cases} Q_p = \dfrac{\omega L_p}{R_p} \\ Q_s = \dfrac{\omega L_s}{R_s} \\ Q_R = \dfrac{\omega L_s}{R_s + R_L} \end{cases} \tag{5.15}$$

Similarly, by solving the following derivative equation:

$$\frac{\partial \eta}{\partial R_L} = 0 \tag{5.16}$$

the optimal value of load resistance for maximizing the transmission efficiency can be derived as

$$R_{L\_\eta} = R_s\sqrt{1 + \kappa^2 Q_s Q_p} \tag{5.17}$$

By substituting (5.17) into (5.14), the maximum transmission efficiency can be given by

$$\eta_{max} = \frac{\kappa^2 Q_p Q_s}{\left(1 + \sqrt{1 + \kappa^2 Q_p Q_s}\right)^2} \tag{5.18}$$

which shows that the maximum efficiency increases with the increasing of Q_p and Q_s.

As shown in Figure 5.3, the tendency of efficiency and normalized output power versus equivalent load resistance are plotted intuitively, where the efficiency is only 50% on the condition of maximum output power. In addition, it can be seen from (5.8) that the system efficiency possesses an extreme point with the variation in the load resistance if the equivalent line resistance in the secondary side is ignored. Furthermore, the small line resistance in the secondary side will generally result in the optimal efficiency point under light load condition.

5.1.2 Controls Ensuring Transmission Performances

Figure 5.4 summarizes the typical control schemes of WPT systems to ensure the transmission performance, which can be classified into three groups: the resonant control, the efficiency control, and the output control. The resonant state is the prerequisite to WPT systems, which should be addressed first. The closed-loop frequency tracking can adaptively adjust the operating frequency to track the split frequencies in the overcoupled region [3–8]. Moreover, dynamic reconfigurable and controllable impedance matching [9–16] circuits switched

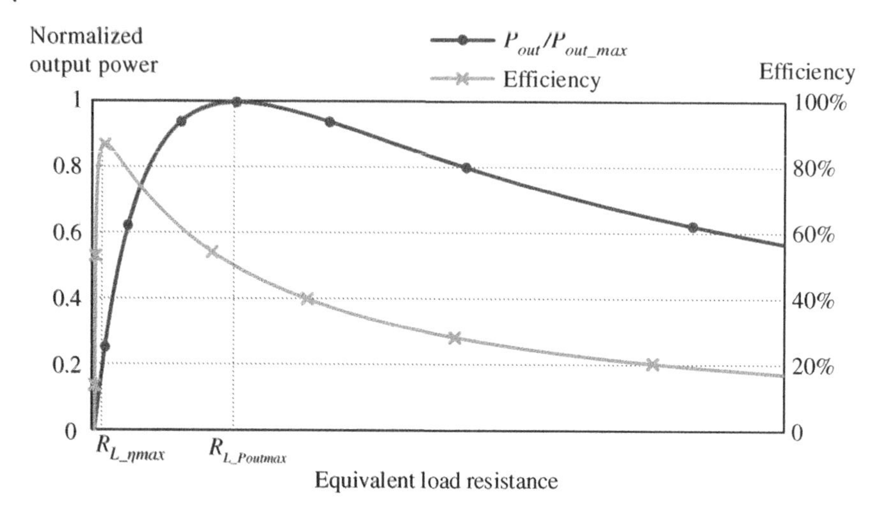

Figure 5.3 Efficiency and normalized output power versus equivalent load resistance.

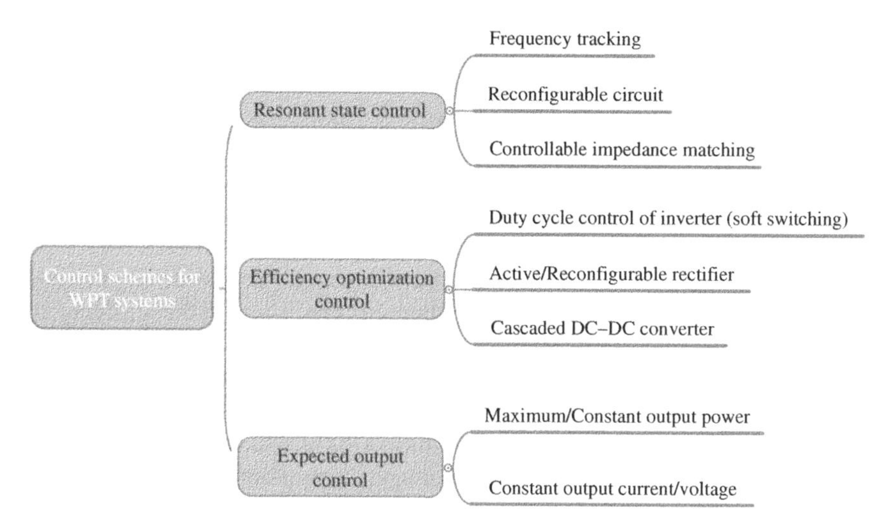

Figure 5.4 Summarization of WPT control schemes.

by relays or semiconductor devices are proposed to ensure the maximum power transfer capability, allowing the system to operate at the resonant state. However, the stability of the resonant operating state cannot ensure the optimal system performances which are highly susceptible to the fluctuations in the mutual coupling state and equivalent load resistance. Thus, the optimizing schemes for system performances are needed, which can be classified into three categories of magnetic coupling mechanism, compensating circuit topology, and closed-loop control, where the design of the magnetic coupling mechanism can effectively improve the transmission distance and the misalignment tolerance which are introduced in detail in Chapter 3. In addition, the efforts for the hybrid topology design of the compensating circuit are mainly contributed on the load-independent output characteristics and soft switching of the inverter, which are described in Chapter 4. In comparison to the above two optimizing schemes, the closed-loop control schemes are more commonly used due to the flexibility and universality.

This chapter introduces the closed-loop control schemes which can be mainly divided into efficiency optimization control and expected output control. Especially in efficiency optimization control schemes, many efforts have been contributed in previous studies, such as the frequency and duty cycle control of inverters for realizing soft switching operation [17, 18], cascaded DC–DC converter [19–27], and active rectifier in pickup side [28–34]. As for the maximum efficiency tracking schemes, it should be noted from Eq. (5.17) that the fluctuations in the mutual coupling coefficient and equivalent load resistance can both be viewed as the deviation in the optimal load value, which means that the efficiency can be controlled by regulating the equivalent load resistant, which is easier to adjust compared with the coupling coefficient. On the other hand, on the basis of the practical demands such as batteries charging [35, 36], the expected output control schemes can be summarized as the output power control [37–39] and constant current or constant voltage control [40–43].

5.2 Maximizing Efficiency Control Schemes

5.2.1 Resonant Control Schemes

Concerning the high-power transfer capability, satisfactory results are only obtained when the system is operating at or close to resonance. However, the resonant operating state may fluctuate due to the dependence on the parameters of resonant components, magnetic coupling coefficient and equivalent load resistance. In order to ensure the resonant operating state, the control schemes can be classified into two types:

(1) ***Frequency tracking*** – The frequency tracking method can adaptively regulate the resonant state via adjusting the resonant operating frequency.
(2) ***Controllable impedance matching*** – The controllable impedance matching method can dynamically adjust the impedance mismatches without changing the resonant frequency.

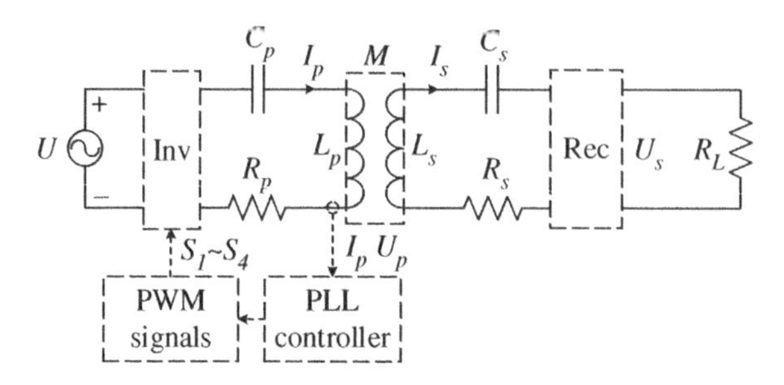

Figure 5.5 Block diagram of ZPA-based frequency tracking control without frequency splitting.

5.2.1.1 Frequency Tracking

5.2.1.1.1 Zero Phase Angle (ZPA) Control For dynamic WPT systems, even if the resonant components are strictly precalculated, the leakage inductances of the magnetic coupling mechanism will be affected by the displacement between the coils, which will result in the optimal resonant frequency varying with the magnetic coupling coefficient. Therefore, the optimal operating frequency is a function of magnetic coupling coefficient, rendering optimal frequency tracking inevitable.

Single ZPA Frequency Without Frequency Splitting Commonly, the optimal frequency tracking is realized by tuning the operating frequency with a zero phase angle (ZPA) state in the primary side [4]. As shown in Figure 5.5, the ZPA frequency is derived in a closed loop, which is performed by sensing the primary current I_p and source voltage U_p. The phase angle θ_p between I_p and U_p can be calculated as

$$\theta_p = \tan^{-1}\left(\frac{\left({R_L}^2 + \left(\omega L_s - \frac{1}{\omega C_s}\right)^2\right)\left(\omega L_p - \frac{1}{\omega C_p}\right) - \omega^2 M^2\left(\omega L_s - \frac{1}{\omega C_s}\right)}{\left({R_L}^2 + \left(\omega L_s - \frac{1}{\omega C_s}\right)^2\right)R_p + \omega^2 M^2 R_L}\right) \tag{5.19}$$

Then, the phase angle θ_p can be measured by zero-cross decoction and controlled to maintain zero by the phase-locked loop (PLL) controller.

Multiple ZPA Frequencies with Frequency Splitting In the overcoupled region, there may exist three frequencies for ZPA operation due to the phenomenon of frequency splitting [5] that will bring instabilities in the control process. The phase angle θ_p varying with the operating frequency is shown in Figure 5.6. An appropriate parameter design method to avoid frequency splitting [6] is the most direct way to overcome the problem. However, this method requires the limitation within a certain range of the coupling coefficient. A primary power flow control method for battery charging is designed to realize the minimum input VA rating exploiting the splitting characteristic [7]. According to the related analysis in Chapter 7,

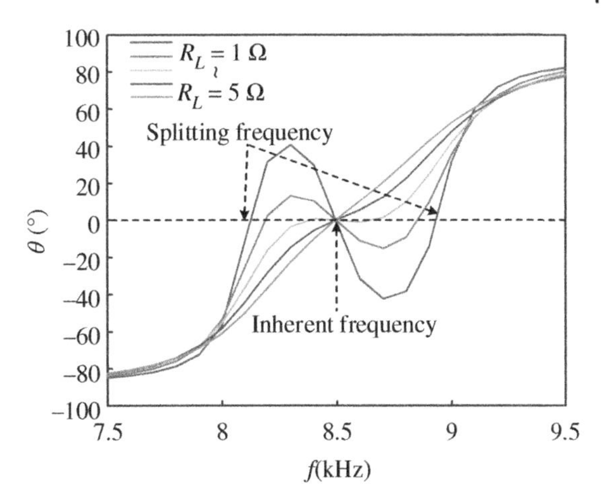

Figure 5.6 Phase angle between the primary current and voltage varying with the operating frequency.

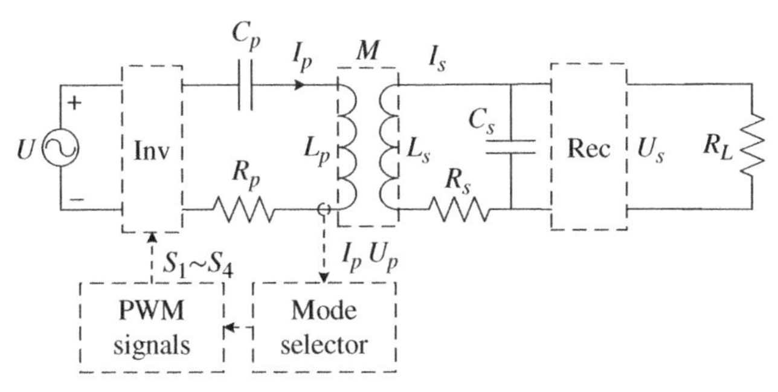

Figure 5.7 Block diagram of ZPA-based frequency tracking control with PLL controller.

the output power of the bifurcation frequencies would be higher than the original resonant frequency under a voltage input source, and as the input is the current source, the result is opposite. Accordingly, the different charging modes of battery can be realized by selecting the three ZPA frequencies according to the input source type, which can be seen in Figure 5.7. Nevertheless, the operating coupling state is fixed.

A PLL control method is proposed to track the optimal frequency under various coupling conditions by zeroing the phase difference between the output current and the source voltage [8]. Based on the forced oscillations theory, the average power supplied by the source U_{in} to the load R of a series L–C–R network gets maximized when the load current i is in phase with the exciting voltage. The phase angle θ_s between I_s and U_p can be calculated as

$$\theta_s = \tan^{-1}\left(\frac{\omega^2M^2 + R_pR_L - \left(\omega L_p - \frac{1}{\omega C_p}\right)\left(\omega L_s - \frac{1}{\omega C_s}\right)}{R_p\left(\omega L_s - \frac{1}{\omega C_s}\right) + R_L\left(\omega L_p - \frac{1}{\omega C_p}\right)}\right) \tag{5.20}$$

Ignoring the primary line resistance R_p, the ZPA frequencies can be calculated by

$$\omega^2M^2 - \left(\omega L_p - \frac{1}{\omega C_p}\right)\left(\omega L_s - \frac{1}{\omega C_s}\right) = 0 \tag{5.21}$$

The maximization of the output power is achieved at ZPA frequencies f_1 and f_2 as

$$f_{1,2} = \sqrt{\frac{\frac{L_p}{C_s} + \frac{L_s}{C_p} \pm \sqrt{\left(\frac{L_s}{C_p} - \frac{L_p}{C_s}\right)^2 + \frac{4M^2}{C_pC_s}}}{8\pi^2(L_pL_s - M^2)}} \tag{5.22}$$

where the corresponding phase angles at frequencies f_1 and f_2 are 0° and 180°, respectively. Compared with the inherent resonant frequency, the maximum output power is achieved at ZPA frequencies f_1 and f_2, while there is no significant reduction in the transmission efficiency. Accordingly, the control process can be realized by controlling the phase angle θ_s to maintain zero by the PLL controller, which can be seen in Figure 5.8. Although, it is worth highlighting that this method avoids the instability issues and restrictions caused by the bifurcation phenomenon, the neglect of secondary line resistance will lead to control errors. Moreover, additional communication is required in this method, which increases the overall system cost.

It should be noted that the above analysis is based on the resistive load, and the mismatch in the resonant state is caused by the fluctuation in mutual inductance. If the above conditions are not met, the ZPA control schemes are no longer applicable. In order to solve the mismatch under broader constraints, a reflection coefficient [9, 10] is introduced as

$$\Gamma(\omega) = \left|\frac{Z_r(\omega) - Z_s(\omega)}{Z_r(\omega) + Z_s(\omega)}\right| \tag{5.23}$$

According to the well-known condition for maximum power transfer of a matching network from the source to the load, the smaller the magnitude of the reflection coefficient, the more power can be transferred from the source to the load.

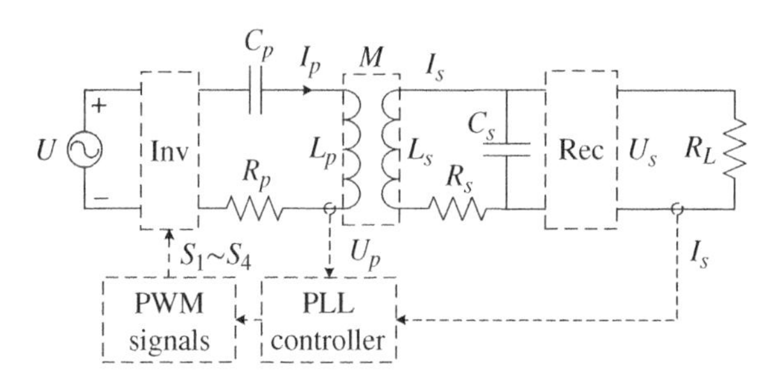

Figure 5.8 Block diagram of ZPA-based frequency tracking control with frequency splitting.

Accordingly, the control of minimizing the magnitude of the reflection coefficient at the desired band will effectively solve the problem of resonant mismatch.

The frequency control scheme for WPT systems is easy to implement, which can regulate the resonant state without using additional circuits. However, frequency control methods also show the following disadvantages:

(1) For a secondary resonant circuit with a fixed inductor and series capacitor, the control of the primary ZPA cannot achieve the optimal output power or system efficiency when the inverter frequency does not match the secondary resonant frequency.
(2) The ZPA control methods are not suitable for dynamic WPT due to the complex control process in the case of frequency splitting.
(3) The wide operating frequency range of the inverter will bring the complicated interference between the magnetic flux due to this varying frequency and the charging devices, and it is difficult to reduce the signal mixing and jamming effects due to this interference.

5.2.1.2 Controllable Impedance Matching

As a potential candidate to increase transmission efficiency, the controllable impedance matching method can effectively deal with the mismatch in resonant state due to the variation in the coupling condition by utilizing a controllable resonant unit to expand the resonant operating range without any auxiliary control circuits. The controllable resonant units mainly include capacitor array [11–13] and reconfigurable loop array [14–16].

As shown in Figure 5.9, the capacitor array can cover a wide range of capacitance values, where the impedance can be automatically adjusted to track the optimal impedance matching point in the case of varying coupling. Meanwhile, a search algorithm for finding the optimum capacitance configuration is needed. Although this method can dynamically adjust the impedance mismatches without changing the resonant frequency, the challenge is that the required complex capacitor matrix in the circuit will add system size, cost, and the controller complexity. In addition, the equivalent series resistance introduced by the series and parallel connection of capacitors, which will influence the transmission efficiency, cannot be overlooked.

In addition to the capacitor array, the reconfigurable loop array can also be used in controllable impedance matching for WPT system. Figure 5.10 depicts the equivalent circuit model of a four-coil WPT system with reconfigurable loop array. Similarly, the reconfigurable loop array can expand the resonant operating range by adaptively switching between different sizes of drive loops and load loops. Compared with the conventional four-coil WPT system, higher efficiency and larger misalignment tolerance can be achieved, while the system size is also inevitably increased. In Ref. [16], the number of turns of the power coil can

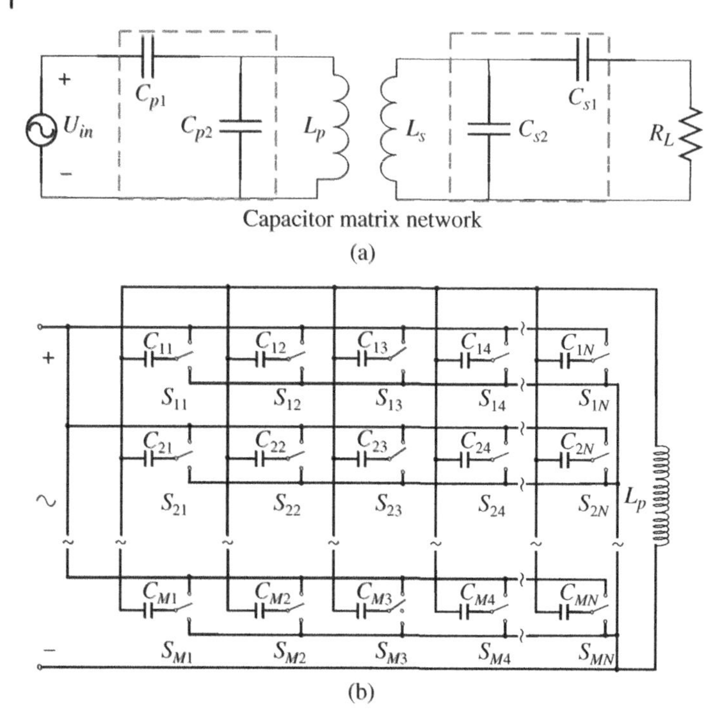

Figure 5.9 Equivalent WPT system with $M \times N$ capacitor matrix: (a) System configuration; (b) Topology of capacitor matrix.

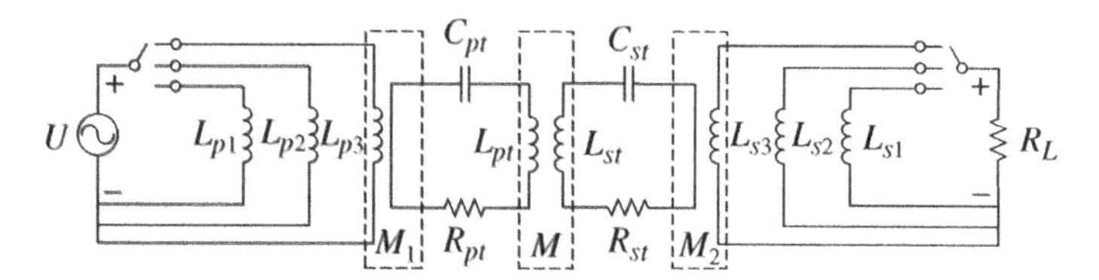

Figure 5.10 Equivalent circuit model of a four-coil WPT system with reconfigurable loop array.

be adaptively changed, which can alter the coupling between two coils in the transmitter or receiver side to realize impedance matching for a large range of load resistance.

5.2.2 Maximizing Efficiency Control Schemes Based on Equivalent Load Resistance Adjustment

The tuning methods of resonance mismatch were analyzed in the Section 5.2.1. However, the system performances will still vary with the coupling state and the load resistance. As can be known from the above analysis, the maximum efficiency can be achieved only at the optimal load resistance point for WPT systems, which depends on the mutual inductance and the line resistance. This section focuses on the maximum efficiency control methods for WPT systems based on matching the optimal value of equivalent load resistance.

5.2.2.1 Equivalent Load Resistance Adjustment Schemes

5.2.2.1.1 Using DC–DC Converters DC–DC converters are designed to adjust the input voltage to the desired output voltage, which are normally used in the secondary side for WPT systems to regulate the output voltage and power. Meanwhile, the DC–DC converters can also be utilized to control the equivalent load resistance to achieve the maximum efficiency by transforming the load resistance to the optimal value. Figure 5.11 depicts a WPT system with the DC–DC converter connected behind the rectifier, where U_{in} and U_{out} are the input and output voltages, respectively, R_e is the equivalent resistance (seen from the output of the rectifier), and R_L is the load resistance. The relationship between R_e and R_L can be obtained assuming that there is no power loss:

$$\frac{U_{in}^{2}}{R_e} = \frac{U_{out}^{2}}{R_L} \tag{5.24}$$

The DC–DC converters can be generally divided into three categories of buck, boost, and buck–boost, and the basic topologies are shown in Figure 5.12.

For buck converter, the relationship between the input voltage U_{in} and the output voltage U_{out} can be deduced as

$$U_{out} = DU_{in} \tag{5.25}$$

where D (0–1) is the duty cycle, and by substituting (5.25) into (5.24),

$$R_e = \frac{R_L}{D^2} \tag{5.26}$$

The equivalent load resistance can be adjusted from R_L to $+\infty$ by controlling the duty cycle.

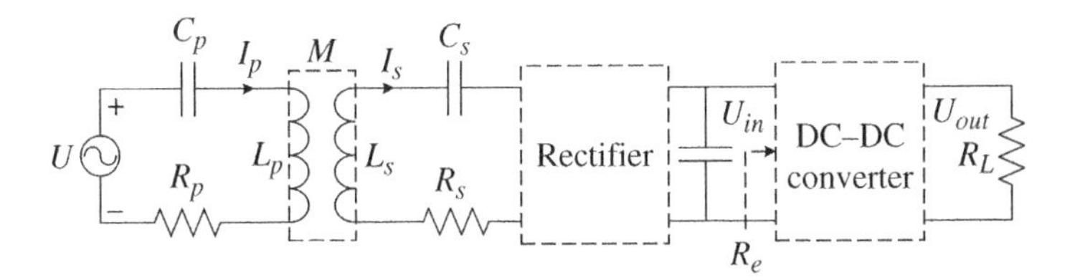

Figure 5.11 Equivalent circuit of a WPT system with the DC–DC converter connecting behind the rectifier.

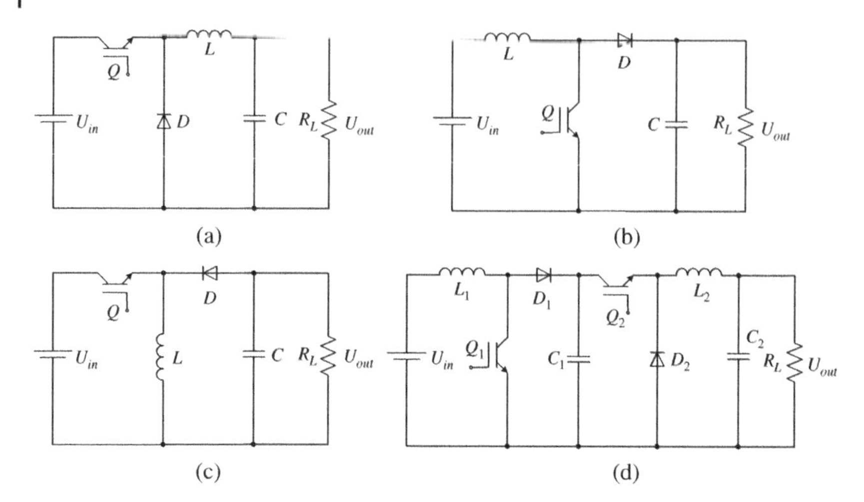

Figure 5.12 Basic DC–DC converter topologies: (a) buck type; (b) boost type; (c) buck–boost type; (d) cascaded boost–buck type.

Similarly, for boost converter:

$$U_{out} = \frac{1}{1-D} U_{in} \tag{5.27}$$

and

$$R_e = (1-D)^2 R_L \tag{5.28}$$

The adjustable range of the equivalent load resistance is from 0 to R_L.

For buck–boost converter

$$U_{out} = \frac{D}{1-D} U_{in} \tag{5.29}$$

and

$$R_e = \frac{(1-D)^2}{D^2} R_L \tag{5.30}$$

The adjustable range of the equivalent load resistance is from 0 to $+\infty$. The transfer characteristics of the basic converters are shown in Table 5.1, where the boost-type converter is more suitable for practical applications as its continuous input current is easier to control. However, the boost-type converter has a limited range of conversion resistance; thus, a cascaded boost–buck-type converter is proposed [19], which is shown in Figure 5.12d. The relationship between the input voltage U_{in} and the output voltage U_{out} can be deduced as

$$U_{out} = \frac{D_2}{1-D_1} U_{in} \tag{5.31}$$

Table 5.1 Comparison of the basic DC–DC converters.

Topology	Output voltage	Equivalent load resistance	Regulating range	Input current
Buck	DU_{in}	R_L/D_2	$R_L \sim +\infty$	Discontinuous
Boost	$U_{in}/(1-D)$	$(1-D_2)R_L$	$0 \sim R_L$	Continuous
Buck–boost	$U_{in}D/(1-D)$	$(1-D_2)R_L/D_2$	$0 \sim +\infty$	Discontinuous
Cascaded boost–buck	$U_{in}D_2/(1-D_1)$	$(1-D_2)R_L/D_2$	$0 \sim +\infty$	Continuous

where D_1 and D_2 (0–1) are the duty cycles of the switches Q_1 and Q_2, respectively. By substituting (5.31) into (5.24),

$$Re = \left(\frac{1-D_1}{D_2}\right)^2 R_L \tag{5.32}$$

The input resistance of the cascaded boost–buck-type converter is easy to control due to its continuous input current feedback. Compared with the basic converters, the cascaded boost–buck converter provides more flexibility in the control process because of the two controllable duty cycles of the boost- and buck-type converters, which can be separately analyzed.

Accordingly, by adjusting the duty cycle of the DC–DC converter, the equivalent load resistance can be transformed to the optimal value in the condition of varying coupling state and load resistance. Normally, the optimal load resistance closed-loop control schemes can be classified into two types as secondary-side control and dual-side control; the block diagrams are shown in Figure 5.13. For secondary-side control, the controllable variable of the voltage conversion ratio *r* can be expressed as

$$r = \frac{U_{out}}{U_s} \tag{5.33}$$

where U_s and U_{out} are the input and output voltages of the DC–DC converter, respectively.

In addition, in order to regulate the output power on the promise of ensuring the system efficiency, the output voltage of the inverter can be regulated by adding DC–DC converter in the primary side, namely dual-side control, which can be seen in Figure 5.13b. Thus, the maximum efficiency can be realized using two controllable variables: the input voltage conversion ratio r_1 and the load resistance conversion ratio r_2, which can be expressed as

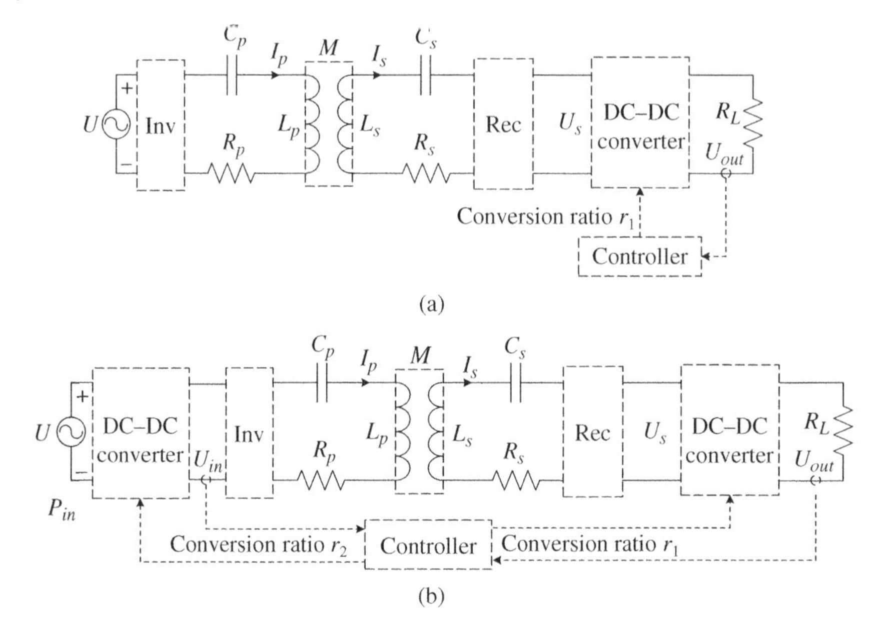

Figure 5.13 Diagram of the optimal load resistance closed-loop control schemes: (a) secondary-side control; (b) dual-side control.

$$\begin{cases} r_1 = \dfrac{U_p}{U_{dc}} \\ r_2 = \dfrac{U_{out}}{U_s} \end{cases} \tag{5.34}$$

where U_p, U_{dc}, U_s, and U_{out} are the input and output voltages of the primary and secondary DC–DC converter, respectively.

5.2.2.1.2 Using Active Rectifier in the Secondary Side There are also a few studies on using active rectifiers to adjust the equivalent load resistance [28–34]. Figure 5.14 shows the typical circuit topology of WPT system with active rectifier in the secondary side. By replacing either two off-diagonals or all of the diodes of the full-bridge diode rectifier with active switches, the output voltage can be controlled by adjusting the turn-on time of the active switches. Normally, the modulation schemes mainly include pulse-width modulation (PWM) and pulse-density modulation (PDM).

For PWM control, the regulation of the output voltage is realized by altering the phase-shift angle between the switching signals of the two rectifier legs. The functionality of the active rectifier is hence comparable to a passive rectifier followed

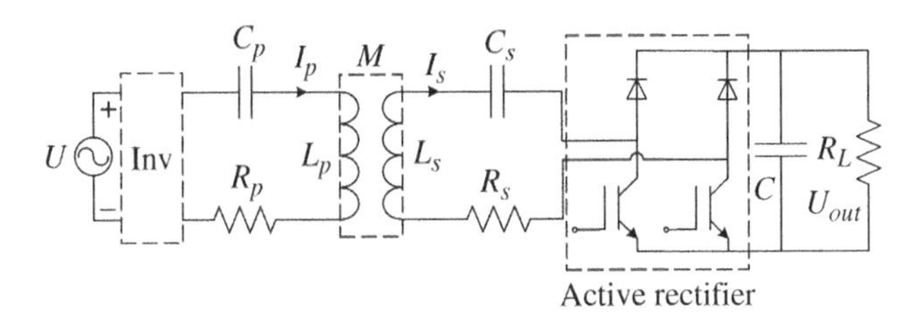

Figure 5.14 Equivalent circuit of a WPT system with the active rectifier in the secondary side.

by a boost converter. Based on Fourier series, the fundamental output voltage can be calculated as

$$U_{in} = \frac{2\sqrt{2}}{\pi} U_{out} \sin\left(\frac{D\pi}{2}\right) \tag{5.35}$$

where D (0–1) is the pulse width of the secondary voltage, and $1 - D$ is the duty cycle of the active switches. By substituting (5.35) into (5.24), the equivalent load resistance can be derived as

$$R_e = \frac{4}{\pi^2} R_L [1 - \cos(D\pi)] \tag{5.36}$$

For PDM control, the ratio of the number of pulse to the number of switching cycles, namely pulse density, is regulated by controlling the switching time of Q_1 and Q_2 in a whole switching cycle. The waveform of the input voltage is shown in Figure 5.15b, and it is worth noting that the waveform is not necessarily periodic. Based on Fourier series, the fundamental output voltage can be calculated as

$$U_{in} = \frac{2\sqrt{2}}{\pi} U_{out} D \tag{5.37}$$

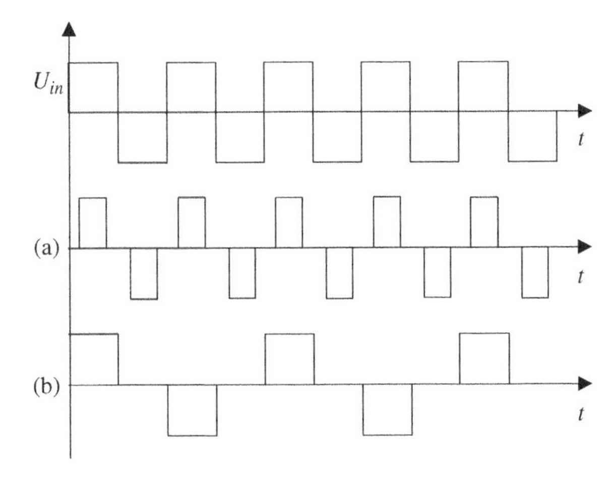

Figure 5.15 Input voltage of active rectifier in the secondary side with different modulation schemes: (a) pulse-width modulation; (b) pulse-density modulation.

where D (0–1) is the pulse width of the secondary voltage, and $1 - D$ is the duty cycle of the active switches. By substituting (5.37) into (5.24), the equivalent load resistance can be derived as

$$R_e = \frac{8}{\pi^2} R_L D^2 \tag{5.38}$$

Accordingly, the equivalent load resistance can be adjusted by controlling the pulse width D. The use of active rectifier can effectively save the system cost, yet the adjustable range of the equivalent load resistance is limited to a smaller area in comparison to using DC–DC converters.

5.2.2.2 Maximizing Efficiency Tracking Schemes

As mentioned above, the equivalent load resistance can be adjusted by additional DC–DC converter and active rectifier. Ideally, the optimal load resistance value is fixed, and the real-time load resistance can be determined by sending the load voltage and current. Nevertheless, for dynamic WPT system, the optimal load resistance will change as the variation of coupling state. Thus, how to track the optimal load resistance under the variation of coupling coefficient is key to realize the maximum system efficiency. Normally, the P&O and estimation of load resistance and coupling coefficient are the methods for tracking the optimal load resistance.

5.2.2.2.1 Perturbation and Observation (P&O) P&O method is normally used to realize the maximum power point tracking for photovoltaic converters, which also has the similar characteristics with the maximum efficiency point tracking for WPT system [20, 23, 24]. Instead of directly calculating the optimal value of load resistance under the variation of coupling coefficient, the maximum efficiency point tracking is realized by adding a series of perturbations. Figure 5.16 depicts the flowchart of the P&O-based maximum efficiency point tracking method; the corresponding block diagram is shown in Figure 5.18a, where the secondary DC–DC converter (which can also be replaced by the active rectifier in the secondary side) is used to perturb the operating point by adjusting the conversion ratio r_1, and the primary DC–DC converter (which can also be replaced by the phase-shift control of the inverter in the primary side) is used to regulate the output power by adjusting the conversion ratio r_2.

(1) The expected system efficiency η_e is defined, and the conversion ratios r_1 and r_2 are initialed.
(2) In order to ensure the accuracy and stability of the control, a diminutive and constant perturbation Δr_1 is applied in a step-by-step manner to change the equivalent load resistance.
(3) The output power P_{out} is measured and regulated to the expected value by adjusting the conversion ratio r_2.

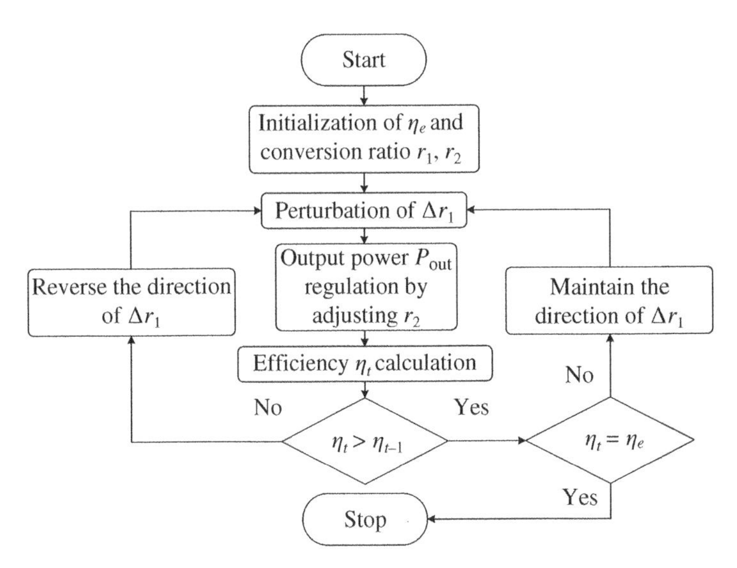

Figure 5.16 Flowchart of P&O-based maximum efficiency point tracking scheme with wireless communication link.

(4) Following each perturbation, the real-time system efficiency η_t is calculated and recorded after the regulation of the output power.
(5) If $\eta_t > \eta_{t-1}$, it means that the adjusted equivalent load resistance approaches the optimal value, and thus, a perturbation with a same sign needs to be applied in the following stage. On the contrary, if $\eta_t > \eta_{t-1}$, it indicates that the adjusted equivalent load resistance deviates from the optimal value, and a perturbation with an opposite sign needs to be applied. Then, the above steps are repeated until the system efficiency is equal to the expected value η_e, and the control procedure is completed.

During the whole process, two variables are needed to be calculated including the output power P_{out} and the system efficiency η_t with three feedback signals including the output current I_{out} and voltage U_{out} in the secondary side and the input power P_{in}. The major difficulty in the above scheme is the wireless communication between the primary and secondary sides. To solve this problem, a simplified method is to regulate the output voltage by the secondary DC–DC converter (which can also be replaced by an active rectifier in the secondary side) and perturb the operating point by the primary DC–DC converter (which can also be replaced by the phase-shift control of the inverter in the primary side). The flowchart and circuit diagram are shown in Figures 5.17 and 5.18b, respectively.

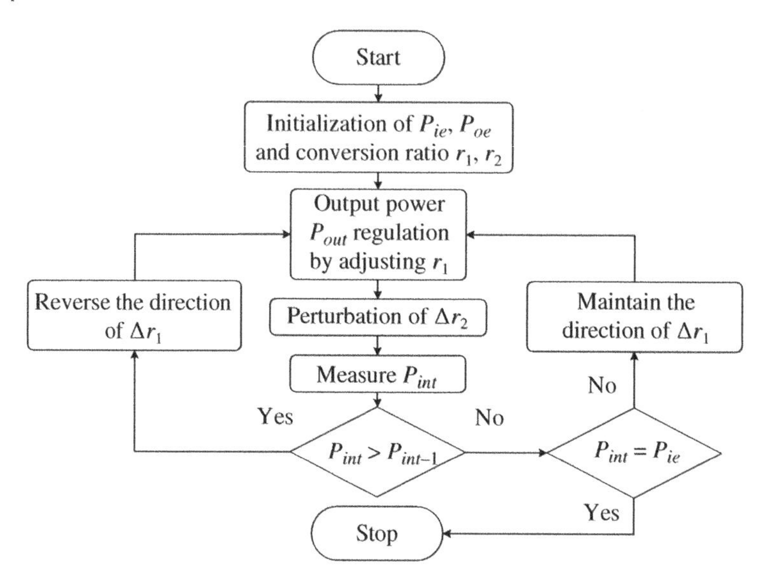

Figure 5.17 Flowchart of P&O-based maximum efficiency point tracking scheme without wireless communication link.

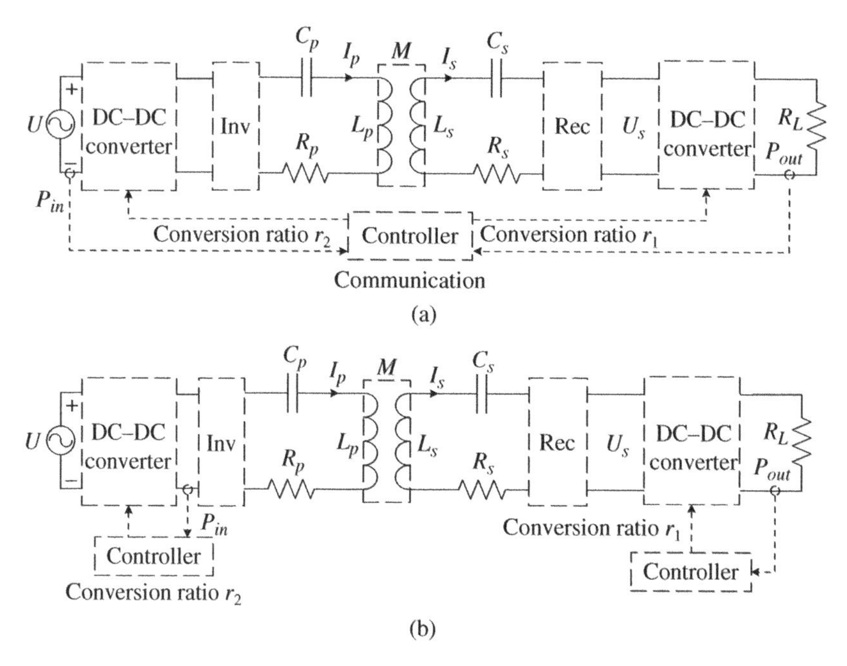

Figure 5.18 Diagram of P&O-based maximum efficiency point tracking control schemes: (a) with wireless communication link; (b) without wireless communication link.

(1) The expected system efficiencies P_{ie}, P_{oe} are defined, and the conversion ratios r_1 and r_2 are initialed.
(2) The output power P_{out} is measured and regulated to the expected value by adjusting the conversion ratio r_1.
(3) In order to ensure the accuracy and stability of the control, a diminutive and constant perturbation Δr_2 is applied in a step-by-step manner to change the equivalent load resistance.
(4) Following each perturbation, the input power P_{int} is measured and recorded.
(5) If $P_t < P_{t-1}$, it means that the adjusted equivalent load resistance approaches the optimal value, and thus a perturbation with a same sign needs to be applied in the following stage. On the contrary, if $P_t > P_{t-1}$, it indicates that the adjusted equivalent load resistance deviates from the optimal value, and a perturbation with an opposite sign needs to be applied. Then, the above steps are repeated until the input power is equal to the expected value P_{ie}, and the control procedure is completed.

This method can save the wireless communication link between the primary and the secondary sides, which can effectively improve the control stability and rapidity.

5.2.2.2.2 Estimation of Load Resistance and Coupling Coefficient For P&O method, there may exist long adjusting time and drastic fluctuations under frequent variations in the coupling coefficient, which may result in the failure of the maximum efficiency point tracking, especially in dynamic WPT system. If the mutual inductance can be accurately estimated in real time, the optimal load resistance can be calculated, and thus the maximum efficiency is worked out. In most of the WPT systems, when the self-inductance is compensated, the resonant state can be considered constant regardless of the variation in the coupling state. Accordingly, the mutual inductance can be calculated by the circuit information in the system.

From the equivalent circuit shown in Figure 5.19, the input impedance Z_{in} can be expressed as

$$Z_{in} = Z_p + \frac{\omega^2 M^2}{Z_s} \tag{5.39}$$

where Z_p and Z_s are defined as the primary and secondary equivalent impedances, respectively, which are given by

$$\begin{cases} Z_p = R_p + j\left(\omega L_p - \dfrac{1}{\omega C_p}\right) \\ Z_s = R_s + Z_L + j\left(\omega L_s - \dfrac{1}{\omega C_s}\right) \end{cases} \tag{5.40}$$

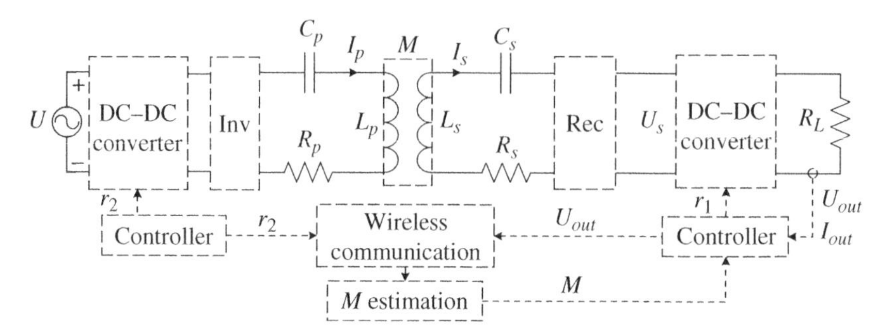

Figure 5.19 Diagram of the maximum efficiency point tracking control scheme based on mutual inductance estimation.

Thus, the mutual inductance can be deduced as

$$M = \frac{1}{\omega}\sqrt{(Z_{in} - Z_p)Z_s} \tag{5.41}$$

where Z_{in} and Z_L can be calculated by

$$\begin{cases} Z_{in} = \dfrac{U_{in}}{I_p}\angle\alpha \\ Z_L = \dfrac{U_{out}}{I_s}\angle\beta \end{cases} \tag{5.42}$$

where α and β are the impedance angles of the equivalent input impedance and the load. When the system operates at the resonant frequency, the self-inductance of the coils can be matched with the compensation capacitances, and considering that the load in the analysis is resistive, the mutual inductance can be simplified as

$$M = \frac{1}{\omega}\sqrt{\frac{(U_{in} - I_pR_p)(U_{out} + R_sI_s)}{I_pI_s}} \tag{5.43}$$

In Eq. (5.43), R_p, R_s, and ω are the known parameters. Accordingly, the mutual inductance can be calculated by measuring the current and voltage of both the primary and secondary sides. A dynamic coupling coefficient estimation method is proposed in Ref. [44] to realize the maximum efficiency tracking. As can be seen in Figure 5.19, the conversion ratio r_1 of the secondary converter is regulated to maintain the equivalent load resistance equal to the optimal value. The real-time load resistance can be detected by measuring the output voltage and current. The conversion ratio r_2 of the primary converter is regulated to maintain the output voltage by transmitting U_{out} to the primary side. Meanwhile, the variation in the conversion ratio r_2 is recorded and transferred to the secondary side. In a short period, the coupling coefficient k is considered to change linearly with respect to the conversion ratio r_2. Accordingly, the mutual inductance can be estimated

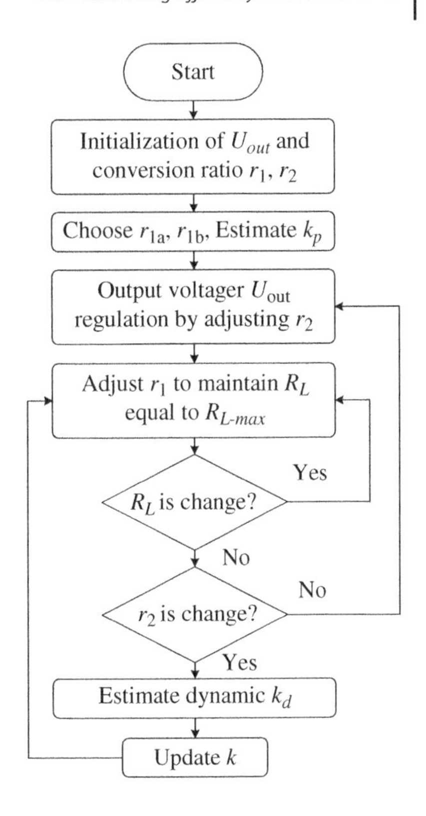

Figure 5.20 Flowchart of the maximum efficiency point tracking control scheme based on mutual inductance estimation.

by the initial value of the coupling coefficient and the variation of conversion ratio r_2. Then, the updated value of the optimal load resistance can be calculated according to the estimated coupling coefficient. The control flowchart is shown in Figure 5.20.

It is believed that this method can be applied to estimate the mutual inductance of a dynamic WPT system. However, one of the issues is that the calculation of the maximum efficiency is derived based on the fundamental components of the waveforms. Higher order harmonics will affect the accuracy of the coupling estimation. In addition, the estimation process requires the measurement to be conducted in two sides simultaneously, which means that the communication is necessary, that is the cost or the increased complexity that the system needs to afford. Then, the question is how to estimate the mutual inductance by only using unilateral data. When the operating frequency deviates slightly from the resonant frequency, the load resistance can be expressed as

$$Z_L = \frac{\omega^2 M^2}{Z_{in} - Z_p} - \left(R_s + j\left(\omega L_s - \frac{1}{\omega C_s}\right)\right) \tag{5.44}$$

Considering the load in the analysis is resistive, the imaginary part of Z_L should be null, and the further deduction is shown as

$$Im\left(\frac{\omega^2M^2}{Z_{in}-Z_p}-j\left(\omega L_s-\frac{1}{\omega C_s}\right)\right)=0 \tag{5.45}$$

Then, the mutual inductance can be derived as

$$M=\sqrt{\frac{\omega L_s-\frac{1}{\omega C_s}}{\omega^2 Im(Z_{in}-Z_p)^{-1}}} \tag{5.46}$$

By substituting (5.46) into (5.44), the load resistance can be calculated as

$$R_L=\frac{\omega L_s-\frac{1}{\omega C_s}}{(Z_{in}-Z_p)(Z_{in}-Z_p)^{-1}}-\left(R_s+j\left(\omega L_s-\frac{1}{\omega C_s}\right)\right) \tag{5.47}$$

Accordingly, the mutual inductance and load resistance can be determined by measuring the amplitude and phase shift of the input voltage and current. However, it should be noted that this estimation method requires the secondary unit to operate at the nonresonant state [45]. Generally, an accurate estimation can be obtained using an operating frequency outside a narrow frequency band (typically 1%) centered on the resonant frequency, which will not sacrifice the system efficiency significantly.

5.2.2.3 Maximizing Efficiency Control Schemes – Design Examples

5.2.2.3.1 A Design Example of Maximum Efficiency Tracking Using P&O Method

To verify the effectiveness of the design example of the maximum efficiency tracking using P&O method, the simulation model is established using MATLAB®/SIMULINK with the key parameters as listed in Table 5.2. The system

Table 5.2 Parameters of the design example.

Item	Symbol	Value
Self-inductance of the primary coil	L_p	60 μH
Self-inductance of the secondary coil	L_s	60 μH
Mutual inductance	M	6 μH
Compensating capacitance of the primary coil	C_p	58.49 nF
Compensating capacitance of the secondary coil	C_s	58.49 nF
Operating frequency	f	85 kHz
Load resistance	R_L	5 Ω
Line resistance of the primary coil	R_p	0.5 Ω
Line resistance of the secondary coil	R_s	0.5 Ω

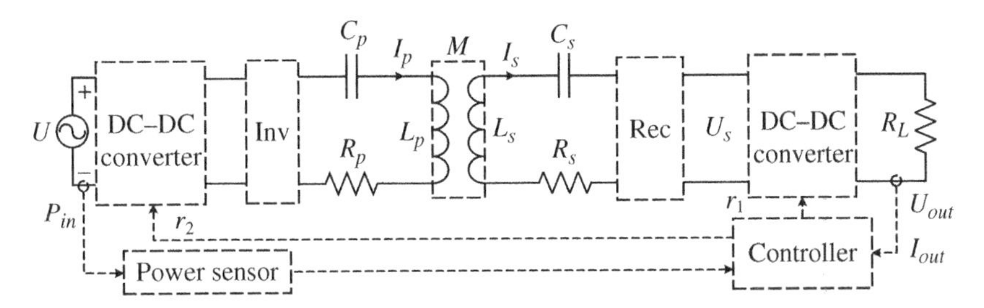

Figure 5.21 System configuration of P&O-based maximum efficiency tracking.

Table 5.3 An example of the experimental system devices.

Item	Symbol	Parameter/type
Power MOSFET	S_1–S_4	Cree/C3M0065090D
Diodes of rectifier	D_1–D_4	SiC Schottky diodes
Compensated capacitor	C_p, C_s	KEMET R41
Inner coil diameter	d_1	10 cm
Outer coil diameter	d_2	20 cm
Power sensor	—	Mini-Circuits ZX47-40LN-S+
Current sensor	—	LEM/LA 25-NP
Electronic load	—	
DC–DC converter	—	
DSP platform	—	DSP28335

configuration for the tracking of the optimal load is shown in Figure 5.21. In the control process, the input power is detected by the power sensor and converted to an analog signal, and the output voltage and current are measured and fed back to the controller. The three feedback signals are used to determine the control variables including the conversion ratios r_1 and r_2.

Table 5.3 gives an example of the experimental system device, which consists of a power sensor and current sensor for the sensing of the input power and load current, two resonance coils and compensated capacitors, a rectifier using SiC Schottky diodes, a cascaded boost–buck DC–DC converter, an electronic load, and a control system based on DSP 28335 platform, where the P&O-based tracking control is implemented based on the feedbacked data. The switching frequency of the DC–DC converter is 20 kHz, and its dimension is 220 mm × 160 mm.

As shown in Figure 5.22a, the experiment of the maximum efficiency tracking with varying coupling coefficient starts from time t_0 with the initial relative

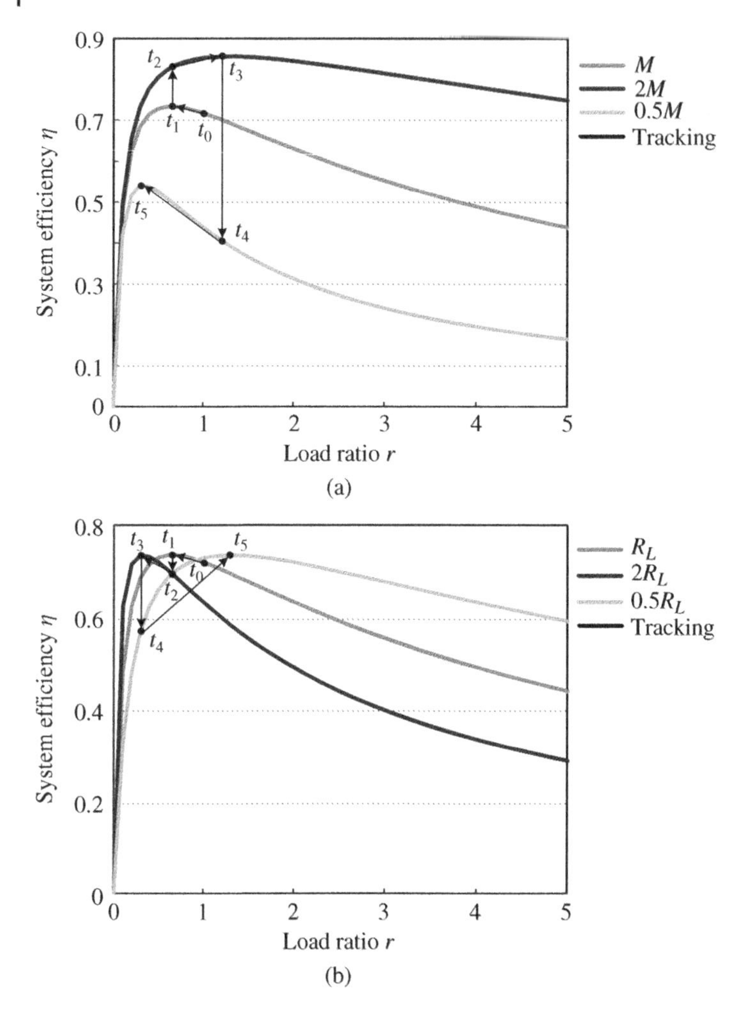

Figure 5.22 Results of the maximum efficiency tracking: (a) tracking of optimal load resistance with varying coupling coefficient; (b) tracking of optimal load resistance with varying load resistance.

position of 12 cm. Then, the relative position is adjusted from 12 to 10 cm at t_2 and then from 10 to 15 cm at t_4. Again, a quick drop in the system efficiency is observed at t_2 because the equivalent load resistance seen by the rectifier deviates from its optimal value due to the variation in k. Through the P&O-based tracking control, the optimal load resistance under the newest coupling coefficient is reached at t_3, which corresponds to the position of 10 cm. A similar procedure is repeated at t_4 and t_5, when the distance is changed again from 10 to 15 cm.

A similar process of the maximum efficiency tracking with varying load resistance is shown in Figure 5.22b. After the initial state with a load resistance of 5 Ω at t_0, the optimal load resistance under the constant k is reached at t_1 by the feedback-based control of the conversion ratio r. At time t_2, the load resistance R_L is switched from 5 to 10 Ω. Since the equivalent load resistance seen by the rectifier deviates from its optimal value, a quick drop in the system efficiency is observed. Through the tracking control using the DC–DC converter (or active rectifier), the optimal load resistance is reached again at t_3; thus, a high system efficiency is achieved. A similar procedure is repeated at t_4 and t_5.

5.2.2.3.2 A Design Example of Maximum Efficiency Tracking Using M Estimation Method

In order to verify the effectiveness of the design example for maximum efficiency tracking using M estimation method, the simulation model is established using MATLAB®/SIMULINK with the key parameters as listed in Table 5.4. The system configuration for the tracking of the optimal load is shown in Figure 5.23. In the control process, the output voltage U_{out} is measured and fed back to the controller by wireless communication, which is regulated by changing the conversion ratio r_2 of the buck–boost converter in the primary side. In addition, the variation of the conversion ratio r_2 is utilized to estimate the dynamical coupling coefficient. The conversion ratio r_1 of the buck–boost converter in the secondary side is adjusted to track the maximum equivalent load resistance to realize the maximum efficiency. Moreover, the load resistance R_L can also be detected by measuring the output dc voltage U_{out} and the current I_{out}, when there exist fluctuations on the load resistance.

Table 5.4 Parameters of the design example.

Item	Symbol	Value
Self-inductance of the primary coil	L_p	60 μH
Self-inductance of the secondary coil	L_s	60 μH
Compensating capacitance of the primary coil	C_p	58.49 nF
Compensating capacitance of the secondary coil	C_s	58.49 nF
Operating frequency	f	85 kHz
Load resistance	R_L	5 Ω
Line resistance of the primary coil	R_p	0.5 Ω
Line resistance of the secondary coil	R_s	0.5 Ω
DC input voltage	U_{in}	10 V
Initial secondary conversion ratio	r_1	0.5
Initial primary conversion ratio	r_2	0.5

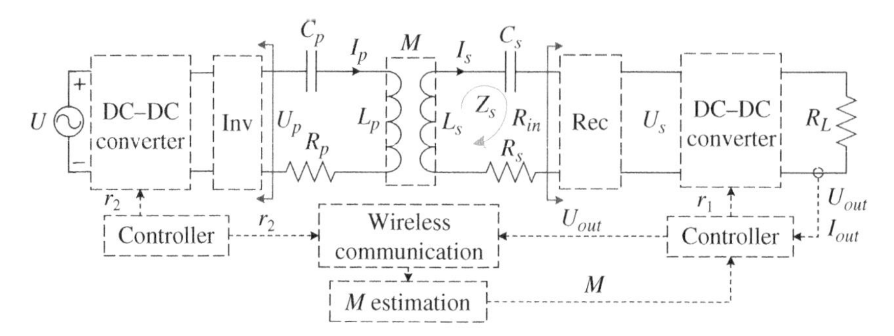

Figure 5.23 Diagram of the maximum efficiency point tracking control scheme based on mutual inductance estimation.

In order to track the maximum efficiency point, the initial value of the coupling coefficient needs to be estimated. The relationship between the coupling coefficient k and the output voltage U_{out} can be expressed as

$$U_{out} = \frac{\sqrt{2}\pi R_{in} r_1 \omega \kappa \sqrt{L_p L_s} U_p Z_s}{4(R_{in} + R_s)(1 - r_1)(\omega^2 \kappa^2 L_p L_s + R_p Z_s)} \tag{5.48}$$

where Z_s is the equivalent impedance of the secondary side, U_p is the output voltage of the inverter in the primary side, and R_{in} is the equivalent impedance of the rectifier, buck–boost converter in secondary side, and load.

As can be seen from Eq. (5.48), the calculation of k will have two solutions based on one instance of output voltage U_{out}. The estimation of the initial value of k can be done only by employing two instances of the output voltage. In the initialization stage, the conversion ratio r_1 is controlled by a secondary controller and maintains $r_1 = r_{1p}$. By choosing initial r_{1a} and r_{1b}, the results of k_{1a}, k_{1b}, k_{2a}, and k_{2a} can be obtained based on the measured output voltage U_{out}. The estimated value of the initial coupling coefficient k_p is obtained by averaging the closest two values among k_{1a}, k_{1b}, k_{2a}, and k_{2a}. The estimation results are shown in Table 5.5

Table 5.5 Simulation results of the initial coupling coefficient estimation.

r_{1a}/r_{1b}	0.3	0.4	0.5	0.6	0.7
0.3	—	0.150	0.149	0.150	0.151
0.4	0.151	—	0.149	0.151	0.150
0.5	0.149	0.150	—	0.149	0.149
0.6	0.150	0.149	0.150	—	0.149
0.7	0.151	0.149	0.149	0.150	—

with different r_{1a} and r_{1b} (the value changes from 0.1 to 0.7), where the reference coupling coefficient is 0.15. It can be seen that the estimation accuracy of the initial coupling coefficient is relatively high and the initial duty cycles have very little effect on the final result of the estimated coupling coefficient.

After the estimation of the initial value of the coupling coefficient, the dynamic value should be estimated in order to track the optimal load resistance. The relationship between the coupling coefficient k and conversion ratio r_2 can be derived as

$$\begin{cases} \kappa = f(r_2) = \dfrac{1}{(1-r_2)\sqrt{L_pL_s}U_r\pi^2\omega r_1}\left(4U_pR_L(1-r_1)r_2 + \sqrt{Ar_2{}^2 + Br_2 + C}\right) \\ A = U_r{}^2R_p\left(r_1{}^2(2\pi^4R_s + 16\pi^2R_L) - 32\pi^2R_Lr_1 + 16\pi^2R_L\right) \\ B = U_r{}^2R_p\left(r_1{}^2(-\pi^4R_s - 8\pi^2R_L) + 16\pi^2R_Lr_1 - 8\pi^2R_L\right) \\ C = U_r{}^2R_p\left(r_1{}^2(-\pi^4R_s - 8\pi^2R_L) + 16\pi^2R_Lr_1 - 8\pi^2R_L\right) \\ \quad + U_p{}^2R_L{}^2\left(16r_1{}^2 - 32r_1 + 16\right) \end{cases} \tag{5.49}$$

As the initial coupling coefficient k_p can be obtained in the initialization process, and the value of r_2 is recorded at the secondary side in real time, the dynamic value of the coupling coefficient k_e can be estimated by

$$\kappa_e = \kappa_p + \Delta r_2 \frac{\partial f(r_2)}{\partial r_2}\Big|_{r_2=r_{2p}} \tag{5.50}$$

The dynamic coupling coefficient estimation can be achieved when the variation in the conversion ratio r_2 and the initial coupling coefficient k_p are determined. Table 5.6 shows the simulation results of the dynamic coupling coefficient estimation when k changes from 0.120 and 0.180. It can be seen that the estimation accuracy is lower when the difference between the dynamic k_d and the previous k_p is larger. Since the estimation of the dynamic coupling coefficient from Eq. (5.50) is the equivalent linearization result, the error will become bigger with the increase in the difference. After the dynamic coupling coefficient is estimated, the maximum efficiency tracking can be achieved by adjusting the equivalent load resistance. The relationship between the conversion ratio r_1 and the optimal equivalent load resistance can be obtained as

$$r_1 = \frac{2(\sqrt{2}\pi\sqrt{R_LR_{\_max}} - 4R_L)}{\pi^2R_{\_max} - 8R_L} \tag{5.51}$$

where $R_{\_max}$ is the optimal load condition to realize the maximum transfer efficiency, which can be expressed as

$$R_{\_max} = R_s\sqrt{1 + \kappa^2\frac{\omega^2L_pL_s}{R_pR_s}} \tag{5.52}$$

Table 5.6 Simulation results of the dynamic coupling coefficient estimation.

Initial k_p	Dynamic k_d	Estimated k_e	Estimation accuracy (%)
0.120	0.090	0.099	90
0.120	0.100	0.107	93
0.150	0.120	0.129	92.5
0.150	0.130	0.136	95.4
0.180	0.150	0.1583	94.5
0.180	0.160	0.165	97

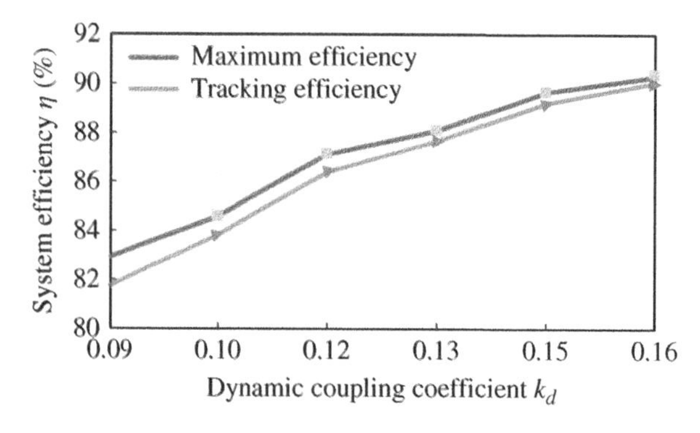

Figure 5.24 Efficiency tracking results with dynamic coupling coefficient.

The maximum efficiency tracking with the variation in coupling coefficient is tested, and the tracking results with the variation of the coupling coefficient is shown in Figure 5.24. It can be seen that the maximum efficiency becomes larger with the increase in the coupling coefficient k. In addition, the tracking efficiency is smaller than the real maximum efficiency due to the estimation error of the dynamic coupling coefficient. The conversion ratios r_1 and r_2 to ensure the maximum efficiency tracking with the variation in coupling coefficient from 0.12, 0.15 to 0.18 are shown in Table 5.7. The output voltage is controlled by the buck–boost converter to be unchanged in the tracking process, so the conversion ratio r_2 gets smaller with the increase in the coupling coefficient. In addition, the conversion ratio r_1 is adjusted to ensure that the equivalent load resistance equals the optimal value.

Table 5.7 Simulation results of the maximum efficiency tracking.

Initial k_p	Dynamic k_d	Ratio r_1	Ratio r_2
0.120	0.090	0.529	0.453
0.120	0.100	0.52	0.467
0.150	0.120	0.497	0.457
0.150	0.130	0.49	0.472
0.180	0.150	0.471	0.462
0.180	0.160	0.466	0.475

References

1. Huang, Z., Wong, S.C., and Tse, C.K. (2017). Design of a single-stage inductive-power-transfer converter for efficient EV battery charging. *IEEE Transactions on Vehicular Technology* 66 (7): 5808–5821.
2. Zhang, W., Wong, S.C., Tse, C.K., and Chen, Q. (2014). Design for efficiency optimization and voltage controllability of series-series compensated inductive power transfer systems. *IEEE Transactions on Power Electronics* 29 (1): 191–200.
3. Sample, A.P., Meyer, D.T., and Smith, J.R. (2011). Analysis, experimental results, and range adaptation of magnetically coupled resonators for wireless power transfer. *IEEE Transactions on Industrial Electronics* 58 (2): 544–554.
4. Wang, C., Covic, G.A., and Stielau, O.H. (2001). General stability criterions for zero phase angle controlled loosely coupled inductive power transfer systems. *IECON* 2: 1049–1054.
5. Wang, C.S., Covic, G.A., and Stielau, O.H. (2004). Power transfer capability and bifurcation phenomena of loosely coupled inductive power transfer systems. *IEEE Transactions on Industrial Electronics* 51 (1): 148–157.
6. Huang, C.Y., Hames, J.E., and Covic, G.A. (2019). Design considerations for variable coupling lumped coil systems. *IEEE Transactions on Power Electronics* 30 (2): 1044–1058.
7. Chao, Y.H. and Shieh, J.J. (2012). Series-parallel loosely coupled battery charger with primary-side control. In: *International Conference on Control, Automation and Information Sciences (ICCAIS)*, 226–230. IEEE.

8 Gati, E., Kampitsis, G., Stavropoulos, I. et al. (2015). Wireless phase-locked loop control for inductive power transfer systems. In: *IEEE Applied Power Electronics Conference and Exposition (APEC)*, 1601–1607. IEEE.

9 Seo, D.W. and Lee, J.H. (2017). Frequency-tuning method using the reflection coefficient in a wireless power transfer system. *IEEE Microwave and Wireless Components Letters* 27 (11): 959–961.

10 Sis, S.A. and Bicakci, S. (2016). A resonance frequency tracker and source frequency tuner for inductively coupled wireless power transfer systems. In: *2016 46th European Microwave Conference (EuMC)*, London, UK, 751–754. IEEE.

11 Lim, Y., Tang, H., Lim, S., and Park, J. (2014). An adaptive impedance-matching network based on a novel capacitor matrix for wireless power transfer. *IEEE Transactions on Power Electronics* 29 (8): 4403–4413.

12 Zhang, Z., Ai, W., Liang, Z., and Wang, J. (2018). Topology-reconfigurable capacitor matrix for encrypted dynamic wireless charging of electric vehicles. *IEEE Transactions on Vehicular Technology* 67 (10): 9284–9293.

13 Zhang, Z. and Pang, H. Continuously-adjustable capacitor for multiple-pickup wireless power transfer under single-power-induced energy field. *IEEE Transactions on Industrial Electronics* https://doi.org/10.1109/TIE.2019.2937056.

14 Dang, Z., Cao, Y., and Qahouq, J.A.A. (2015). Reconfigurable magnetic resonance-coupled wireless power transfer system. *IEEE Transactions on Power Electronics* 30 (11): 6057–6069.

15 Cao, Y. and Qahouq, J.A.A. (2019). Modelling and control design of reconfigurable wireless power transfer system for transmission efficiency maximization and output voltage regulation. *IET Power Electronics* 12 (8): 1903–1916.

16 Lee, G., Waters, B.H., Shin, Y.G. et al. (2016). A reconfigurable resonant coil for range adaptation wireless power transfer. *IEEE Transactions on Microwave Theory and Techniques* 64 (2): 624632.

17 Li, H., Chen, S., Fang, J. et al. (2019). A low-subharmonic, full-range, and rapid pulse density modulation strategy for ZVS full-bridge converters. *IEEE Transactions on Power Electronics* 34 (9): 8871–8881.

18 Sun, Y., Zhang, H., Hu, A.P. et al. (2017). The recognition and control of nonideal soft-switching frequency for wireless power transfer system based on waveform identification. *IEEE Transactions on Power Electronics* 32 (8): 6617–6627.

19 Fu, M., Ma, C., and Zhu, X. (2014). A cascaded boost-buck converter for high-efficiency wireless power transfer systems. *IEEE Transactions on Industrial Electronics* 10 (3): 1972–1980.

20 Li, H., Li, J., Wang, K. et al. (2015). A maximum efficiency point tracking control scheme for wireless power transfer systems using magnetic resonant coupling. *IEEE Transactions on Power Electronics* 30 (7): 3998–4008.

21 Orekan, T., Zhang, P., and Shih, C. (2018). Analysis, design, and maximum power-efficiency tracking for undersea wireless power transfer. *IEEE Journal of Emerging and Selected Topics in Power Electronics* 6 (2): 843–854.

22 Yeo, T.D., Kwon, D., Khang, S.T., and Yu, J.W. (2017). Design of maximum efficiency tracking control scheme for closed-loop wireless power charging system employing series resonant tank. *IEEE Transactions on Power Electronics* 32 (1): 471–478.

23 Zhong, W.X. and Hui, S.Y.R. (2015). Maximum energy efficiency tracking for wireless power transfer systems. *IEEE Transactions on Power Electronics* 30 (7): 4025–4043.

24 Fu, M., Yin, H., Zhu, X., and Ma, C. (2015). Analysis and tracking of optimal load in wireless power transfer systems. *IEEE Transactions on Power Electronics* 30 (7): 3952–3963.

25 Zhang, W., Wu, X., Xia, C., and Liu, X. (2019). Maximum efficiency point tracking control method for series–series compensated wireless power transfer system. *IET Power Electronics* 12 (10): 2534–2542.

26 Liu, Y., Mai, R., Liu, D. et al. (2019). Optimal load ratio control for dual-receiver dynamic wireless power transfer maintaining stable output voltage. *IET Power Electronics* 12 (10): 2669–2677.

27 Huang, Z., Wong, S.C., and Tse, C.K. (2018). Control design for optimizing efficiency in inductive power transfer systems. *IEEE Transactions on Power Electronics* 33 (5): 4523–4534.

28 Huwig, D. and Wambsganß, P. (2013). Digitally controlled synchronous bridge-rectifier for wireless power receivers. In: *Twenty-Eighth Annual IEEE Applied Power Electronics Conference and Exposition (APEC)*, California, USA, 2598–2603. IEEE.

29 Berger, A., Agostinelli, M., Vesti, S. et al. (2015). Phase-shift and amplitude control for an active rectifier to maximize the efficiency and extracted power of a wireless power transfer system. In: *IEEE Applied Power Electronics Conference and Exposition (APEC)*, 1620–1624. IEEE.

30 Ozalevli, E., Femia, N., Capua, G.D. et al. (2018). A cost-effective adaptive rectifier for low power loosely coupled wireless power transfer systems. *IEEE Transactions on Circuits and Systems I: Regular Papers* 65 (7): 2318–2329.

31 Diekhans, T. and Doncker, R.W.D. (2015). A dual-side controlled inductive power transfer system optimized for large coupling factor variations and partial load. *IEEE Transactions on Power Electronics* 30 (11): 6320–6328.

32 Colak, K., Asa, E., Bojarski, M. et al. (2015). A novel phase-shift control of semibridgeless active rectifier for wireless power transfer. *IEEE Transactions on Power Electronics* 30 (11): 6288–3297.

33 Mai, R., Liu, Y., Li, Y. et al. (2018). An active-rectifier-based maximum efficiency tracking method using an additional measurement coil for wireless power transfer. *IEEE Transactions on Power Electronics* 33 (1): 716–728.

34 Tang, X., Zeng, J., Pun, K. et al. (2018). Low-cost maximum efficiency tracking method for wireless power transfer systems. *IEEE Transactions on Power Electronics* 33 (6): 5317–5329.

35 Jang, Y. and Jovanovic, M.M. (2003). A contactless electrical energy transmission system for portable telephone battery chargers. *IEEE Transactions on Industrial Electronics* 50 (3): 520–527.

36 Choi, B., Nho, J., Cha, H. et al. (2004). Design and implementation of low-profile contactless battery charger using planar printed circuit board windings as energy transfer device. *IEEE Transactions on Industrial Electronics* 51 (1): 140–147.

37 Tse, K.K., Ho, B.M.T., Chung, H.S.H., and Hui, S.Y.R. (2004). A comparative study of maximum-power-point trackers for photovoltaic panels using switching-frequency modulation scheme. *IEEE Transactions on Industrial Electronics* 51 (2): 410–418.

38 Yin, J., Lin, D., Lee, C.K., and Hui, S.Y.R. (2015). A systematic approach for load monitoring and power control in wireless power transfer systems without any direct output measurement. *IEEE Transactions on Power Electronics* 30 (3): 1657–1667.

39 Fan, S., Wei, R., Zhao, L. et al. (2018). An ultralow quiescent current power management system with maximum power point tracking (MPPT) for batteryless wireless sensor applications. *IEEE Transactions on Power Electronics* 33 (9): 7326–7337.

40 Li, Y., Hu, J., Li, X. et al. (2020). Analysis, design, and experimental verification of a mixed high-order compensations-based WPT system with constant current outputs for driving multistring LEDs. *IEEE Transactions on Power Electronics* 67 (1): 203–213.

41 Song, K., Li, Z., Jiang, J., and Zhu, C. (2018). Constant current/voltage charging operation for series–series and series–parallel compensated wireless power transfer systems employing primary-side controller. *IEEE Transactions on Power Electronics* 33 (9): 8065–8080.

42 Liu, F., Chen, K., Zhao, Z. et al. (2018). Transmitter-side control of both the CC and CV modes for the wireless EV charging system with the weak communication. *IEEE Journal of Emerging and Selected Topics in Power Electronics* 6 (2): 955–965.

43 Li, Y., Hu, J., Liu, M. et al. (2019). Reconfigurable intermediate resonant circuit based WPT system with load-independent constant output current and voltage for charging battery. *IEEE Transactions on Power Electronics* 34 (3): 1988–1992.

44 Dai, X., Li, X., Li, Y., and Hu, A.P. (2018). Maximum efficiency tracking for wireless power transfer systems with dynamic coupling coefficient estimation. *IEEE Transactions on Power Electronics* 33 (6): 5005–5015.

45 Yin, J., Lin, D., Parisini, T., and Hui, S. (2016). Front-end monitoring of the mutual inductance and load resistance in a series–series compensated wireless power transfer system. *IEEE Transactions on Power Electronics* 31 (10): 7339–7352.

6

Control Scheme for Multiple-pickup WPT System

6.1 Introduction

Currently, the conventional 1-to-1 transmission is incapable of meeting the complex application and various requirements from the customers, for example, the nonunified wireless power transfer (WPT) standards for consumer electronics, the dynamic wireless charging for electric vehicles (EVs), and contactless charging for sensor networks. Consequently, how to tackle 1-to-*n* transmission has become the research focus in recent years, specifically, the control scheme for multiple-pickup WPT systems.

Hence, this chapter firstly introduces two forms of 1-to-n transmission strategy, namely, the single-frequency time-sharing transmission and multifrequency simultaneous transmission, as well as various multifrequency excitations in details. Then, the impedance matching strategy for 1-to-*n* transmission, which holds multiple resonating frequencies, is recommended, and the compensation approach aiming to eliminate cross-coupling among the pickup coils is elaborated. Finally, other distinctive control schemes, which are different from the 1-to-1 transmission, are discussed with emphasis on power allocation, maximum efficiency for multitransmitter, and constant voltage control.

6.2 Transmission Strategy

As previously depicted in Section 2.1.3 and shown in Figure 6.1, the 1-to-*n* WPT systems are classified into two operating modes, that is, the single-frequency excitation and multifrequency excitation. As for single-frequency excitation, there are two power distribution patterns, namely, the simultaneous transmission pattern where only one resonate frequency exists, and the time-sharing pattern where various operating frequencies are utilized.

Wireless Power Transfer: Principles and Applications, First Edition. Zhen Zhang and Hongliang Pang.

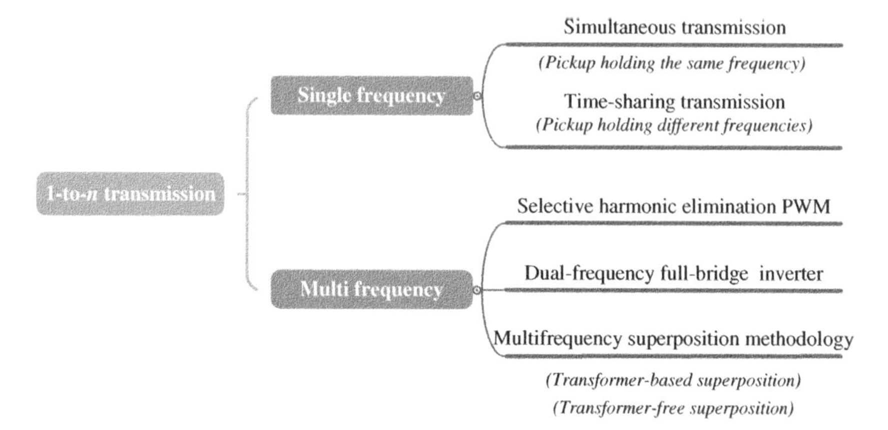

Figure 6.1 Structure diagram of 1-to-n transmission.

Regarding the single-frequency simultaneous transmission, all the pickups hold the same resonate frequency, embrace within the same open induced electromagnetic field, and acquire the wirelessly transmitted power simultaneously. Accordingly, single-frequency simultaneous transmission holds many similar control characteristics compared to 1-to-1 transmission, which has been detailed in other chapters. Thus, a specific introduction of single-frequency time-sharing transmission as well as multifrequency excitation transmission will follow up in the following parts of this section.

6.2.1 Single-frequency Time-sharing Transmission

6.2.1.1 Modeling and Analysis

Unlike the conventional 1-to-1 inductive power transfer (IPT) system where only one designed resonant frequency is adopted, in the previously proposed 1-to-n IPT system, different operating frequencies are adopted internally to excite the induced electromagnetic field, for example, the selective WPT systems [1–3]. Nevertheless, at one specific point of time, there only exists one operating frequency, and the idea of 1-to-1 transmission has been expressed more concretely in Section 2.3.1.3. Hence, the selective WPT should be categorized as a single-frequency time-sharing WPT system, especially not transferring multifrequency power simultaneously.

Aiming to facilitate the following theoretical analysis, a general two-pickup WPT system under single-frequency time-sharing operating mode, shown in Figure 6.2, is depicted to further explore the related transmission characteristics.

Denote f, f_{s1}, f_{s2} as the operating frequency of the primary side and the resonating frequencies of pickup one and pickup two, respectively, which are given as

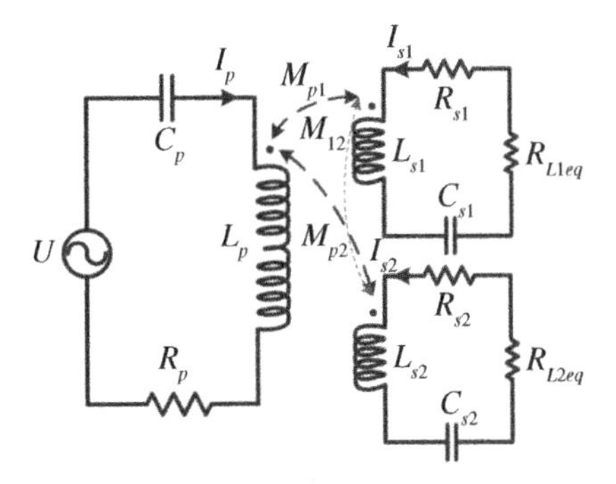

Figure 6.2 Simplified two-pickup WPT system under single-frequency time-sharing operating mode.

$$\begin{cases} f = \dfrac{1}{2\pi\sqrt{L_p C_p}} \\ f_{s1} = \dfrac{1}{2\pi\sqrt{L_{s1} C_{s1}}} \\ f_{s2} = \dfrac{1}{2\pi\sqrt{L_{s2} C_{s2}}} \end{cases} \tag{6.1}$$

Based on Kirchhoff's voltage law (KVL), the secondary pickup one current $\overrightarrow{I_{s1}}$ and pickup two current $\overrightarrow{I_{s2}}$ can be, respectively, deduced as

$$\begin{cases} \overrightarrow{I_{s1}} = -\dfrac{\omega^2 M_{p2} M_{12} + j\omega M_{p1}(Z_{s2} + R_{L2eq})}{\omega^2 M_{12}^2 + (Z_{s1} + R_{L1eq})(Z_{s2} + R_{L2eq})} \overrightarrow{I_p} \\ \overrightarrow{I_{s2}} = -\dfrac{\omega^2 M_{p1} M_{12} + j\omega M_{p2}(Z_{s1} + R_{L1eq})}{\omega^2 M_{12}^2 + (Z_{s1} + R_{L1eq})(Z_{s2} + R_{L2eq})} \overrightarrow{I_p} \end{cases} \tag{6.2}$$

The similar calculation procedure has already been clarified in Section 2.3.4, where $\overrightarrow{Z_p}$, $\overrightarrow{Z_{s1}}$, and $\overrightarrow{Z_{s2}}$ represent the impedance of the primary side, secondary pickup one, and secondary pickup two, respectively, which are given by

$$\begin{cases} \overrightarrow{Z_p} = R_p + j\omega L_p + \dfrac{1}{j\omega C_p} \\ \overrightarrow{Z_{s1}} = R_{s1} + j\omega L_{s1} + \dfrac{1}{j\omega C_{s1}} \\ \overrightarrow{Z_{s2}} = R_{s2} + j\omega L_{s2} + \dfrac{1}{j\omega C_{s2}} \end{cases} \tag{6.3}$$

Subsequently, the expression can be rearranged as

$$\begin{cases} \overrightarrow{I_{s1}} = -\dfrac{\dfrac{\omega^2 M_{p2} M_{12}}{Z_{s2} + R_{l2eq}} + j\omega M_{p1}}{\dfrac{\omega^2 M_{12}^2}{Z_{s2} + R_{l2eq}} + Z_{s1} + R_{L1eq}} \overrightarrow{I_p} \\ \overrightarrow{I_{s2}} = -\dfrac{\dfrac{\omega^2 M_{p1} M_{12}}{Z_{s1} + R_{L1eq}} + j\omega M_{p2}}{\dfrac{\omega^2 M_{12}^2}{Z_{s1} + R_{L1eq}} + Z_{s2} + R_{L2eq}} \overrightarrow{I_p} \end{cases} \tag{6.4}$$

Assume that there is a large difference between f_{s1} and f_{s2}, when the operating frequency f equals f_{s1}. According to Eqs. (6.1) and (6.3), we can obtain that $\overrightarrow{Z_{s1}} \approx R_{s1}$, and $\overrightarrow{Z_{s2}} \to \infty$. Then, Eq. (6.4) can be simplified as

$$\begin{cases} \overrightarrow{I_{s1}} = -\left(\dfrac{\left(\dfrac{\omega^2 M_{p2} M_{12}}{Z_{s2}(\to \infty) + R_{l2eq}} \right) \to 0 + j\omega M_{p1}}{\left(\dfrac{\omega^2 M_{12}^2}{Z_{s2}(\to \infty) + R_{l2eq}} \right) \to 0 + R_{s1} + R_{L1eq}} \right) \overrightarrow{I_p} \left(\overrightarrow{I_{s1}} \approx -\frac{j\omega M_{p1}}{R_{s1} + R_{L1eq}} \overrightarrow{I_p} \right) \\ \overrightarrow{I_{s2}} = -\dfrac{\dfrac{\omega^2 M_{p1} M_{12}}{R_{s1} + R_{L1eq}} + j\omega M_{p2}}{\dfrac{\omega^2 M_{12}^2}{R_{s1} + R_{L1eq}} + Z_{s2}(\to \infty) + R_{L2eq}} \overrightarrow{I_p} (\overrightarrow{I_{s2}} \approx 0) \end{cases} \tag{6.5}$$

Hence, $|\overrightarrow{I_{s1}}| >> |\overrightarrow{I_{s2}}|$. Accordingly, when the frequency is operating at the resonant frequency for pickup one, most of the power will distribute to pickup one rather than pickup two, and vice versa. Consequently, just by changing the operating frequency to the resonate frequency of pickup one and pickup two, the wirelessly transmitted power can be selectively distributed to the desired pickup.

Similarly, for WPT systems with n pickups, the closer the operating frequency is to the desired secondary pickups, the more power the particular pickup will acquire. By altering the operating frequency, time-sharing transmission for multiple pickups will be achieved.

6.2.1.2 Verification

The verification was conducted under two pickups whose inherent resonant frequencies are 181 and 203 kHz (Figure 6.3a) and 191 and 203 kHz (Figure 6.3b).

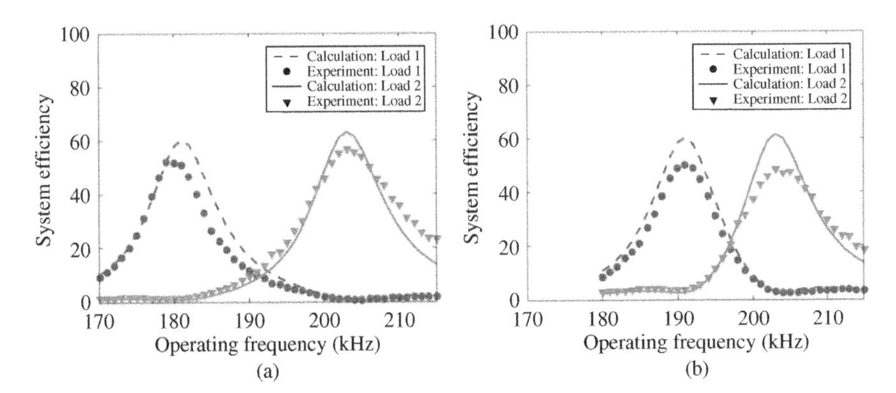

Figure 6.3 Efficiency versus operating frequency under two-pickup selective wireless power transmission. (a) Resonate at 181 and 203 kHz; (b) Resonate at 191 and 203 kHz. Source: Adapted from Kim et al. [2].

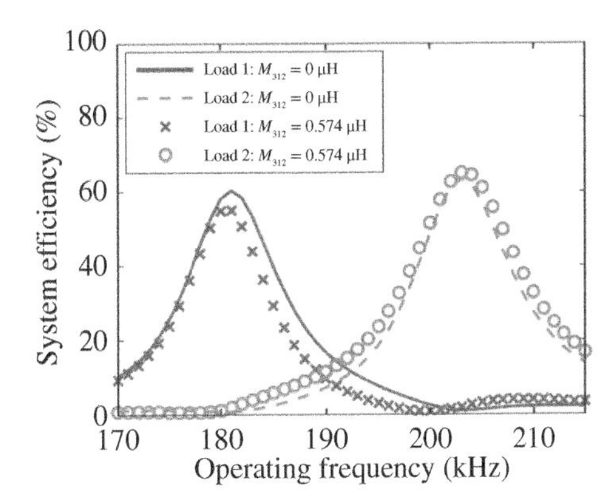

Figure 6.4 Efficiency versus operating frequency with and without cross-coupling between two pickups. Source: Adapted from Kim et al. [2].

As shown above, the efficiency of each pickup peaks at its inherent resonant frequency, irrespective of the resonant frequency of another pickup and the operating frequency. When operating at each resonant frequency, another pickup can hardly acquire any power from the primary side, which explains the so-called selective wireless power transmission. Moreover, in Figure 6.3b, when setting the operating frequency at approximately 198 kHz, both pickups can acquire the wirelessly transmitted power. However, the efficiency is relatively low compared to the resonant frequency. Besides, when considering the cross-coupling effect, the verification of the relevant system efficiency is as plotted in Figure 6.4. From the results below, the efficiency curves are roughly the same, which indicates that the cross-coupling effect has a small influence on the selective wireless power transfer system.

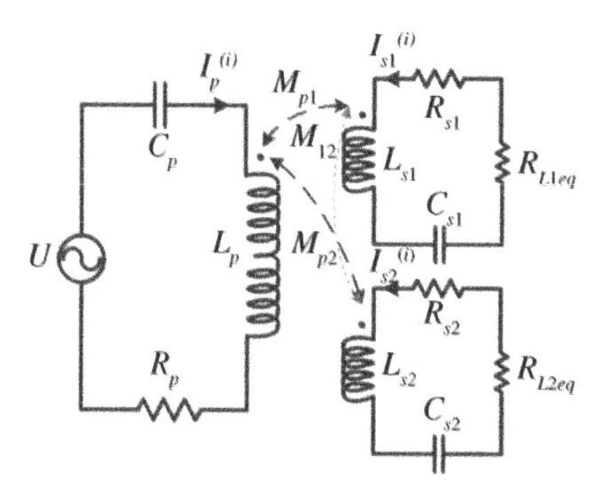

Figure 6.5 Simplified two-pickup WPT system under multifrequency simultaneous operating mode.

6.2.2 Multifrequency Simultaneous Transmission

With the various demands from customers and different standards among manufacturers, multifrequency simultaneous transmission should be taken into consideration. Being different from the single-frequency simultaneous/time-sharing transmission, which contains only one transmitting frequency at the same time, in multifrequency transmission, multiple frequencies are superposed to excite the induced electromagnetic field, namely, the excitation source contains various operating frequencies at the same time. In this section, the circuit model for multifrequency simultaneous transmission is established, and the fundamental harmonic analysis (FHA) method is adopted to analyze the transmission characteristics. Subsequently, some common methods for the excitation of multifrequency are elaborated. Finally, the comparison among the aforementioned approaches is put forward as a guidance for practical WPT application scenarios.

6.2.2.1 Modeling and Analysis

Figure 6.5 depicts a two-pickup WPT system to illustrate the working principle of multifrequency simultaneous transmission, where the superscript i ($i = 1$ or 2) represents the transmission channel at the angular frequency of ω_1 and ω_2, respectively. Being different from the concept of time-sharing multiplexing, the crux of the presented multifrequency transmission is to pick up the power with various frequencies.

According to KVL, the corresponding primary side current $\overrightarrow{I_p}^{(i)}$, secondary pickup one current $\overrightarrow{I_{s1}}^{(i)}$, and secondary pickup two current $\overrightarrow{I_{s2}}^{(i)}$ can be, respectively, deduced as

$$\begin{cases} \overrightarrow{I_p^{(i)}} = \dfrac{U^{(i)}}{Z_p^{(i)} + \dfrac{\omega_i^2 M_{p1}^2 \left(Z_{s2}^{(i)} + R_{L2eq}\right) + \omega_i^2 M_{p2}^2 \left(Z_{s1}^{(i)} + R_{L1eq}\right) - 2j\omega_i^3 M_{p1} M_{p2} M_{12}}{\omega_i^2 M_{12}^2 + \left(Z_{s1}^{(i)} + R_{L1eq}\right)\left(Z_{s2}^{(i)} + R_{L2eq}\right)}} \\ \overrightarrow{I_{s1}^{(i)}} = -\dfrac{U^{(i)} \left(\omega_i^2 M_{p2} M_{12} + j\omega M_{p1} \left(Z_{s2}^{(i)} + R_{L2eq}\right)\right)}{Z_p^{(i)} \left(Z_{s1}^{(i)} + R_{L1eq}\right)\left(Z_{s2}^{(i)} + R_{L2eq}\right) + Z_p^{(i)} \omega_i^2 M_{12}^2 + \omega_i^2 M_{p1}^2 \left(Z_{s2}^{(i)} + R_{L2eq}\right) + \omega_i^2 M_{p2}^2 \left(Z_{s1}^{(i)} + R_{L1eq}\right) - 2j\omega_i^3 M_{p1} M_{p2} M_{12}} \\ \overrightarrow{I_{s2}^{(i)}} = -\dfrac{U^{(i)} \left(\omega_i^2 M_{p1} M_{12} + j\omega_i M_{p2} \left(Z_{s1}^{(i)} + R_{L1eq}\right)\right)}{Z_p^{(i)} \left(Z_{s1}^{(i)} + R_{L1eq}\right)\left(Z_{s2}^{(i)} + R_{L2eq}\right) + Z_p^{(i)} \omega_i^2 M_{12}^2 + \omega_i^2 M_{p1}^2 \left(Z_{s2}^{(i)} + R_{L2eq}\right) + \omega_i^2 M_{p2}^2 \left(Z_{s1}^{(i)} + R_{L1eq}\right) - 2j\omega_i^3 M_{p1} M_{p2} M_{12}} \end{cases} \tag{6.6}$$

Then, the corresponding total input power $P_{in}^{(i)}$, load power of pickup one $P_{L1}^{(i)}$, and load power of pickup two $P_{L2}^{(i)}$ can be, respectively, obtained as

$$P_{in}^{(i)} = \overrightarrow{I_p^{(i)}} U^{(i)}, P_{L1}^{(i)} = \left|\overrightarrow{I_{s1}^{(i)}}\right|^2 R_{L1eq}, P_{L2}^{(i)} = \left|\overrightarrow{I_{s2}^{(i)}}\right|^2 R_{L2eq} \tag{6.7}$$

Hence, the corresponding transmission efficiency can be gained as

$$\eta = \frac{P_{L1}^{(1)} + P_{L1}^{(2)} + P_{L2}^{(1)} + P_{L2}^{(2)}}{P_{in}^{(1)} + P_{in}^{(2)}} \tag{6.8}$$

According to the above calculation analysis, it can be revealed that the transmission characteristics of the multifrequency simultaneous system are intensively related to the operating frequencies that contain various frequency components and desired voltage amplitude at the corresponding values. Consequently, how to implement the multifrequency excitation becomes the prime technical concern.

6.2.2.2 Method of Multifrequency Excitation

This section mainly presents some common methods for multifrequency excitation, with emphasis on the principle, theoretical analysis, and verifications. In this section, all the proposed methods are utilized under the circumstance of only one power supply.

6.2.2.2.1 Multifrequency Programmed Pulse Width Modulation The programmed pulse width modulation (PWM) signal is widely adopted for generating the square waveforms with a fundamental switching frequency equal to the designed resonate frequency, where all switching angles are predetermined and optimized by the desired output [4]. In industrial applications, full control switches are widely adopted as high-frequency control components. Hence, the converter outputs pulse voltages with equal amplitude and unequal width. Only higher harmonics with a low amplitude which are relatively easy to remove are left, which greatly improves the output characteristics of the inverter. Evolved from the programmed PWM, the multifrequency programmed PWM was proposed in [5, 6], where two frequencies, the fundamental element and an arbitrary harmonic element, are selected as the output. The detailed principle is given below.

Multifrequency programmed PWM adopts Fourier analysis to synthesize a pulse sequence, which consists of a number of discrete switching angles. Figure 6.6 depicts the unipolar/bipolar multifrequency programmed PWM, where $\theta_1, \theta_2, \theta_3, \theta_4$, and θ_n represent every switching angle during a quarter cycle of the pulse wave. Based on the Fourier series analysis, the waveform of unipolar PWM, shown in Figure 6.6a, can be unfolded as

$$U_{uni}(\omega t) = \sum_{n=1}^{\infty}(a_n \cos(n\omega t) + b_n \sin(n\omega t)) \tag{6.9}$$

where

$$\begin{cases} a_n = \frac{1}{\pi}\int_0^{2\pi} U_{uni}\cos(n\omega t)d(\omega t) \\ b_n = \frac{1}{\pi}\int_0^{2\pi} U_{uni}\sin(n\omega t)d(\omega t) \end{cases} \tag{6.10}$$

During $[0-\pi]$, U_{uni} regards $\frac{\pi}{2}$ as the axis symmetry; meanwhile, during $[0-2\pi]$, U_{uni} regards $(\pi,0)$ as the point symmetry. Hence, the following equation can be obtained:

$$\begin{cases} U_{uni}(t) = -U_{uni}(t+\pi) \\ U_{uni}(t) = U_{uni}(\pi - t) \end{cases} \tag{6.11}$$

By substituting (6.11) into (6.10), the cosine component and even sine component of the Fourier series equal zero, which is shown below:

$$\begin{cases} a_n = 0, n = 0,1,2,3,\cdots \\ b_n = 0, n = 0,2,4,6,\cdots, b_n = \sum_{n=1,3,5\cdots}^{\infty} \frac{4U}{n\pi}\left[\sum_{k=1}^{n}(-1)^{k+1}\cos(n\theta_k)\right], \\ n = 1,3,5,7\cdots \end{cases} \tag{6.12}$$

Then, Eq. (6.9) can be rearranged as

$$U_{uni}(\omega t) = \sum_{n=1,3,5\cdots}^{\infty} \frac{4U}{n\pi}\left[\sum_{k=1}^{n}(-1)^{k+1}\cos(n\theta_k)\right]\sin(n\omega t)) \tag{6.13}$$

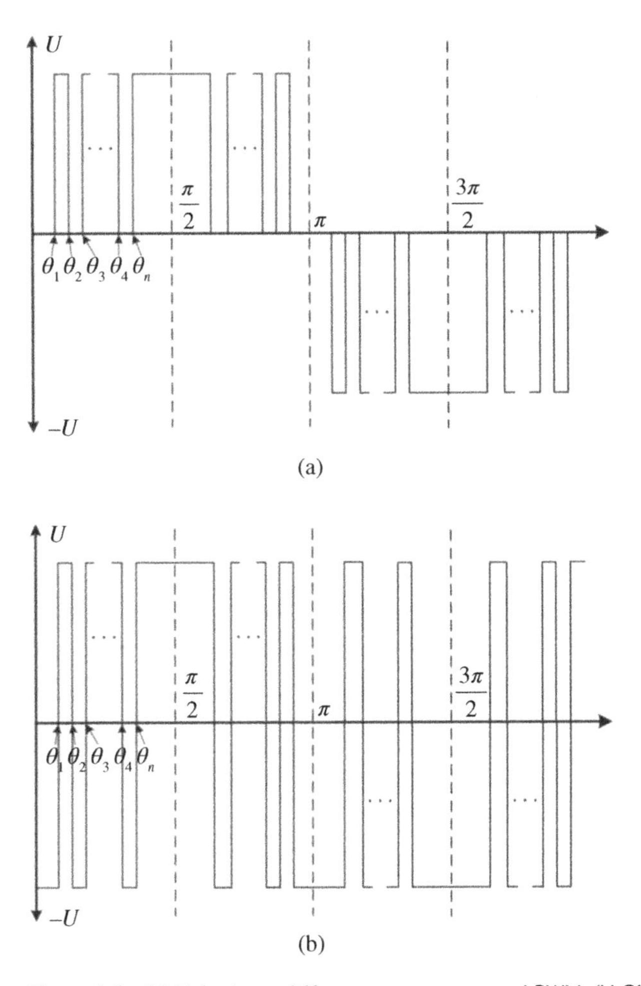

Figure 6.6 (a) Unipolar multifrequency programmed PWM; (b) Bipolar multifrequency programmed PWM.

Similarly, the Fourier expansion of the bipolar PWM, shown in Figure 6.6b, can be expressed as

$$U_{bi}(\omega t) = \sum_{n=1,3,5\cdots}^{\infty} \frac{4U}{n\pi} \left[1 + 2\sum_{k=1}^{n}(-1)^k \cos(n\theta_k) \right] \sin(n\omega t)) \tag{6.14}$$

As for the detailed regulation of the multifrequency programmed PWM, according to Eq. (6.13), the unipolar transcendental equations are firstly set up as

$$\begin{cases} \frac{4U}{\pi}(\cos(1\theta_1) - \cos(1\theta_2) + \cos(1\theta_3) - \cdots + \cos(1\theta_k)) = U_1 \\ \frac{4U}{3\pi}(\cos(3\theta_1) - \cos(3\theta_2) + \cos(3\theta_3) - \cdots + \cos(3\theta_k)) = U_3 \\ \frac{4U}{5\pi}(\cos(5\theta_1) - \cos(5\theta_2) + \cos(5\theta_3) - \cdots + \cos(5\theta_k)) = U_5 \\ \cdots\cdots \\ \frac{4U}{n\pi}(\cos(n\theta_1) - \cos(n\theta_2) + \cos(n\theta_3) - \cdots + \cos(n\theta_k)) = U_k \end{cases} \tag{6.15}$$

where U_1, U_3, U_5,$\cdots$ U_k represent the desired amplitudes of fundamental or specified odd harmonics frequencies, which can be arbitrary set when satisfying the following condition:

$$\left\{ \frac{U_k}{U} \le 1, k = 1,3,5,\cdots \right. \tag{6.16}$$

Then, a set of specific values for θ_1, θ_3, θ_5, $\cdots\theta_k$ is obtained based on the Newton–Raphson iteration algorithm [7], or other related methods [8–10], where most of the approaches can only be solved offline. Finally, the calculated angles are imported into the microcontroller unit (MCU) to control the switches. Hence, the desired output contains various frequencies with different amplitudes. Remarkably, in order to obtain the *kth* odd harmonics frequency, minimum $(k+1)/2$ objective functions are required [6].

Take an example, set $U_1 = 0.6U$, $U_5 = 0.8U$, $U_3 = U_7 = U_k = 0$ and then substitute them into Eq. (6.15) to figure out the related value of angles. Hence, the output waveform only includes the fundamental and the fifth harmonic frequencies with amplitude values of $0.6U$ and $0.8U$, respectively.

Similarly, the bipolar objective equations can be derived based on Eq. (6.14):

$$\begin{cases} \frac{4U}{\pi}(1 - 2\cos(1\theta_1) + 2\cos(1\theta_2) - 2\cos(1\theta_3) + \cdots + 2\cos(1\theta_k)) = U_1 \\ \frac{4U}{3\pi}(1 - 2\cos(3\theta_1) + 2\cos(3\theta_2) - 2\cos(3\theta_3) + \cdots + 2\cos(3\theta_k)) = U_3 \\ \frac{4U}{5\pi}(1 - 2\cos(5\theta_1) + 2\cos(5\theta_2) - 2\cos(5\theta_3) + \cdots + 2\cos(5\theta_k)) = U_5 \\ \cdots\cdots \\ \frac{4U}{n\pi}(1 - 2\cos(n\theta_1) + 2\cos(n\theta_2) - 2\cos(n\theta_3) + \cdots + 2\cos(n\theta_k)) = U_k \end{cases} \tag{6.17}$$

Then, the same procedure as for unipolar transcendental equations can be adapted to acquire the desired outputs.

However, there exist trade-offs between uni- and bipolar multifrequency programmed PWM modulation for WPT applications. Both methods have various total harmonic distortion (THD) and switching losses. For instance, regarding the bipolar multifrequency programmed PWM, in order to prevent the short circuit phenomenon caused by the simultaneously turning on of two switches in the same bridge, dead zone time should be added; thus, the calculated angles need to be reconsidered. Meanwhile, the aforementioned dead zone time need not be considered in unipolar multifrequency programmed PWM. Consequently, the selection of a certain scheme is restrained by the output power range, requirements of control strategy, system efficiency, etc. [5].

Accordingly, by eliminating the unwanted odd harmonics (set corresponding U_k as zero), arbitrary combinations of any specified odd harmonics frequencies with various amplitudes can be selected to compose the multifrequency excitation. The aforementioned approach is also called selective harmonic elimination PWM in power electronics [8–10].

The verification was conducted under 87 and 205 kHz, which are the low and high ends of the Qi standard. However, these two frequencies exceed the Qi standard if adopting the fundamental component (87 kHz) and its third harmonic (261 kHz). Hence, a flexible selection of frequencies and the relevant objective functions for the unipolar multifrequency programmed PWM are reconstructed as

$$\begin{cases} \frac{4U}{\pi}(\cos(1\theta_1) - \cos(1\theta_2) + \cos(1\theta_3) - \cdots + \cos(1\theta_k)) = U_{ac29} \\ \frac{4U}{3\pi}(\cos(3\theta_1) - \cos(3\theta_2) + \cos(3\theta_3) - \cdots + \cos(3\theta_k)) = U_{ac87} \\ \frac{4U}{5\pi}(\cos(5\theta_1) - \cos(5\theta_2) + \cos(5\theta_3) - \cdots + \cos(5\theta_k)) = 0 \\ \frac{4U}{7\pi}(\cos(7\theta_1) - \cos(7\theta_2) + \cos(7\theta_3) - \cdots + \cos(7\theta_k)) = U_{ac205} \\ \cdots\cdots \\ \frac{4U}{n\pi}(\cos(n\theta_1) - \cos(n\theta_2) + \cos(n\theta_3) - \cdots + \cos(n\theta_k)) = 0 \end{cases} \tag{6.18}$$

As seen from Eq. (6.18), the fundamental component U_{ac29} cannot equal zero. This phenomenon is determined by the fundamental component naturally formed in the shape of the unipolar multifrequency programmed PWM. And if there remains no fundamental component, there will be no more output power. Accordingly, the values of U_{ac87} and U_{ac205} cannot surpass the value of the fundamental component U_{ac29}. In this case, shown in Figure 6.7, the fundamental component U_{ac29} is set to 0.6, and the third and fifth harmonics U_{ac87} and U_{ac205} are set to 0.35, respectively. Then, the numeric iteration algorithm is utilized to solve the transcendental equation sets and gain the related control value of θ_k.

The unipolar multifrequency programmed PWM requires an existing fundamental component; however, the fundamental frequency can be set to 0 in the bipolar case shown in Figure 6.8. The related objective functions for the bipolar multifrequency programmed PWM are rewritten as

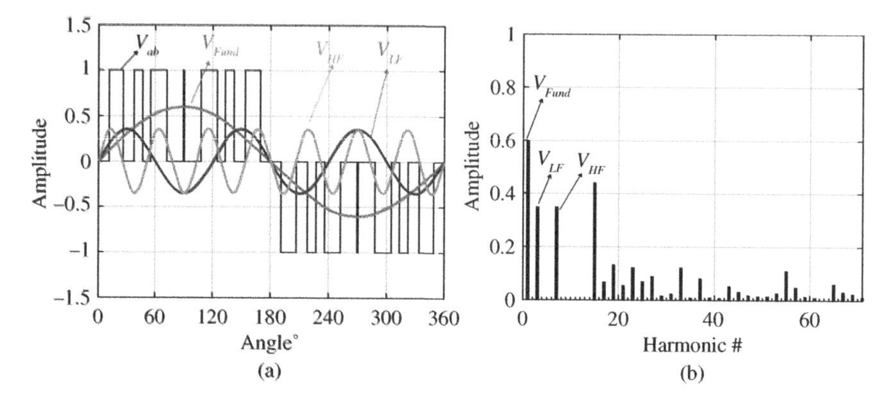

Figure 6.7 Unipolar multifrequency programmed PWM under $U_{ac29} = 0.6$, $U_{ac87} = 0.35$, and $U_{ac205} = 0.35$. (a) Time-domain spectrum. (b) FFT results. Source: Zhao and Costinett [6].

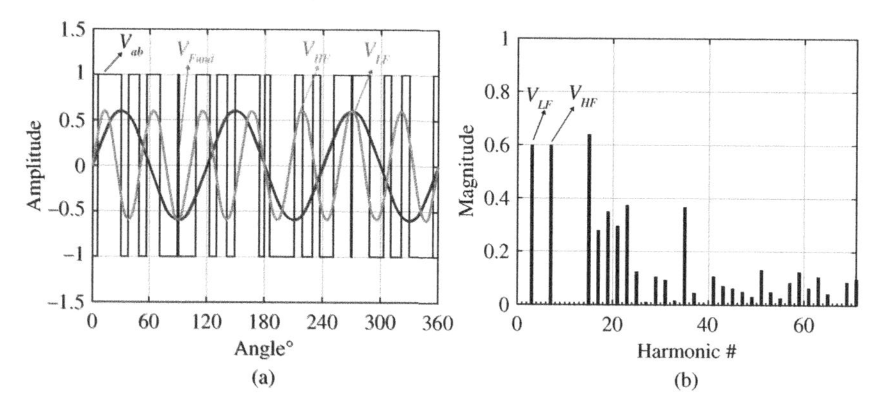

Figure 6.8 Bipolar multifrequency programmed PWM under $U_{ac29} = 0$, $U_{ac87} = 0.6$, and $U_{ac205} = 0.6$. (a) Time-domain spectrum. (b) FFT results. Source: Zhao and Costinett [6].

$$\begin{cases} \frac{4U}{\pi}(1 - 2\cos(1\theta_1) + 2\cos(1\theta_2) - 2\cos(1\theta_3) + \cdots + 2\cos(1\theta_k)) = U_{ac29} = 0 \\ \frac{4U}{3\pi}(1 - 2\cos(3\theta_1) + 2\cos(3\theta_2) - 2\cos(3\theta_3) + \cdots + 2\cos(3\theta_k)) = U_{ac87} \\ \frac{4U}{5\pi}(1 - 2\cos(5\theta_1) + 2\cos(5\theta_2) - 2\cos(5\theta_3) + \cdots + 2\cos(5\theta_k)) = 0 \\ \frac{4U}{7\pi}(1 - 2\cos(5\theta_1) + 2\cos(7\theta_2) - 2\cos(7\theta_3) + \cdots + 2\cos(7\theta_k)) = U_{ac205} \\ \cdots\cdots \\ \frac{4U}{n\pi}(1 - 2\cos(n\theta_1) + 2\cos(n\theta_2) - 2\cos(n\theta_3) + \cdots + 2\cos(n\theta_k)) = 0 \end{cases} \tag{6.19}$$

As shown in Figure 6.8, the results verify the aforementioned analysis. Harmonics higher than the ninth order can be regarded as high-order harmonics, which can be eliminated by the filtering characteristics of the resonant circuit itself.

6.2.2.2.2 Dual-frequency Full-bridge Resonant Inverter The configuration of a dual-frequency full-bridge resonant inverter is shown in Figure 6.9, where one leg of the inverter is switched at low frequency (D_1, D_2), and the other leg is switched at high frequency (D_3, D_4) [11]. Usually, the switching frequency on the high-frequency leg (f_h) is set at three or five times of that on the low-frequency leg (f_l). For example, set $f_h = 3f_l$ and duty ratio of 50%, the control signal for each switch and output voltage U_{ac} of the dual-frequency modulation is as given in Figure 6.10.

The output voltage and current of the two pickups are shown in Figure 6.11, where two frequencies are composed in the exciting circuit. This idea is similar to combining two half-bridge inverters that operate at two different frequencies.

6.2.2.2.3 Multifrequency Superposition Methodology

Transformer-based Superposition As illustrated in Figure 6.12, the transformer-based superposition method contains multiple half-bridge voltage-fed inverters that share a common DC voltage source and operate at multiple switching frequencies, namely $f_1, f_2, \cdots, f_n$. In order to combine the output of each inverter with various frequencies, a superposition scheme that adopts series-connected transformers is proposed in [12, 13], where the primary side is driven by a mixed-frequency square voltage. As an alternative method to the half-bridge inverter, the full-bridge inverters can also be adopted in this scheme. Due to the topology being almost the same, the transformer-based full-bridge voltage fed inverter will not be further discussed in this section.

Verification Key waveforms of the transformer-based dual-frequency system are depicted in Figure 6.13, where v_{in1} and v_{in2} indicate the midpoint voltages of

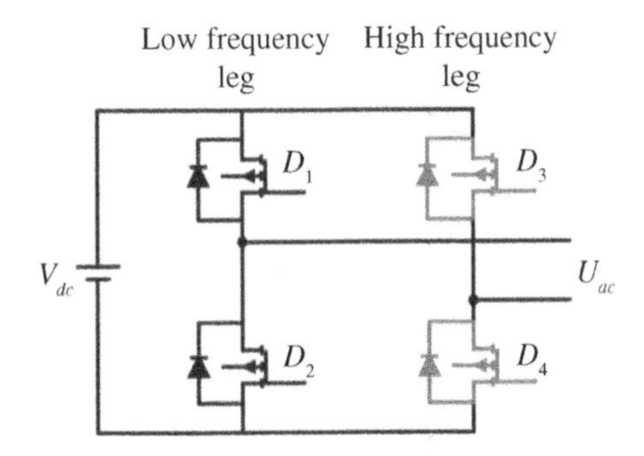

Figure 6.9 Configuration of dual-frequency full-bridge resonant inverter.

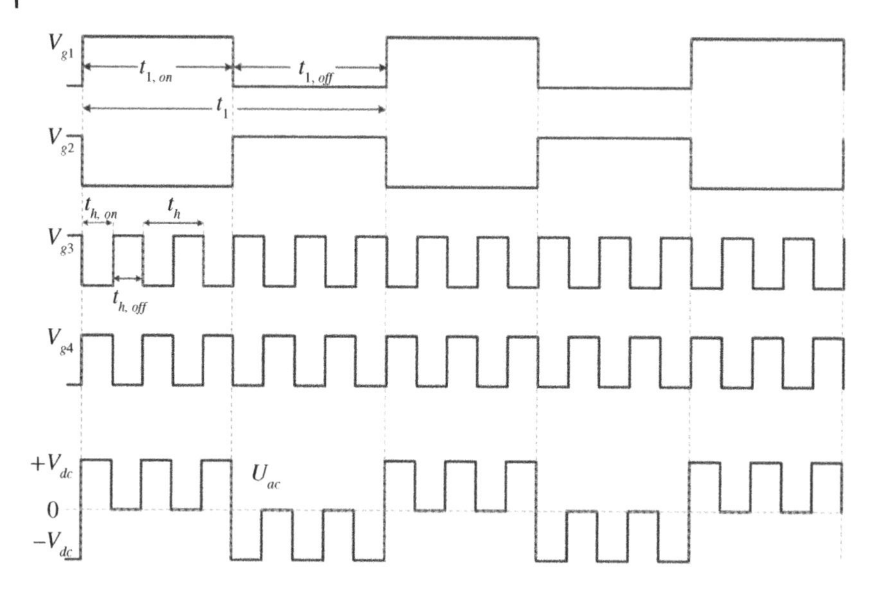

Figure 6.10 Control signal and output voltage of the dual-frequency modulation under the circumstance of $f_h = 3f_l$ and duty ratio of 50% ($t_{l,on} = t_{l,off}$, $t_{h,on} = t_{h,off}$).

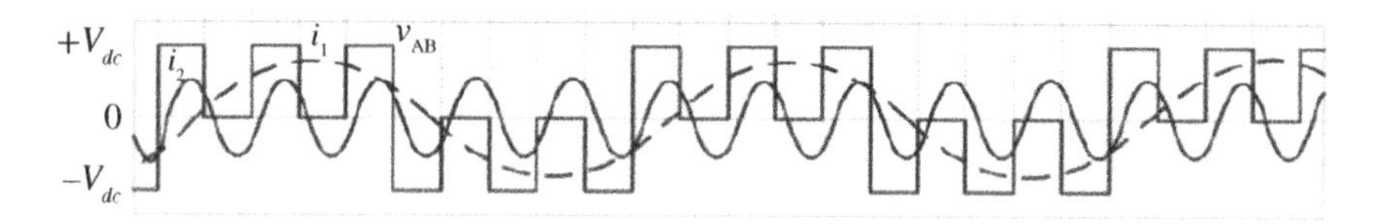

Figure 6.11 Inverter output voltage and two load currents. Source: Papani et al. [11]/with permission of The Institution of Engineering and Technology.

half-bridge inverters operating at the frequencies of f_1 and f_2, respectively. i_p, i_{s1}, and i_{s2} represent the currents of transmitting coils and two-pickup coils. v_{01} and v_{02} are the midpoint voltages of rectifiers. As can be seen from Figure 6.13, the transmitting currents consist of the multifrequency components that are driven by a mixed-frequency square voltage. Meanwhile, the dominant frequency components of each output are f_1 and f_2, which indicates that the power is wirelessly transferred to the desired loads selectively. However, in real application, the power loss in the primary transformer cannot be ignored, which limits the popularity of the transformer-based superposition method.

Transformer-free Superposition The transformer-free superposition methodology is illustrated in Figure 6.14, where superscript (i) represents the superposed ith order fundamental or harmonic current. Notably, a synthetic half-cycle

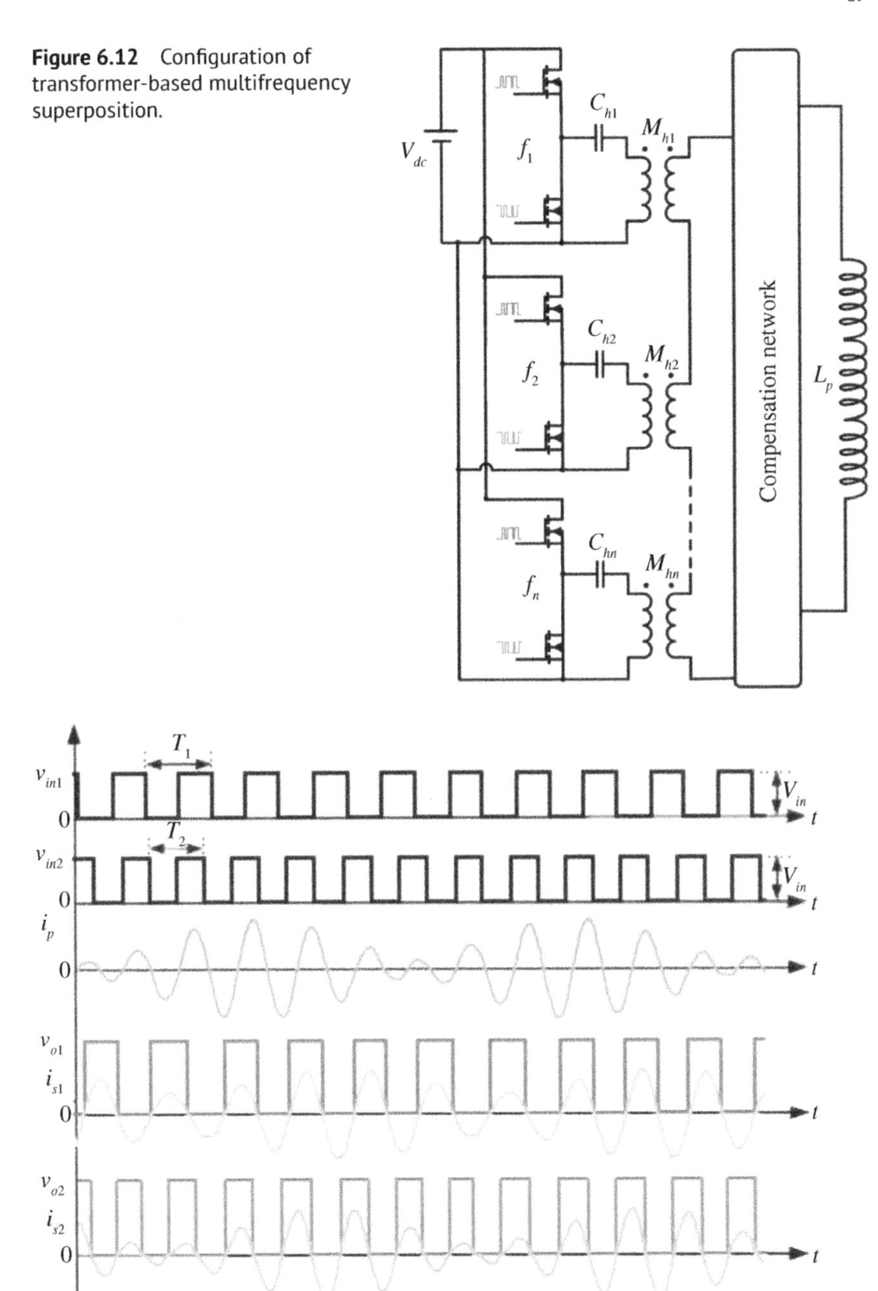

Figure 6.12 Configuration of transformer-based multifrequency superposition.

Figure 6.13 Waveform of dual-frequency transformer-based multifrequency superposition. Source: Liu et al. [12].

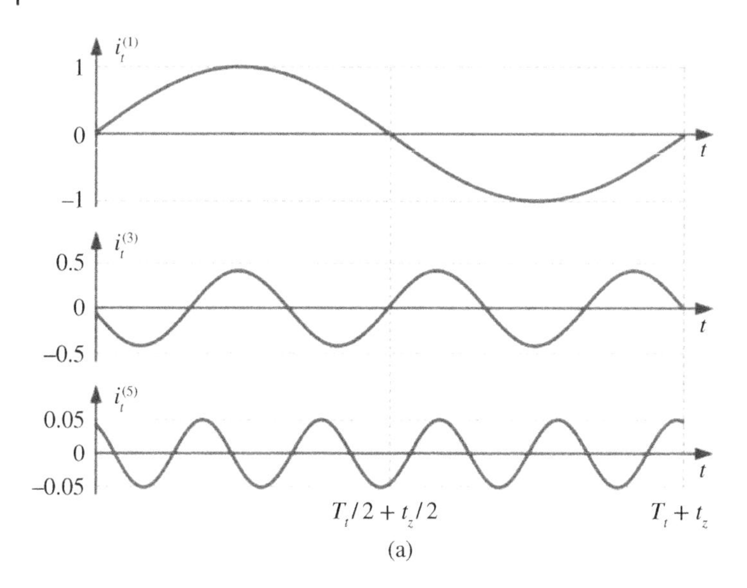

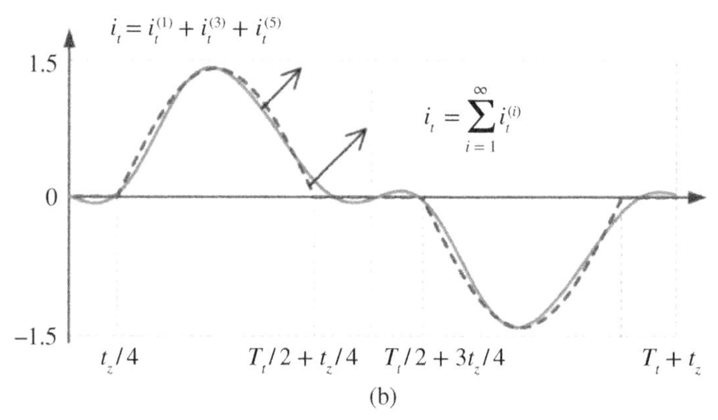

Figure 6.14 Methodology for transformer-free multifrequency superposition. (a) Superposed fundamental, third, and fifth harmonic currents. (b) Synthetic transmitting current. Source: Liu et al. [14].

sinusoidal transmitting current can be superposed by the fundamental, third-, and fifth-harmonic current components. Theoretically, a more ideal half-cycle sinusoidal transmitting current will be constructed by adopting more high-order harmonics. The synthetic half-cycle sinusoidal can be deduced as

$$i_t(t) = \begin{cases} 0, & t \in t_{izero} \\ A_t \sin(\omega_t(t - t_Z/4)), & t \in [t_Z/4,\ \ T_t/2 + t_Z/4] \\ A_t \sin(\omega_t(t - 3t_Z/4)), & t \in [T_t/2 + 3t_Z/4,\ \ T_t + 3t_Z/4] \end{cases} \tag{6.20}$$

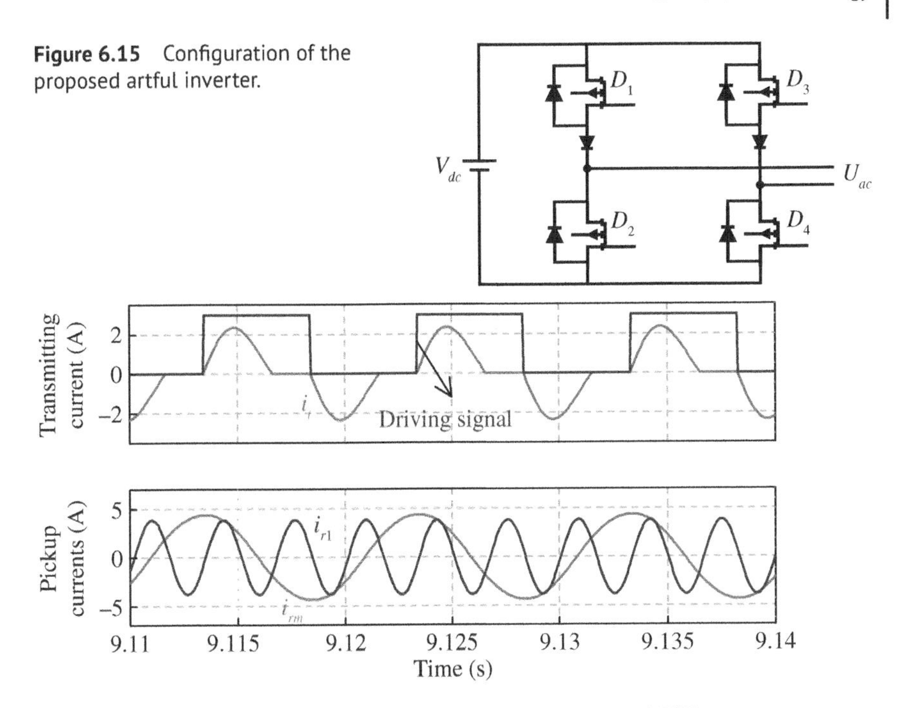

Figure 6.15 Configuration of the proposed artful inverter.

Figure 6.16 Waveforms of transformer-free superposition-based WPT system.

where t_Z is the duration of t_{izero} during one switching period. Owing to the odd function characteristic of its periodic extension, $i_t(t)$ is only superposed by odd harmonic components.

In practical implementation, an improved transmitter is introduced by connecting one diode in series to each leg, as shown in Figure 6.15. The transmitting side can approximately generate a synthetic half-cycle sinusoidal current as expressed in Eq. (6.20). Hence, it holds the competence to transmit the fundamental and high-order harmonics. Under the exemplified transformer-free superposition-based WPT, the switching regulates at 100.73 kHz. Figure 6.16 depicts the simulation waveforms of the driving signal, superposed transmitting current, and induced currents of two pickups. It is worth noting that the superposed transmitting current is ideal half-cycle sinusoidal owing to the utilization of ideal power devices.

6.2.2.3 Discussion

According to the aforementioned analysis and verifications, it can be concluded that all the proposed methods can realize multifrequency excitation. It is feasible for scholars to choose any models for conducting the research. However, limitations and particularities lie among various methods. Accordingly, their merits and demerits are summarized in Table 6.1.

Table 6.1 Comparison of four methods.

Method	Advantages	Disadvantages
Multifrequency programmed pulse width modulation	(1) Single inverter and multiple frequencies (2) Arbitrarily design the voltage amplitude for each frequency (3) High quality of output waveform	(1) Offline calculation of switching angles (unable to adjust in real time) (2) Only contains a random combination of fundamental and odd harmonics
Dual-frequency full-bridge resonant inverter	(1) Single inverter (2) Easy to operate and control	(1) Only contains two frequencies (2) Contains high-order harmonics
Transformer-based superposition	(1) Multiple frequencies (2) Arbitrarily design the voltage amplitude for each frequency (can be higher than the source voltage)	(1) Consumes multiple inverters and transformers (2) Power loss for transformers would be large
Transformer-free superposition	(1) Single inverter and multiple frequencies (2) Easy to control	(1) The amplitude of each frequency cannot be arbitrarily set (2) Only contains a random combination of fundamental and odd harmonics

6.3 Impedance Matching Strategy for Multifrequency Transmission

As aforementioned in Chapters 2 and 4, the compensation circuit that is applicable for multifrequency transmission becomes the prerequisite for minimizing the voltage–ampere (VA) rating of the converter/power supply in multiple-pickup WPT systems. Consequently, how to design an artful topology that can simultaneously nullify the reactance under the circumstances of multiple frequencies turns into the prime technical concern.

6.3.1 Compensation Network for Multifrequency

6.3.1.1 Dual-frequency Compensation Network

As shown in Figure 6.17, a dual-frequency resonating compensation network is presented, which contains transmitting coil L_{p1}, capacitors C_{p1}, parallel-connected inductor L_{p2}, and capacitors C_{p2} [15, 16]. The separate resonant frequencies can be obtained as

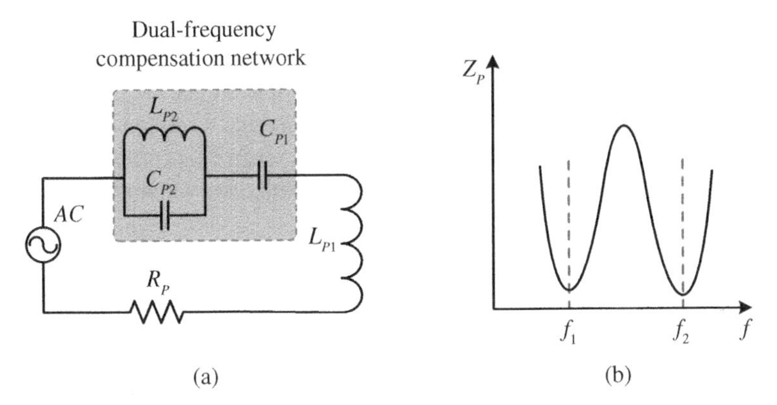

Figure 6.17 Dual-frequency resonating compensation network: (a) Circuit diagram; (b) Impedance characteristic.

$$\begin{cases} f_{p1} = \dfrac{1}{2\pi\sqrt{L_{p1}C_{p1}}} \\ f_{p2} = \dfrac{1}{2\pi\sqrt{L_{p2}C_{p2}}} \end{cases} \tag{6.21}$$

Based on KVL, the total impedance shown in Figure 6.17 can be expressed as

$$\begin{aligned} Z_p &= R_p + j\omega_i L_{p1} + \frac{1}{j\omega_i C_{p1}} + \frac{1}{\frac{1}{j\omega_i L_{p2}} - j\omega_i C_{p2}} \\ &= R_p + j\frac{-L_{p1}L_{p2}C_{p1}C_{p2}\omega_i^{\,4} + (L_{p1}C_{p1} + L_{p2}C_{p2} + L_{p2}C_{p1})\omega_i^{\,2} - 1}{\omega_i C_{p1}\left(1 - \omega_i^{\,2}L_{p2}C_{p2}\right)}, \\ & i = 1 \text{ or } 2 \end{aligned} \tag{6.22}$$

where R_p represents the internal resistance of the transmitting coil, and the superscript i ($i = 1, 2$) represents the transmission channel at the frequencies of ω_1 and ω_2, respectively. Accordingly, letting Eq. (6.22) equal zero yields dual resonant frequencies:

$$\begin{cases} f_1 = \dfrac{\sqrt{\dfrac{L_{p1}C_{p1} + L_{p2}C_{p2} + L_{p2}C_{p1} - \sqrt{(L_{p1}C_{p1} + L_{p2}C_{p2} + L_{p2}C_{p1})^2 - 4L_{p1}L_{p2}C_{p1}C_{p2}}}{2L_{p1}L_{p2}C_{p1}C_{p2}}}}{2\pi} \\ f_2 = \dfrac{\sqrt{\dfrac{L_{p1}C_{p1} + L_{p2}C_{p2} + L_{p2}C_{p1} + \sqrt{(L_{p1}C_{p1} + L_{p2}C_{p2} + L_{p2}C_{p1})^2 - 4L_{p1}L_{p2}C_{p1}C_{p2}}}{2L_{p1}L_{p2}C_{p1}C_{p2}}}}{2\pi} \end{cases} \tag{6.23}$$

Hence, the impedance characteristic can be drawn as shown in Figure 6.17b, which indicates two separate resonant frequencies.

In practical dual-frequency multiple-pickup WPT applications, two different frequencies are predefined according to various requirements. Consequently, the design procedure of the parameters in a dual-frequency resonating compensation network should be elaborated. Firstly, define three intermediate variables:

$$\begin{cases} r_L = \frac{L_{p1}}{L_{p2}} \\ r_C = \frac{C_{p1}}{C_{p2}} \\ r_f = \frac{f_1}{f_2} \end{cases} \tag{6.24}$$

By substituting Eqs. (6.21) and (6.24) into (6.23), the dual resonant frequencies can be rearranged as

$$\begin{cases} f_1 = f_{p1}\sqrt{\frac{1}{2} + \frac{r_c}{2}\left(1 + r_L - \sqrt{\left(\frac{r_L r_c - 1}{r_c}\right)^2 + 2\left(\frac{r_L r_c + 1}{r_c}\right) + 1}\right)} \\ f_2 = f_{p2}\sqrt{\frac{1}{2} + \frac{r_c}{2}\left(1 + r_L + \sqrt{\left(\frac{r_L r_c - 1}{r_c}\right)^2 + 2\left(\frac{r_L r_c + 1}{r_c}\right) + 1}\right)} \end{cases} \tag{6.25}$$

Then, the ratio r_c can be expressed as

$$r_c = \frac{\left[\left(r_f^4 + 1\right) r_L - 2r_f^2\right] \pm \sqrt{\left(r_f^4 - 1\right)^2 r_L^2 - 4r_f^2\left(r_f^4 + 2r_f^2 + 1\right) r_L}}{2\left(r_f^2 r_L^2 + 2r_f^2 r_L + r_f^2\right)} \tag{6.26}$$

In order to ensure the discriminant above zero, the ratio r_f should satisfy the inequality as

$$r_L > \frac{4r_f^2\left(r_f^4 + 2r_f^2 + 1\right)}{\left(r_f^4 - 1\right)^2} \tag{6.27}$$

Finally, the practical design procedure of the dual-frequency resonating compensation network is addressed as

Step 1: Predefine two operating frequencies f_1 and f_2, calculate the value of r_f, and wind the transmitting coil L_{p1}.

Step 2: Determine the value of r_L based on formula (6.27), solve L_{p2} based on Eq. (6.24), and then calculate the value of r_c based on formula (6.26).

Step 3: Substitute f_1, r_L, and r_c into Eq. (6.25) and then obtain f_{p1}.

Step 4: Solve C_{p1} based on Eq. (6.21) and then solve C_{p2} based on (6.24).

Employing the proposed dual-frequency resonating compensation network, the imaginary part of Z_p equals zero under two resonant frequencies f_1 and f_2, thus successfully reducing the reactive power loss and maintaining the transmission performance.

6.3.1.1.1 ***Verification*** Figure 6.18 depicts the verification results of the impedance Z_p for the primary circuit. As seen from Figure 6.18a, the modulus value of Z_p equals almost zero at both 100 and 300 kHz as the imaginary part of the Z_p can be eliminated at these two frequencies by adopting the dual-frequency compensation network. The impedance angle curve shown in Figure 6.18b further verifies that the dual-frequency compensation network can work at the resonate state concerning multiple frequencies, thus forming an ideal multichannel transmission passageway for multiple-pickup WPT systems.

6.3.1.2 Analysis for Multifrequency Compensation Network

As shown in Figure 6.19, a triple-frequency resonating compensation network is presented, which contains transmitting coil L_{p1}, capacitors C_{p1}, parallel-connected inductor L_{p2} and capacitors C_{p2}, and parallel-connected inductor L_{p3} and capacitors C_{p3}. Based on KVL, the corresponding total impedance shown in Figure 6.19a can be expressed as

$$
\begin{aligned}
Z_p &= R_p + j\omega_i L_{p1} + \frac{1}{j\omega_i C_{p1}} + \frac{j\omega_i L_{p2}}{1-\omega_i^2 L_{p2}C_{p2}} + \frac{j\omega_i L_{p3}}{1-\omega_i^2 L_{p3}C_{p3}} \\
&= R_p + j\frac{\begin{matrix}\omega_i^6 L_{p1}L_{p2}L_{p3}C_{p1}C_{p2}C_{p3} - \omega_i^4 \\ \begin{pmatrix} L_{p1}L_{p2}C_{p1}C_{p2} + L_{p1}L_{p3}C_{p1}C_{p3} + L_{p2}L_{p3}C_{p2}C_{p3} \\ +L_{p2}L_{p3}C_{p1}C_{p2} + L_{p2}L_{p3}C_{p1}C_{p3} \end{pmatrix} \\ +\omega_i^2(L_{p1}C_{p1} + L_{p2}C_{p2} + L_{p3}C_{p3} + L_{p2}C_{p1} + L_{p3}C_{p2}) - 1\end{matrix}}{\omega_i C_{p1}\left(1-\omega_i^2 L_{p2}C_{p2}\right)\left(1-\omega_i^2 L_{p3}C_{p3}\right)}
\end{aligned}
\tag{6.28}
$$

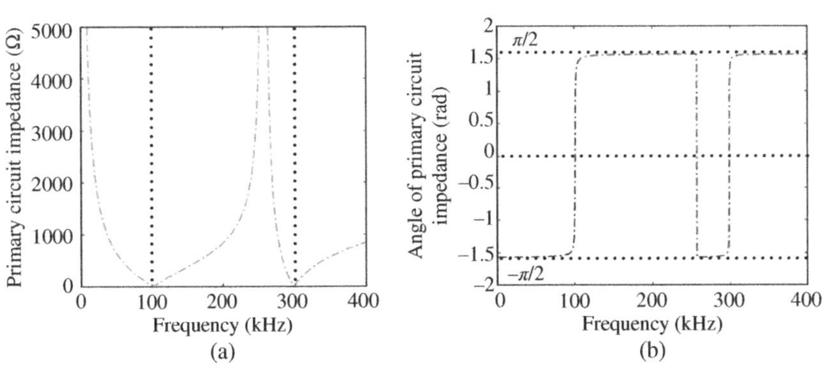

Figure 6.18 Verification results of Z_p: (a) $|Z_p|$; (b) Impedance angle of Z_p.

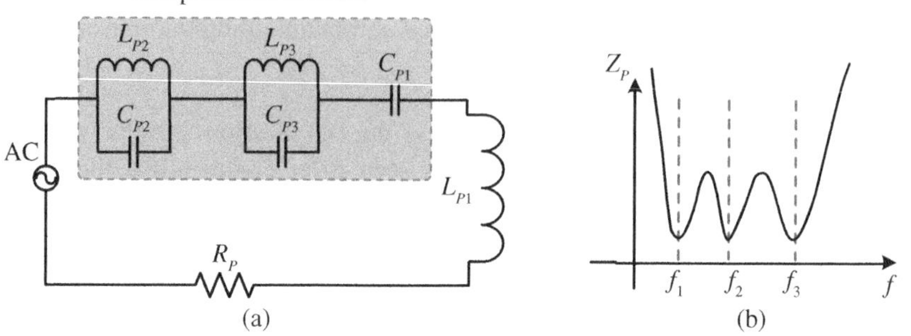

Figure 6.19 Triple-frequency resonating compensation network: (a) Circuit diagram; (b) Impedance characteristic.

where $i = 1, 2, \text{or}\, 3$, and R_p denotes the internal resistance of the transmitting coil.

Set $x = \omega_i^2$, then the numerator of Eq. (6.28) can be regarded as a univariate cubic equation as given in Eq. (6.29), where the corresponding intermediate variable parameters a, b, and c are addressed as

$$ax^3 + bx^2 + cx - 1 = 0$$

$$\begin{cases} a = L_{p1}L_{p2}L_{p3}C_{p1}C_{p2}C_{p3} \\ b = -(L_{p1}L_{p2}C_{p1}C_{p2} + L_{p1}L_{p3}C_{p1}C_{p3} + L_{p2}L_{p3}C_{p2}C_{p3} \\ \quad +L_{p2}L_{p3}C_{p1}C_{p2} + L_{p2}L_{p3}C_{p1}C_{p3}) \\ c = L_{p1}C_{p1} + L_{p2}C_{p2} + L_{p3}C_{p3} + L_{p2}C_{p1} + L_{p3}C_{p2} \end{cases} \tag{6.29}$$

According to the root formula of the univariate cubic equation, setting $x = y - \frac{b}{3a}$, Eq. (6.29) can be rearranged as

$$y^3 + \frac{3ac - b^2}{3a^2}y + \frac{-27a^2 - 9abc + 2b^3}{27a^3} - 1 = 0 \tag{6.30}$$

Accordingly, solving Eq. (6.30) yields three roots as follows:

$$\begin{cases} y_1 = \sqrt[3]{-\frac{q}{2} + \sqrt{\Delta}} + \sqrt[3]{-\frac{q}{2} - \sqrt{\Delta}} \\ y_2 = \frac{-1 + \sqrt{3}i}{2} \cdot \sqrt[3]{-\frac{q}{2} + \sqrt{\Delta}} + \frac{-1 - \sqrt{3}i}{2} \cdot \sqrt[3]{-\frac{q}{2} - \sqrt{\Delta}} \\ y_3 = \frac{-1 - \sqrt{3}i}{2} \cdot \sqrt[3]{-\frac{q}{2} + \sqrt{\Delta}} + \frac{-1 + \sqrt{3}i}{2} \cdot \sqrt[3]{-\frac{q}{2} - \sqrt{\Delta}} \end{cases} \tag{6.31}$$

where $p = \frac{3ac - b^2}{3a^2}$, $q = \frac{-27a^2 - 9abc + 2b^3}{27a^3}$, $\Delta = \frac{q^2}{4} + \frac{p^3}{27}$.

Correspondingly, the three resonant frequencies can be calculated as

$$\begin{cases} f_1 = \dfrac{\sqrt{\sqrt[3]{-\frac{q}{2}+\sqrt{\Delta}}+\sqrt[3]{-\frac{q}{2}-\sqrt{\Delta}}-\frac{b}{3a}}}{2\pi} \\ f_2 = \dfrac{\sqrt{\frac{-1+\sqrt{3}i}{2}\cdot\sqrt[3]{-\frac{q}{2}+\sqrt{\Delta}}+\frac{-1-\sqrt{3}i}{2}\cdot\sqrt[3]{-\frac{q}{2}-\sqrt{\Delta}}-\frac{b}{3a}}}{2\pi} \\ f_3 = \dfrac{\sqrt{\frac{-1-\sqrt{3}i}{2}\cdot\sqrt[3]{-\frac{q}{2}+\sqrt{\Delta}}+\frac{-1+\sqrt{3}i}{2}\cdot\sqrt[3]{-\frac{q}{2}-\sqrt{\Delta}}-\frac{b}{3a}}}{2\pi} \end{cases} \tag{6.32}$$

Hence, the impedance characteristic can be drawn as shown in Figure 6.19b, which indicates three separate resonant frequencies. Meanwhile, in order to ensure three unequal real roots, the value Δ should be controlled to less than zero. Thus, the optimal selection of the related parameters should be taken into consideration. In addition, it should be noted that, based on the relationship between roots and the coefficient of Eq. (6.29), where $x_1 + x_2 + x_3 = -\frac{b}{a} > 0$, and $x_1\, x_2\, x_3\, \frac{1}{a} > 0$, the actual curve of $W = ax^3 + bx^2 + cx - 1$ can be drawn, which is shown in Figure 6.20 with two different distribution of roots.

Calculating the derivative function of W, namely $V = 3ax^2 + 2bx + c$, the result implies that the value of $V = c$, which is positive when $x = 0$. Consequently, the distribution of all three roots should be larger than zero as described in Figure 6.20b. As a result, this three-frequency resonating network can be validated to have three resonating frequencies as shown in Eq. (6.32) and can be applied to provide a simultaneous transmission path in triple-frequency transmission.

6.3.1.2.1 *Verification* Figure 6.21 illustrates the magnitude and phase characteristics of the input impedance Z_p, wherein the primary circuit utilizes triple-resonating compensation with the resonating frequency of 100/300/500 kHz. It can be found that the phase angle is kept at zero to facilitate the zero phase angle (ZPA) and the unity power factor. Owing to the full

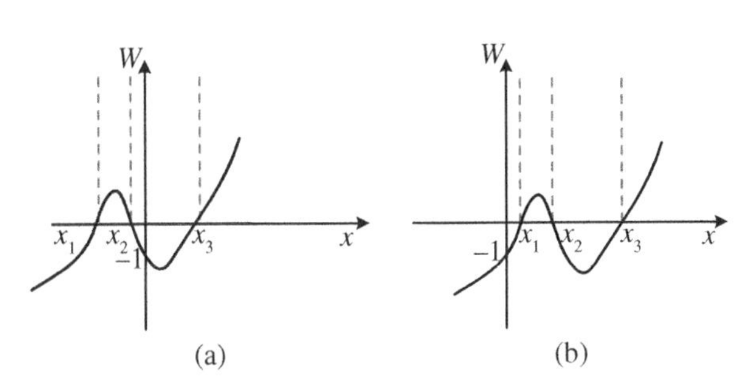

Figure 6.20 Roots distribution: (a) $x_1, x_2 < 0$, $x_3 > 0$; (b) $x_1, x_2, x_3 > 0$.

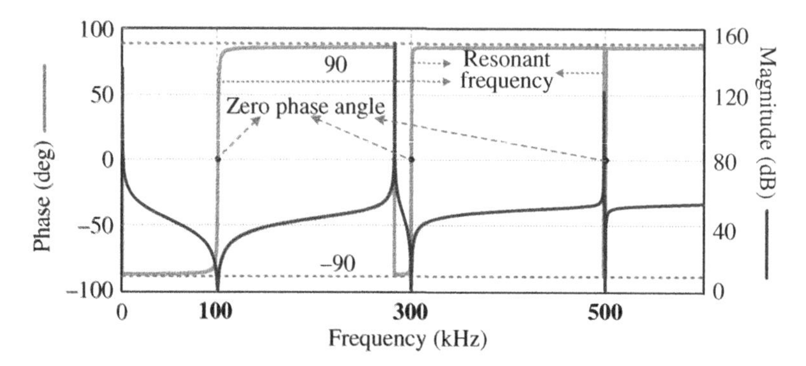

Figure 6.21 Impedance characteristics of triple-resonating compensation under resonating frequencies of 100, 300, and 500 kHz.

elimination of the imaginary part, the magnitude of Z_p is zero at each resonating frequency.

According to the aforementioned analysis of dual- and triple-frequency resonating compensation networks, the circuit diagram of multifrequency resonating compensation can be concluded as shown in Figure 6.22. The detailed derivation process of the proposed topology is quite similar and thus left as an exercise for readers and can be referred to in [17].

6.3.2 Compensation for Cross-coupling on the Pickup Side

Owing to the random placement of pickups, the cross-coupling effect inevitably appears in practical applications. It addresses more challenging analysis and optimization issues in the operation of the system and significantly decreases the system efficiency [18, 19]. Hence, fruitful achievements have been made regarding how to eliminate the cross-coupling effect [20–22].

Notably, if two-pickup WPT systems hold the same set of the primary current and secondary currents as well as identical equivalent resistance, its important characteristics, the total equivalent impedance seen from the power supply,

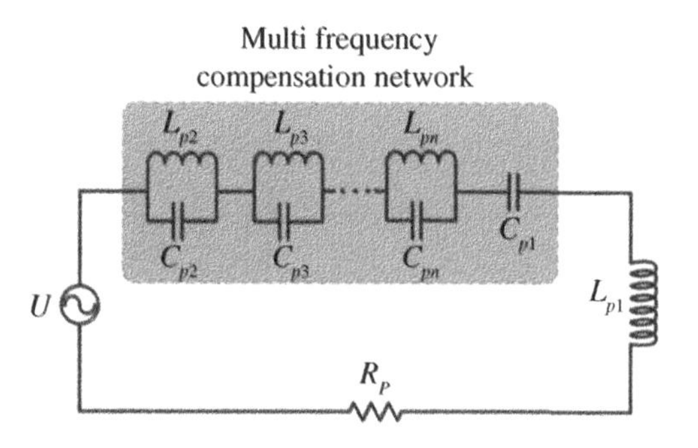

Figure 6.22 Circuit diagram of multifrequency resonating compensation network.

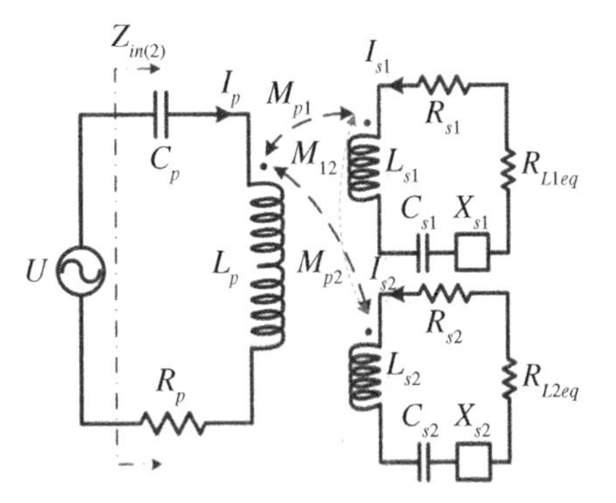

Figure 6.23 Simplified two-pickup IPT system with cross-coupling and extra reactance.

transmission efficiency, input power, and load power will be equal to each other whether there remains cross-coupling effect or not. The above conditions can be realized by adopting properly designed extra reactance as shown in Figure 6.23.

Denote X_{s1} and X_{s2} as the extra reactance of pickup one and pickup two, respectively. Accordingly, based on KVL, the circuit as shown in Figure 6.23 can be described as

$$\begin{bmatrix} \vec{U} \\ 0 \\ 0 \end{bmatrix} = \begin{bmatrix} \vec{Z_p} & j\omega M_{p1} & j\omega M_{p2} \\ j\omega M_{p1} & \vec{Z_{s1}} + R_{L1eq} + jX_{s1} & j\omega M_{12} \\ j\omega M_{p2} & j\omega M_{12} & \vec{Z_{s2}} + R_{L2eq} + jX_{s2} \end{bmatrix} \begin{bmatrix} \vec{I_p} \\ \vec{I_{s1}} \\ \vec{I_{s2}} \end{bmatrix} \tag{6.33}$$

By substituting (2.39) into (6.33) gives

$$\begin{cases} X_{s1} = -\dfrac{\omega M_{p2} M_{12}(R_{s1} + R_{L1eq})}{M_{p1}(R_{s2} + R_{L2eq})} \\ X_{s2} = -\dfrac{\omega M_{p1} M_{12}(R_{s2} + R_{L2eq})}{M_{p2}(R_{s1} + R_{L1eq})} \end{cases} \tag{6.34}$$

The negative value of X_{s1} and X_{s2} means the capacitive impedance, which signifies that this method can be implemented by adjusting the series-connected capacitors. Adopting this approach, the numerical and theoretical analyses of the total equivalent impedance, transmission efficiency, input power, and load power can be well preserved. Accordingly, all the significant conclusions addressed in Section 2.3.3.4 to Discussion (under the circumstances without cross-coupling) remain valid. Regarding the aforementioned approach, capacitance impedance needs to be calculated in advance, which hampers the practical application in which some parameters fluctuate due to the working condition and temperature change. Accordingly, a self-adaptive synchronous series compensator instead of a capacitor series connected with the pickup circuit was proposed [22]. It can automatically cancel the reactance caused by the inductor, nonresistive load, or frequency variation, hence increasing the transmission performance. However, the applicable conditions are based on low coupling and thus can be a solution for low-power medical implant devices.

6.4 Others

6.4.1 Power Allocation

As a unique characteristic of multiple-pickup WPT systems, the power distribution issues among multipickup become critical technical concerns. Fruitful power allocation approaches have been proposed [2, 12, 14, 23–32]. Regarding the time-sharing transmission, it can easily achieve power distribution by changing the transmission time durations to multipickup [2]. Regarding multifrequency simultaneous transmission, the most common type is utilizing the impedance adjustment transformation of DC–DC converters to regulate the individual output voltages [23–25], wherein the zero-voltage switching in the transmitting side can be realized. However, additional DC–DC modules will result in system increment, control complexity, and unexpected power loss, which will further decrease the transmission efficiency. Besides, the output power can be distributed by changing the operating frequency of an inverter which synthesizes a half-cycle sinusoidal current in the transmitting coil [14, 26]. However, in order to generate such a half-cycle sinusoidal current, one diode needs to be series-connected to the upper leg of the inverter, which will bring additional turn-on losses. Meanwhile, the power cannot be arbitrarily distributed by the customers. To achieve an individual and continuous power distribution, a multifrequency modulation method based on a look-up table and delta-sigma modulation scheme is introduced, wherein the components of the synthesized driving voltage pulse have continuously and individually varying amplitudes of multiple specific frequencies [27]. The attention that the control method will be relatively complex should be paid when adopting this approach.

In practice, some of the pickups may transiently require more power than their normal operating state. Thus, an energy injection method that adopts the synchronization method of the primary and reverse injection pickups was proposed [28]. A bidirectional pickup topology is utilized to act as a transceiver in multiple-pickup WPT systems. A similar concept of wireless energy trading through routers on the pickup side was also introduced in Ref. [29]. Both the aforementioned approaches employ a complementary form of energy on the pickup side, regardless of the transmitting circuit.

In the near future, there will exist more portable wireless charging tables in different scenarios. Thus, an objective-based optimal power distribution method was put forward [30], wherein the charging capacity of devices, waiting time, and the number of charging times are taken into consideration, which offers a complete commercial charging solution. Besides, a demand-customized selective wireless power transfer system was proposed in which the system can simultaneously deliver power to meet different output requirements with different electrical appliances for authorized receivers, while preventing unauthorized pickups from stealing the power even when they are placed in the charging area [31, 32].

6.4.2 Maximum Efficiency for Multitransmitter

Efficiency maximization based on load optimization and frequency tracking has been thoroughly investigated for the conventional 1-to-1 transmission. Recently, an increasing number of researches have been conducted to further investigate the maximum efficiency for multiple-input multiple-output (MIMO) wireless power transmission [33–35]. Regarding MIMO WPT systems, efficiency maximization requires a joint optimization of transmitting current and pickup loads. The system in [33] contains a primary current controller to stabilize the voltage and an optimal resistance controller to track the maximum efficiency point. Besides, two controllers are independent without communication. Similarly, a two-step tracking method that firstly optimizes the transmitting current ratios according to the magnetic coupling ratios and then optimizes the pickup load impedance was proposed in [34]. *Q*-Thang presents a maximum efficiency expression at any frequency, with any number of transmitters and pickups, and under any coil arrangement, wherein the optimal complex values of input currents and loads are derived by applying the first-order necessary condition followed by the second partial derivative test [35].

6.4.3 Constant Voltage Control

How to maintain constant output voltage for each pickup with the coupling conditions changing and load varying has always been a challenge for researchers. Bo proposed a system in which each load can maintain constant power and constant efficiency when the coupling coefficient changes over a large range. The proposed approach combines the parity-time symmetry principle and the time-sharing control strategy. Besides, arbitrary power distribution among multiple loads is achieved only by adjusting the duty cycle α of each controllable resonant capacitor [36]. A multiple-pickup WPT platform that is capable of controlling the load voltages to satisfy individually rated values and stabilizing the load voltages against the load variations was presented in [37], wherein the voltage control is performed by effortlessly adjusting the arrangement of the transmitting coils inside the charging platform. Similarly, a load-independent LCC-compensated WPT system for multipickup with a decoupled compact coupler design was addressed in [38, 39].

For multiple-pickup WPT systems, there remain plenty of points, for example energy trading among pickups, cancelation of cross-coupling at both sides, and a more flexible and intelligent power distribution strategy for readers to investigate.

References

1 Zhang, Y., Lu, T., Zhao, Z. et al. (2015). Selective wireless power transfer to multiple loads using receivers of different resonant frequencies. *IEEE Transactions on Power Electronics* 30 (11): 6001–6005.
2 Kim, Y.-J., Ha, D., Chappell, W.J., and Irazoqui, P.P. (2016). Selective wireless power transfer for smart power distribution in a miniature-sized multiple-receiver system. *IEEE Transactions on Industrial Electronics.* 63 (3): 1853–1862.
3 Lee, K., and Chae, S. H. (2020). Comparative analysis of frequency-selective wireless power transfer for multiple-Rx systems, *IEEE Transactions on Power Electronics*, 35 (5): 5122–5131.
4 Huynh, P. S., Vincent, D., Patnaik, L., and Williamson, S.S. (2018). FPGA-based PWM implementation of matrix converter in inductive wireless power transfer systems In: *Proceedings of the IEEE PELS Workshop on Emerging Technologies: Wireless Power Transfer (Wow)*, Montréal, QC, Canada.
5 Zhao, C., Costinett, D., Trento, B., and Friedrichs, D. (2016). A single-phase dual frequency inverter based on multi-frequency selective harmonic elimination. In: *Proceedings of the IEEE Applied Power Electronics Conference and Exposition*, Long Beach, CA, USA.
6 Zhao, C. and Costinett, D. (2017). GaN-based dual-mode wireless power transfer using multifrequency programmed pulse width modulation. *IEEE Transactions on Industrial Electronics* 64 (11): 9165–9176.
7 Patel, H.S. and Hoft, R.G. (1973). Generalized technique of harmonics elimination and voltage control in thyristor inverts: part I harmonic elimination. *IEEE Transactions on industrial application* IA-9 (3): 310–317.
8 Ahmadi, D., Zou, K., Li, C. et al. (2011). A universal selective harmonic elimination method for high-power inverters. *IEEE Transactions on Industrial Electronics* 26 (10): 2743–2752.
9 Chiasson, J.N., Tolbert, L.M., McKenzie, K.J., and Du, Z. (2004). A complete solution to the harmonic elimination problem. *IEEE Transactions on Power Electronics* 19 (2): 491–499.
10 Agelidis, V. G., Balouktsis, A., Balouktsis, I., and Cossar, C., (2006). Multiple sets of solutions for harmonic elimination PWM bipolar waveforms: analysis and experimental verification. *IEEE Transactions on Power Electronics*, 21 (2): 415–421.
11 Papani, S.K., Neti, V., and Murthy, B.K. (2015). Dual frequency inverter configuration for multiple-load induction cooking application. *IET Power Electronics.* 8 (4): 591–601.
12 Liu, F., Yang, Y., Ding, Z. et al. (2018). A multifrequency superposition methodology to achieve high efficiency and targeted power distribution for a multiload MCR WPT system. *IEEE Transactions on Power Electronics* 33 (10): 9005–9016.

13 Liu, F., Yang, Y., Ding, Z. et al. (2018). Eliminating cross interference between multiple receivers to achieve targeted power distribution for a multi-frequency multi-load MCR WPT system. *IET Power Electronics* 11 (8): 1321–1328.

14 Liu, W., Chau, K.T., Lee, C.H.T. et al. (2019). Multi-frequency multi-power one-to-many wireless power transfer system. *IEEE Transactions on Magnetics* 55 (7): 1, 8001609–9.

15 Kung, M. and Lin, K. (2015). Enhanced analysis and design method of dual-band coil module for near-field wireless power transfer systems. *IEEE Transactions on Microwave Theory and Techniques* 63 (3): 821–832.

16 Zhang, Z., Li, X., Pang, H. et al. (2020). Multiple-frequency resonating compensation for multichannel transmission of wireless power transfer. *IEEE Transactions on Power Electronics* 36 (5): 5169–5180.

17 Pang, H., Chau, K.T., Liu, W., and Tian, X. Multi-resonating-compensation for multi-channel multi-pickup wireless power transfer. *IEEE Transactions on Magnetics* https://doi.org/10.1109/TMAG.2022.3145878.

18 Lee, C.K., Zhong, W., and Hui, S. (2012). Effects of magnetic coupling of nonadjacent resonators on wireless power domino-resonator systems. *IEEE Transactions on Power Electronics.* 27 (4): 1905–1916.

19 Fu, M., Zhang, T., Ma, C., and Zhu, X. (2015). Efficiency and optimal loads analysis for multiple-receiver wireless power transfer systems. *IEEE Transactions on Microwave Theory and Techniques* 63 (3): 801–812.

20 Fu, M., Zhang, T., Zhu, X. et al. (2016). Compensation of cross coupling in multiple-receiver wireless power transfer systems. *IEEE Transactions on Industrial Informatics* 12 (2): 474–482.

21 Narayanamoorthi, R., Juliet, A.V., and Chokkalingam, B. (2019). Cross interference minimization and simultaneous wireless power transfer to multiple frequency loads using frequency bifurcation approach. *IEEE Transactions on Power Electronics* 34 (11): 10898–10909.

22 Shi, L., Alou, P., Oliver, J.Á. et al. (2021). A self-adaptive wireless power transfer system to cancel the reactance. *IEEE Transactions on Industrial Electronics* 68 (12): 12141–12151.

23 Fu, M., Yin, H., Liu, M. et al. (2018). A 6.78MHz multiple-receiver wireless power transfer system with constant output voltage and optimum efficiency. *IEEE Transactions on Power Electronics* 36 (6): 5330–5340.

24 Kim, J., Kim, D.-H., and Park, Y.-J. (2016). Free-positioning wireless power transfer to multiple devices using a planar transmitting coil and switchable impedance matching networks. *IEEE Transactions on Microwave Theory and Techniques* 64 (11): 3714–3722.

25 Huang, Y., Liu, C., Zhou, Y. et al. (2019). Power allocation for dynamic dual-pickup wireless charging system of electric vehicle. *IEEE Transactions on Magnetics* 55 (7): 8600106.

26 Liu, W., Chau, K.T., Lee, C.H.T. et al. (2020). Wireless energy-on-demand using magnetic quasi-resonant coupling. *IEEE Transactions on Power Electronics* 35 (9): 9057–9069.

27 Qi, C., Huang, S., Chen, X., and Wang, P. (2021). Multifrequency modulation to achieve an individual and continuous power distribution for simultaneous MR-WPT system with an inverter. *IEEE Transactions on Power Electronics* 36 (11): 12440–12455.

28 Dai, X., Wu, J., Jiang, J. et al. (2021). An energy injection method to improve power transfer capability of bidirectional WPT system with multiple pickups. *IEEE Transactions on Power Electronics* 36 (5): 5095–5107.

29 Liu, W., Chau, K.T., Chow, C.C.T., and Lee, C.H.T. (2022). Wireless energy trading in traffic internet. *IEEE Transactions on Power Electronics* 37 (4): 4831–4841. https://doi.org/10.1109/TPEL.2021.3118458.

30 Zhang, Z., Pang, H., and Wang, J. (2018). Multiple objective-based optimal energy distribution for wireless power transfer. *IEEE Transactions on Magnetics* 54 (11): 1, 8600205–5.

31 Tian, X., Chau, K.T., Hua, Z., and Han, W. Design and analysis of demand-customized selective wireless power transfer system. *IEEE Transactions on Industrial Electronics* https://doi.org/10.1109/TIE.2022.3142392.

32 Tian, X., Chau, K.T., Liu, W., and Lee, C.H.T. (2021). Selective wireless power transfer using magnetic field editing. *IEEE Transactions on Power Electronics* 36 (3): 2710–2719.

33 Xu, J., Li, X., Li, H. et al. (2022). Maximum efficiency tracking for multitransmitter multireceiver wireless power transfer system on the submerged buoy. *IEEE Transactions on Industrial Electronics* 69 (2): 1909–1919.

34 Kim, D.-H. and Ahn, D. (2020). Maximum efficiency point tracking for multiple-transmitter wireless power transfer. *IEEE Transactions on Power Electronics* 35 (11): 11391–11400.

35 Duong, Q. and Okada, M. (2018). Maximum efficiency formulation for multiple-input multiple-output inductive power transfer systems. *IEEE Transactions on Microwave Theory and Techniques* 66 (7): 3463–3477.

36 Luo, C., Qiu, D., Gu, W. et al. (2022). Multiload wireless power transfer system with constant output power and efficiency. *IEEE Transactions on Industry Applications* 58 (1): 1101–1114.

37 Vo, Q.-T., Duong, Q.-T., and Okada, M. (2019). Load-independent voltage control for multiple-receiver inductive power transfer systems. *IEEE Access* 7: 139450–139461.

38 Cheng, C., Lu, F., Zhou, Z. et al. (2020). A load-independent LCC-compensated wireless power transfer system for multiple loads with a compact coupler design. *IEEE Transactions on Industrial Electronics* 67 (6): 4507–4515.

39 Cheng, C., Li, W., Zhou, Z. et al. (2020). A load-independent wireless power transfer system with multiple constant voltage outputs. *IEEE Transactions on Power Electronics* 35 (4): 3328–3331.

7

Energy Security of Wireless Power Transfer

This chapter aims to raise the attentions on the consideration of energy security for wireless power transfer (WPT) systems, especially the issues on multiple-pickup applications. Firstly, the frequency characteristics of WPT systems are elaborated with emphasis on the impact of the transmitting power, including the frequency sensitivity and the frequency splitting, where the frequency characteristic is the theoretical basis for the presented energy encryption strategy. After revealing the working principle of energy encryption, simulated and experimental verifications are both given to illustrate the feasibility and the applicability. Lastly, the opportunities and challenges of the energy encryption will be drawn for WPT systems.

7.1 Introduction

The wireless power transfer (WPT) systems have been widely studied from now on. As the development of WPT systems, there have been increasing requirements that the traditional 1-to-1 WPT systems cannot fulfill, for example the dynamic wireless charging for electric vehicles (EVs). Accordingly, the WPT systems with one single transmitter for multiple pickups (STMPs) have been increasingly studied and employed for both academic researchers and the industrial engineers in recent years. Compared to multiple 1-to-1 WPT systems, the STMP-WPT systems have many advantages. For example, the number of inverters is reduced, which can save costs and increase the power density. Then, the induced electromagnetic fields of multiple 1-to-1 WPT systems affect each other, while only using one single transmitter can avoid this problem. In addition, what we called one single transmitter means only one induced electromagnetic field instead of only one transmitting coil. Accordingly, by increasing the number of transmitting coils, the charging area can be expanded, which has significant meanings for some specific applications, for example the dynamic wireless charging for EVs.

Wireless Power Transfer: Principles and Applications, First Edition. Zhen Zhang and Hongliang Pang.

As aforementioned, the STMP-WPT systems have many advantages. However, everything is a double-edged sword. For example, the security performance of STMP-WPT systems becomes a key challenge and has hardly been studied by researchers, which causes the limitation of some specific applications. Maybe an idea comes to your mind, that is we can turn on–off the power supply to ensure the security. However, simply turning on and off the power supply is useless because for multiple pickups, they work in the same induced electromagnetic field. Accordingly, the energy may be undesired to be transmitted to some pickups. Consequently, in this chapter, the energy security is introduced, and the corresponding energy encryption scheme is carried out to ensure the energy security.

7.2 Characteristics of Frequency

As mentioned in Chapter 2, WPT system adopts the high-frequency-induced electromagnetic field to realize the power transmission under the loosely coupled conditions. It can be seen that the frequency characteristic is the key to WPT technologies. In order to realize the advanced control for WPT systems such as the energy encryption, the impact of frequency on the transmission performance should be particularly paid attention. Accordingly, the frequency sensitivity and the frequency splitting are analyzed in this section, respectively.

7.2.1 Frequency Sensitivity

The transmission performance of WPT systems is determined by various factors, especially for the operating frequency. According to the theoretical analysis in Chapter 2, it shows that the power transmission performance, including the power, the efficiency, and the distance, is significantly affected by the operating frequency. Taking the series–series compensation topology as an exemplification, the output power P_{out} can be expressed as

$$P_{out} = \frac{\omega^2 M^2 R_L U_{in}^2}{(R_P R_L + \omega^2 M^2)^2} \tag{7.1}$$

Based on (7.1), the power curve can be drawn with respect to the operating frequency, as depicted in Figure 7.1, which shows that the transmitting power is significantly affected by the operating frequency. Figure 7.1 takes the EVs charging system as an example, where the load power and operating frequency are chosen around 3.3 kW and 20 kHz. The corresponding result means that when the WPT system works at the optimal frequency, the power can be efficiently transferred to the pickup. On the contrary, when the operating frequency deviates from

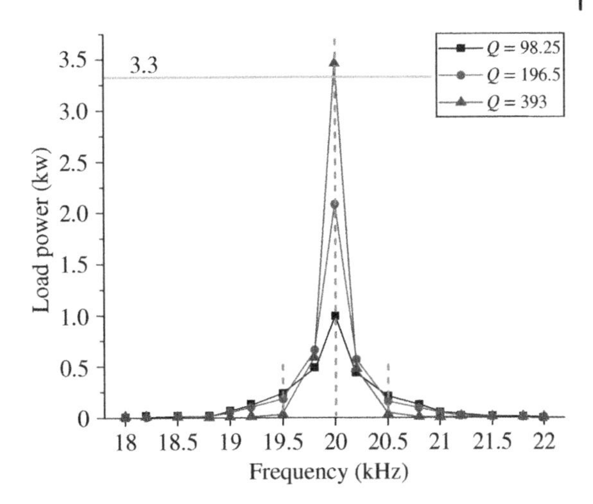

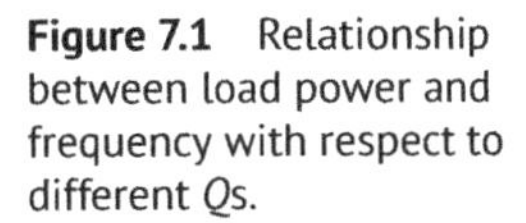
Figure 7.1 Relationship between load power and frequency with respect to different Qs.

the optimal value, the transmitting power will be suppressed to an extremely low level. The phenomenon that the transmitting power of WPT systems varies with the change in the operating frequency is called the frequency sensitivity. Accordingly, the frequency sensitivity increases the difficulty for tracking the maximum transmission power point of WPT systems.

To further verify the frequency sensitivity, the electromagnetic field analysis is performed by taking the two-coil WPT systems as an example. Figure 7.2 shows the distribution of the induced electromagnetic field when the WPT system works at the optimal frequency and the frequency deviating from the optimal value, respectively. By comparison, it shows that the magnetic flux line of the optimal operating frequency distributes more intensive and uniform than that of the nonoptimal frequency. Thus, Figure 7.2 illustrates the impact of frequency sensitivity on the flux density and even the transmitting power of WPT systems.

7.2.2 Frequency Splitting

As an inherent characteristic of WPT systems, the frequency splitting refers to the phenomenon that the output power of WPT systems has multiple peaks when the coupling coefficient is large enough. As shown in Figure 7.3, multiple peaks appear on the frequency curve when the WPT system has a short transmission distance. When the WPT system works at the resonant frequency (20 kHz), the output power drops down to the valley instead of rising to the top. However, two peaks appear symmetrically on both sides of the original resonant frequency. In [1], it elaborates the corresponding reason, that is the increased coupling coefficient results in the increasing of the reflected impedance from the secondary side to the primary side. Then, the input impedance represents a nonpure resistance

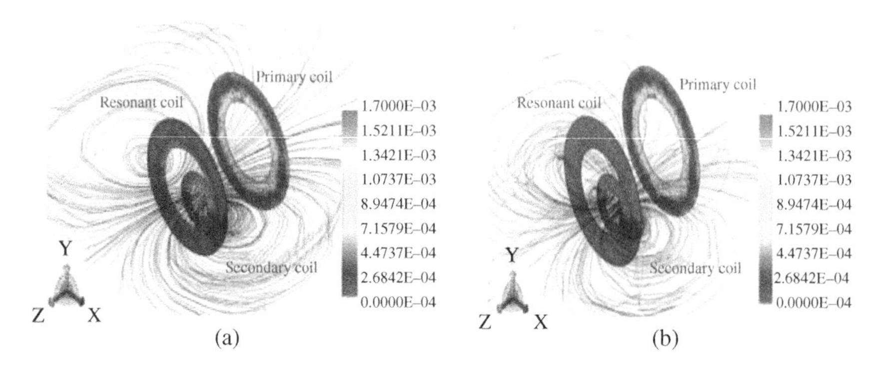

Figure 7.2 Distribution of the induced electromagnetic field. (a) Optimal operating frequency; (b) Nonoptimal operating frequency.

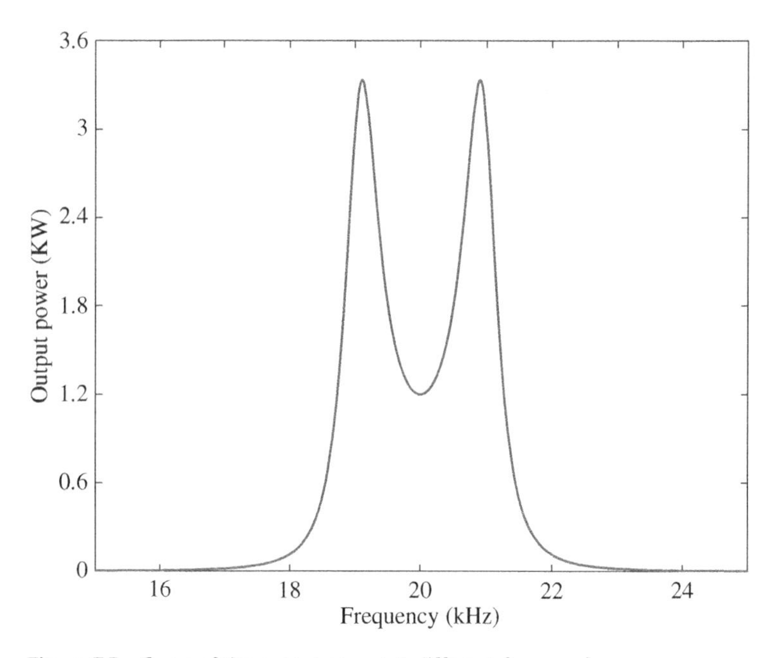

Figure 7.3 Curve of the output power at different frequencies.

state, which means that the imaginary part exists in the equivalent circuit of WPT systems, thus resulting in more than one resonant frequency in the system. The frequency-splitting phenomenon has been reported in previous studies [2, 3]. It has shown that the frequency splitting results in multiple peaks for the output power rather than the efficiency. Accordingly, this chapter focuses on revealing the impact on the output power of WPT systems by giving the theoretical analysis of the frequency splitting.

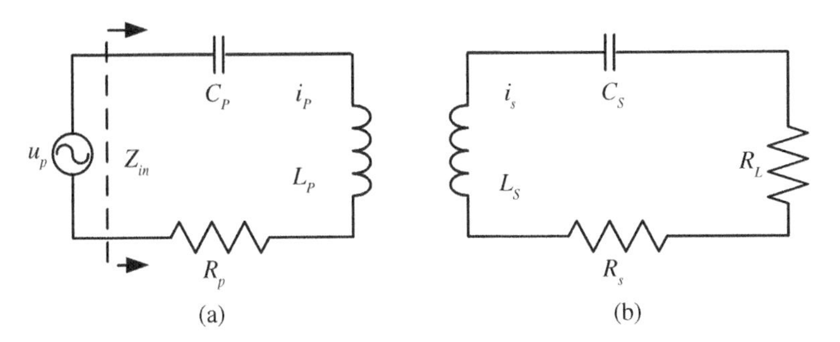

Figure 7.4 Equivalent circuit of two-coil WPT systems. (a) Primary side; (b) Secondary side.

Taking two-coil WPT systems as the exemplification [4-7], Figure 7.4 shows the equivalent circuit of two-coil WPT systems, where C_p and C_s are the primary and the secondary capacitances, L_p and L_s are the inductances of the transmitting and the pickup coils, R_p and R_s are the internal resistances of the primary and the secondary sides, R_L is the load, Z_{in} is the input impedance, and M is the mutual inductance between the transmitting and the pickup coils, respectively. The natural resonant frequencies of the primary and the secondary sides, namely f_p and f_s, can be expressed as

$$f_p = \frac{1}{2\pi\sqrt{L_pC_p}} \tag{7.2}$$

$$f_s = \frac{1}{2\pi\sqrt{L_SC_S}} \tag{7.3}$$

In order to facilitate the analysis, we assume that the parameters of the primary and the secondary sides are identical, that is $R_p = R_s = R_0$, $L_p = L_s = L_0$, $C_p = C_s = C_0$. When the primary and secondary units are both in the resonate state, the system resonant frequency f_0 can be given as

$$f_0 = f_p = f_s \tag{7.4}$$

In the primary side, the input impedance Z_{in} can be calculated as

$$Z_{in} = R_0 + j\left(\omega L_0 - \frac{1}{\omega C_0}\right) + \frac{(\omega M)^2}{R_0 + j\left(\omega L_0 - \frac{1}{\omega C_0}\right) + R_L} \tag{7.5}$$

where ω is the resonant angular frequency. Additionally, the current of the primary side i_p can be expressed as

$$i_p = \frac{u_p}{Z_{in}} \tag{7.6}$$

In the secondary side, the loop impedance Z_s can be obtained as

$$Z_s = R_0 + j\left(\omega L_0 - \frac{1}{\omega C_0}\right) + R_L \tag{7.7}$$

Thus, the current of the secondary side i_s can be calculated based on the voltage induced by the primary side, which is given by

$$i_s = \frac{j\omega M i_p}{Z_s} \tag{7.8}$$

Hereby, three factors are introduced, including the load-matching factor (L_M), the transfer quality factor (T_Q), and the frequency deviation factor (F_D), which are, respectively, defined by:

$$L_M = \frac{R_L}{R_0} \tag{7.9}$$

$$T_Q = \frac{\omega M}{L_0} \tag{7.10}$$

$$F_D = \frac{1}{R_0}\left(\omega L_0 - \frac{1}{\omega C_0}\right) \tag{7.11}$$

Using (7.5)–(7.11), the output power P_{out} can be derived as

$$\begin{aligned} P_{out} &= i_s^2 R_L \\ &= \frac{u_p^2}{R_L} \frac{T_Q^2 L_M^2}{\left(1 + L_M + T_Q^2\right)^2 + F_D^4 + \left[1 + (1 + L_M)^2 - 2T_Q^2\right] F_D^2} \end{aligned} \tag{7.12}$$

At the original resonant frequency, $F_D = 0$. Hence, the P_{out} has the maximum value when

$$T_Q = \sqrt{1 + L_M} \tag{7.13}$$

According to (7.9)–(7.11), L_M is independent of the frequency ω, while T_Q and F_D vary with the change in frequency. Due to the inequality $\partial T_Q/\partial\omega \ll \partial F_D/\partial\omega$, T_Q is usually regarded as a constant value ignoring the variation of ω. Thus, let $\partial P_{out}/\partial F_D = 0$, it can be obtained as

$$F_D\left(2F_D^2 + \delta\right) = 0 \tag{7.14}$$

where $\delta = 1 + (1 + {L_M}^2) - 2{T_Q}^2$. Based on (7.14), we can harbor the idea that the critical point where P_{out} splits is determined by δ. When $\delta > 0$, the output power P_{out} decreases with the increasing F_D. As a result, P_{out} has only one peak when the operating frequency equals the original resonant frequency. When $\delta < 0$, however, P_{out} peaks multiply at the left- and right-hand sides of the original resonant frequency as depicted in Figure 7.4. The two frequency deviation factors corresponding to these two peaks are given by

$$F_D = \pm\sqrt{-\frac{\delta}{2}} \tag{7.15}$$

Accordingly, the critical point of the frequency splitting can be found at $\delta = 0$.

As aforementioned, the mechanism of frequency splitting has been elaborated by adopting the circuit theory. Similarly, the circuit theory can be utilized to reveal the frequency splitting for multiple-coil WPT systems, such as the three- and four-coil structures. Due to the frequency-splitting characteristic of the WPT systems, it is difficult to obtain the maximum output power at the resonance frequency when the coupling coefficient is extremely large. In order to inhibit the frequency splitting, a number of efforts have been made in previous studies. In two-coil WPT systems, the nonidentical resonant coil is utilized to avoid overcoupling and thus suppress the frequency splitting [8]. In four-coil WPT systems, the frequency splitting can be suppressed by tracking the frequency [9]. In addition, the frequency splitting can be suppressed by adjusting the load resistor [10].

To sum up, no matter the frequency sensitivity and the frequency splitting can directly affect the transmitting power of WPT systems, from the perspective of traditional and regular views, undoubtedly, these two frequency characteristics are taken as negative factors to deteriorate the transmission performance of WPT systems. However, everything has pros and cons. Are these two negative characteristics really useless? Then, one question comes into our mind, that is what can we do using these flaws?

7.3 Energy Encryption

The energy security is a common issue for various utilizations in power systems, but it is a brand-new concept for WPT systems. In previous studies, a great number of efforts have been made to improve the transmission performance. However, the energy security has not been paid enough attentions compared with the other research fields. This concept was first proposed and discussed by the research team of The University of Hong Kong in 2014 [11] and then improved and verified in 2015 [12, 13]. It has obtained increasing attentions from the academia and the industry since then [14].

In this section, firstly, some basic concepts of cryptography are briefly introduced. Then, the chaotic sequence cipher, which is chosen to be the security key to encrypt the transferred energy, is introduced as the exemplification. After introducing the encryption mechanism and giving the implementation considerations, the energy encryption scheme is discussed.

7.3.1 Cryptography

Cryptography is a science that studies making and breaking the codes and ciphers, including the ciphering and the deciphering. The ciphering is to study the objective law of cryptographic changes, which is utilized in the coding of passwords to ensure the security and privacy of the communication. Corresponding to the ciphering, the process that aims to decode passwords for obtaining the communication information is called deciphering. The ciphering and deciphering are collectively called cryptography. There are many branches of cryptography, including the symmetric cipher, the hash function, the asymmetric cipher, and the signature algorithms, which are all utilized for the security of communication in current applications. Along with the development of WPT technologies, the "*thing*" that is wirelessly transmitted over the air is not only the information but also the energy. In other words, the concept of security is gradually expanding from the signal level to the power level. Thus, the wirelessly transmitted energy is also capable of being encrypted by utilizing the scheme of cryptography to ensure the energy security of WPT systems.

In the symmetric cryptosystems, the same password is used to encrypt and decrypt the message. Accordingly, the sender and the receiver are both aware of the secret key and the process of encryption/decryption. Figure 7.5 shows the process of the encryption and the decryption. As a branch of the symmetric cipher, the sequence cipher is often used in the application field including the wireless communication and the diplomatic communication, which is also called stream cipher. It has salient advantages of simple implementation, easy hardware implementation, fast encryption and decryption processing, and no or limited error propagation. Accordingly, the sequence cipher is widely utilized by a number of practical applications, which is an ideal technical solution to realize the energy encryption for WPT systems.

The security of cryptosystems depends on the generation of random numbers, which are just like the keys in the data encryption standard algorithm. The

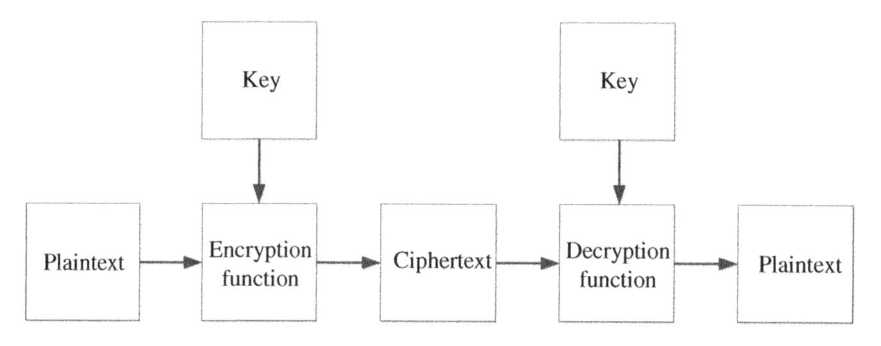

Figure 7.5 Process of encryption and decryption.

confidentiality of a sequence cipher completely depends on the randomness of the key. If the key is a truly random number, the system is theoretically indecipherable. However, the amount of keys required by the true random number is surprisingly large, which is actually not feasible in practical application. Accordingly, random-like sequences are generally adopted as key sequences instead of random sequences. Employing the random-like characteristic, the unpredictability of orbits, and the sensitivity of control parameter and initial state, which coincides with cryptography, the chaotic system is adopted for the sequence cipher. Thus, the chaotic cryptography emerges as the times require, which has been widely studied in recent years. Up to now, there have been many chaotic systems adopted in cryptography including Logistic map, Chebyshev map, Tent map, and piecewise linear chaotic map. In addition, some applications utilize various chaotic systems for cryptography to enhance the security performance.

Analogous to Figure 7.5, chaotic encryption mainly adopts the sequence iteratively generated by the chaotic system as a key sequence to encrypt the plaintext. Subsequently, the ciphertext is transferred through the channel. Lastly, the receptor will decrypt the ciphertext using the corresponding security key. Accordingly, the chaotic encryption shows characteristics as follows:

- The extreme sensitivity of initial state, which significantly enhances the confidentiality of the system.
- The randomness of encryption, which guarantees the security of the system.
- The relatively high accuracy, which can be used to generate a large number of candidate keys.

To sum up, combining with the advantages of the sequence cipher and the chaotic system, the chaos-based encryption shows the promising prospect for the energy encryption of WPT systems. Accordingly, the corresponding encryption mechanism and implementation are discussed in detail in Section 7.3.2.

7.3.2 Energy Encryption Scheme

In Section 7.2.1, the frequency sensitivity has been discussed. We have already known that the transmitting power of the WPT systems varies with the change in the operating frequency. Accordingly, this phenomenon which is called frequency sensitivity can be positively utilized to realize energy encryption. That is, the operating frequency is regularly adjusted by the power supply unit. The confidentiality and unpredictability of the frequency regulation significantly enhance the security of energy which is transferred wirelessly to the secondary side. By this way, the transferred energy is divided into many parts by the time and simultaneously packed with various frequencies. Thus, the key for the secondary side to get the energy is to know the regulation law of the changing frequency. It means that when the frequency is regularly adjusted, the unauthorized secondary side which does not have the knowledge of the regulation cannot retrieve the expected energy due to the frequency sensitivity. On the contrary, the authorized secondary side can obtain enough energy by adjusting the compensation circuit with the security key acquired from the primary side. Accordingly, the energy is effectively transferred to the authorized secondary side by utilizing the frequency sensitivity of WPT systems.

Then, the question comes into our mind, that is how to regularly adjust the frequency? As stated in Section 7.3, chaotic encryption is chosen due to its implementation ability and the random-like feature. As a simple 1-dimension and quadratic map, the logistic map can be utilized as an exemplified chaotic system to generate the security key, which is given by

$$X_{n+1} = f(X_n, A) = AX_n(1 - X_n) \tag{7.16}$$

Denote the sequence as $X_{n,}$ and the bifurcation parameter as A. Figure 7.6 shows the end value of X_n after 400 iterations concerning different values of A. Accordingly, the bifurcation parameter can determine the end value of the logistic map, periodically or chaotically. When $A > 3.57$, the value of X_n significantly relies on the initial state $X_{0,}$ and its value is chaotic. That is, a little change in X_0 can result in a remarkable variation of the trajectory. Consequently, $A = 3.9$ is chosen to generate the security key $X_n \in (0,1)$ to encrypt the operating frequency.

Figure 7.7 shows the basic circuit of the exemplified WPT systems, where C_p and C_s are the primary and the secondary capacitances, L_p and L_s are the inductances of the transmitting and the pickup coils, and R_p and R_s are the internal resistances of the primary and the secondary sides, respectively. Additionally, denote R_L as the load.

In order to reduce the volt–ampere (VA) rating of the power supply in the primary side and maximize the transmission power capability, the reactive power of the circuit should be suppressed as much as possible and even eliminated, which is elaborated in Section 7.4. Then, the primary and the secondary sides both work at the resonant state, which are given by

$$\begin{cases} \omega L_p = \dfrac{1}{\omega C_p} \\ \omega L_s = \dfrac{1}{\omega C_s} \end{cases} \tag{7.17}$$

where ω is the operating frequency. Accordingly, the transferred power can be maintained at the maximum value by ensuring the equations as

$$\omega_0 = \frac{1}{\sqrt{L_p C_p}} = \frac{1}{\sqrt{L_s C_s}} \tag{7.18}$$

where ω_0 is the resonant frequency of both the primary and the secondary sides.

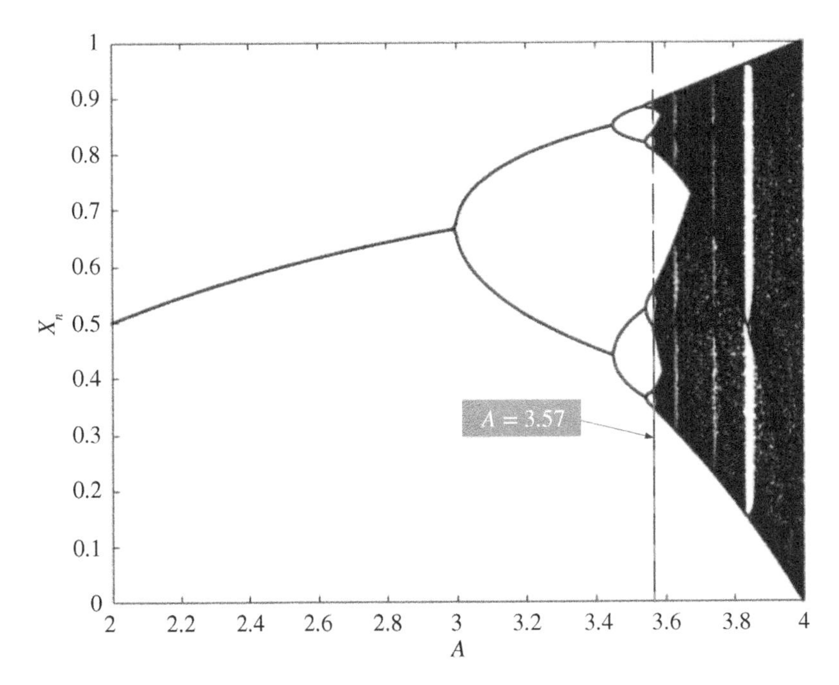

Figure 7.6 X_n concerning different values of A.

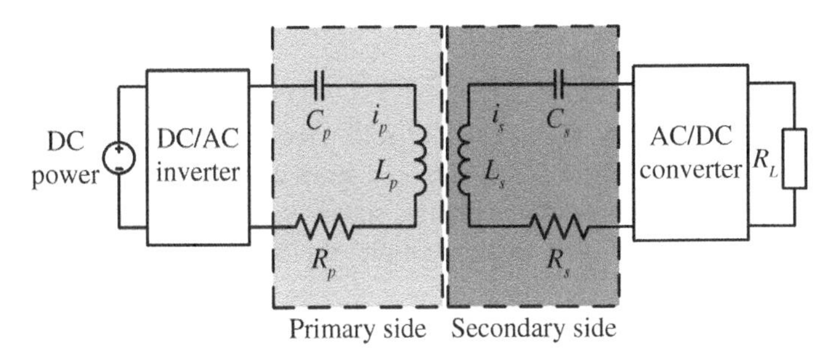

Figure 7.7 Basic circuit of two-coil WPT systems.

As aforementioned, the logistic map is chosen to encrypt the frequency, where the sequence $X_n \in (0,1)$, and the bifurcation parameter $A = 3.9$. Accordingly, the encrypted operating frequency can be expressed as

$$\omega = Y_n \omega_0 \tag{7.19}$$

where Y_n is the security key which is generated by the logistic map. Y_n is given by

$$Y_n = i + (j - i)X_n, 0 < i < j \tag{7.20}$$

The reason why Y_n is adopted instead of X_n is to further improve the security performance and extend the regulation range of the operating frequency. In such a way, the operating frequency is chaotically encrypted. According to the encrypted operating frequency, the capacitors of the circuit are synchronously adjusted to maintain the transferred power at the maximum value, which are, respectively, given by

$$\begin{cases} C_p = \dfrac{1}{Y_n^2 \omega_0^2 L_p} \\ C_s = \dfrac{1}{Y_n^2 \omega_0^2 L_s} \end{cases} \tag{7.21}$$

It means that the capacitors need to be adjusted accordingly in both the primary and the secondary sides to realize the energy encryption and decryption. Commonly, the capacitor array/matrix is utilized to realize the capacitance adjustment, which can also be adopted for the energy encryption scheme. In Figure 7.8, for example, the capacitor array consists of three capacitors in the primary side and six capacitors in the secondary side. Accordingly, by switching on and off the corresponding capacitors, the capacitance can be adjusted to ensure the resonant working state for both the primary and the secondary sides.

Since the primary side adopts three capacitors to form the capacitor array, the WPT system is set up at five resonant states as exemplification, which can be chosen by

$$Z_n = floor(5X_n) \tag{7.22}$$

where *floor* is a round function, Z_n is used to select section, and X_n is the chaotic sequence. Accordingly, the WPT system can work at the resonant state determined by Z_n. The corresponding switching states of the capacitor array are listed in Table 7.1. Taking $Z_n = 3$ as an example, the states of the primary

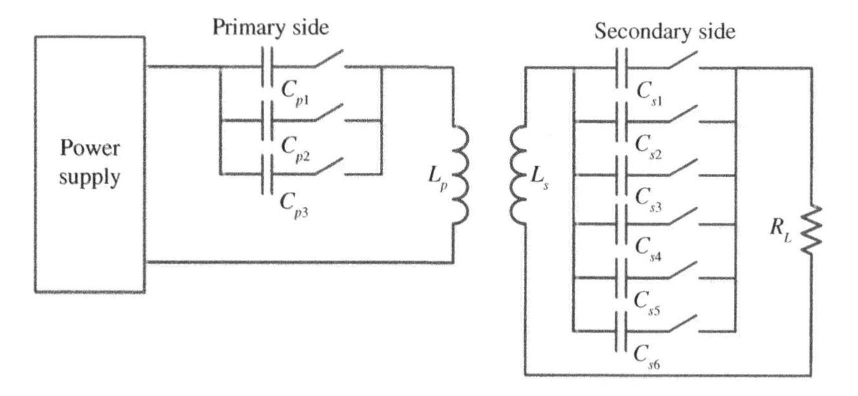

Figure 7.8 Capacitor arrays of the primary and secondary sides.

Table 7.1 Switching states of the capacitor array.

Z_n	Frequency (kHz)	Switching state of C_{p1}–C_{p3}	Switching state of C_{s1}–C_{s6}
0	—	—	—
1	85	1 1 1	0 0 1 1 1 1
2	91	0 1 1	1 1 0 0 1 1
3	100	1 0 1	0 1 0 1 0 1
4	110	0 0 1	0 1 0 0 0 1
5	135	1 1 0	0 1 0 1 1 0

capacitor array are 1(ON), 1(ON), and 0(OFF), which represent the switching states of C_{p1}, C_{p2}, and C_{p3}, respectively. The corresponding states of the secondary capacitor array are 0(OFF), 1(ON), 0(OFF), 1(ON), 1(ON), and 0(OFF), which represent the switching states of C_{s1}, C_{s2}, C_{s3}, C_{s4}, C_{s5}, and C_{s6}, respectively. The operating frequency is correspondingly adjusted to 100 kHz. Besides, if $Z_n = 0$, the switching state and operating frequency have no change.

According to the presented encryption mechanism above, the entire procedure of the exemplified energy encryption scheme is shown in Figure 7.9, which can be summarized as follows:

- ***Step 1*** – The initial value of the logistic map is randomly chosen to generate the chaotic sequence by Eq. (7.16). Then, the chaotic sequence is used based on Eqs. (7.19) and (7.20) to generate the security key to pack and unpack the energy, which is the prerequisite to realize the energy encryption for WPT systems.
- ***Step 2*** – The operating frequency of the primary side is chaotically adjusted using the security key. It means that the adjustment regulation is random-like and even unpredictable. Accordingly, the secondary side cannot acquire the energy without knowing the value and regulation of the operating frequency. Meanwhile, the primary capacitor is adjusted to correspond to the operating frequency, according to Eq. (7.21), which aims to ensure that the primary side works at the resonant state. Then, the volt–ampere (VA) rating of the power supply in the primary side can be reduced, accordingly.
- ***Step 3*** – The primary side works at the standby state as mentioned above to await the request of acquiring the energy from the secondary side. Once receiving the request, the primary side verifies the identification first. If the identification is not accepted, namely the secondary side is unauthorized, the request is rejected, and the primary side comes back to the standby state. On the contrary, the request from the authorized secondary side will be accepted, and the security key will be simultaneously transmitted to the authorized secondary side. For the secondary side, it awaits the response until receiving the security key.

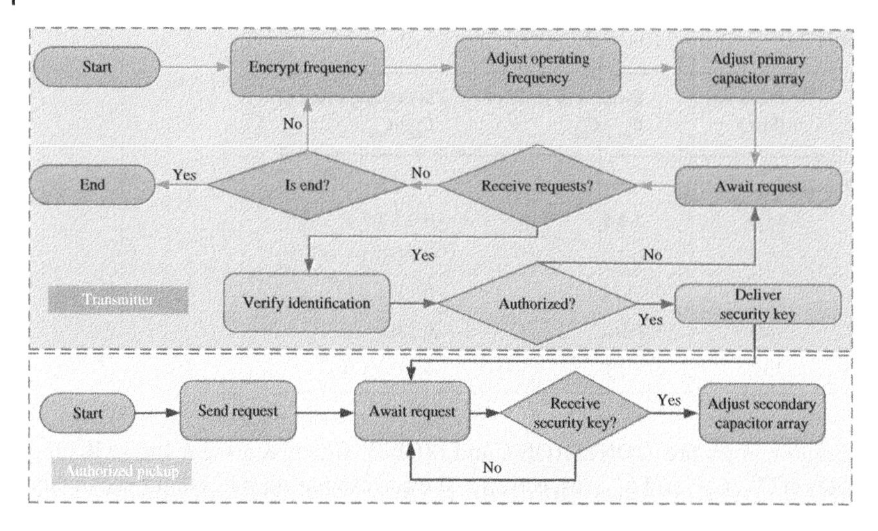

Figure 7.9 Procedure of exemplified energy encryption scheme.

- ***Step 4*** – After receiving the security key, the capacitance of the secondary side will be synchronously adjusted based on Eq. (7.21) to ensure the resonant state of the secondary side. Accordingly, the transferred power can be effectively maintained at the maximum value.

As aforementioned, by decrypting the energy through the security key encrypted and transmitted from the primary side, the authorized secondary side acquires the energy. In addition, unknowing the security key, the unauthorized secondary side cannot obtain the energy illegally due to the frequency sensitivity. Accordingly, the security performance of WPT systems is improved significantly.

7.4 Verifications

To verify the energy encryption scheme, the related simulation and the experimentation are both carried out in this section. The simulation verification gives the electromagnetic field analysis and the circuit simulation using MATLAB®/SIMULINK and finite element analysis software, respectively. The key parameters of the transmitting and the pickup coils are given in Table 7.2, and the parameters of the capacitor arrays are listed in Table 7.3. Also, experimental verification is carried out to further verify the feasibility of the energy encryption, where the key parameter of the prototype is the same as that listed in Tables 7.2 and 7.3. The digital-signal-process (DSP 28335) microcontroller is utilized to realize the energy encryption scheme as shown in Figure 7.9.

Table 7.2 Key parameters of the transmitting and pickup coils.

Items	Value (mH)
Transmitting coil inductance (L_p)	0.096
Pickup coil inductance (L_s)	0.0093
Mutual inductance (L_{ps})	0.003586
Mutual inductance (L_{ss})	0.0000465

Table 7.3 Parameters of the capacitor array.

Items	Value (nF)					
Primary capacitance	C_{p1}		C_{p2}		C_{p3}	
	4.7		10		22	
Secondary capacitance	C_{s1}	C_{s2}	C_{s3}	C_{s4}	C_{s5}	C_{s6}
	1	2.2	4.7	47	100	220

7.4.1 Simulation

Based on the energy encryption scheme as introduced in Section 7.3.2, the simulation is carried out to demonstrate the energy encryption scheme, which consists of one STMP. The operating frequency is encrypted around 100 kHz using a chaotic sequence. The capacitor array is adopted to adjust the compensation capacitance, accordingly. In order to comprehensively verify the feasibility of the energy encryption as aforementioned, case studies are carried out to cover various application scenarios which require paying attention regarding the energy security. In the STMP-WPT system, there is another challenge, namely power distribution for multiple pickups, where the energy is required to be transmitted to pickups separately or simultaneously. In such a case, the energy security should be ensured not only in single pickup but also in multiple pickups. Additionally, in some practical applications, there may be a pickup that is not expected to obtain the energy, or even tries to steal the energy. Accordingly, two main cases are given, respectively, in simulation as follow:

- ***Case 1*** – The WPT system with one single transmitter for multiple authorized pickups.
- ***Case 2*** – The WPT system with one STMP consists of authorized pickup and unauthorized pickup.

The simulation results are given as follows, respectively:

7.4.1.1 Case 1 – One Single Transmitter with Authorized Pickups

In this case, the WPT system with one single transmitter for two authorized pickups is taken as an exemplification. Figure 7.10 shows the simulation results of the two authorized pickups, which include the load current, the load voltage, and the active time of the authorized signal. From Figure 7.10a,b, it can be observed that during 0.3–0.4 seconds and 0.5–0.6 seconds, the load voltage and load current of Pickup A can reach about 8.5 V and 1.3 A, respectively. For Pickup B, it reaches the same value of load voltage and load current as Pickup A during 0.1–0.2 seconds. In particular, it can be observed that during 0.2–0.3 seconds, the load voltage and the load current of Pickup A and B are both reduced since both pickups share the transmitted power. Besides, it shows that the waveforms of load voltage and current are fluctuating within a slight range during the active time. The reason is that when the operating frequency is adjusted to encrypt the transmission channel, the inductances of transmitting and pickup coils inevitably vary with the variation in frequency within a slight range. More importantly, in order to compensate the nonresonance caused by the variation in operating frequency, the compensation capacitance should be synchronously regulated, accordingly. The discrete adjustment way of the adopted capacitor array mainly results in the fluctuation of the transmitting power. However, it should be noted that such a fluctuation has little influence on the transmission ability of the WPT systems. In Figure 7.10c, the shaded area represents the active time of the authorized signal, which shows that the active time of Pickup A are 0.2–0.4 seconds and 0.5–0.6 seconds, while the active time of Pickup B is 0.1–0.3 seconds, which well corresponds to Figure 7.10a,b.

Furthermore, the curve of the load power is also given in Figure 7.11, which illustrates that the load power of each pickup can reach 8 W when both of them work at resonant state. If only one of the pickups works at the resonant state, the average power of load can rise to about 14 W. The simulation result of the load power well agrees with that shown in Figure 7.10, which can verify that the energy encryption scheme can ensure the power transmitted to multiple authorized pickups separately or simultaneously. Accordingly, the energy encryption scheme provides a solution for the problem of power distribution and also ensures the energy security in practical applications, such as the selective charging of the portable devices putting on the same pad or selective charging of the EVs in the same area.

7.4.1.2 Case 2 – One Single Transmitter with Authorized Pickup and Unauthorized Pickup

In this case, the simulation results are given to verify two working conditions, which are: (i) one single transmitter with two pickups consists of authorized and unauthorized pickups and (ii) one single transmitter with three pickups consists of

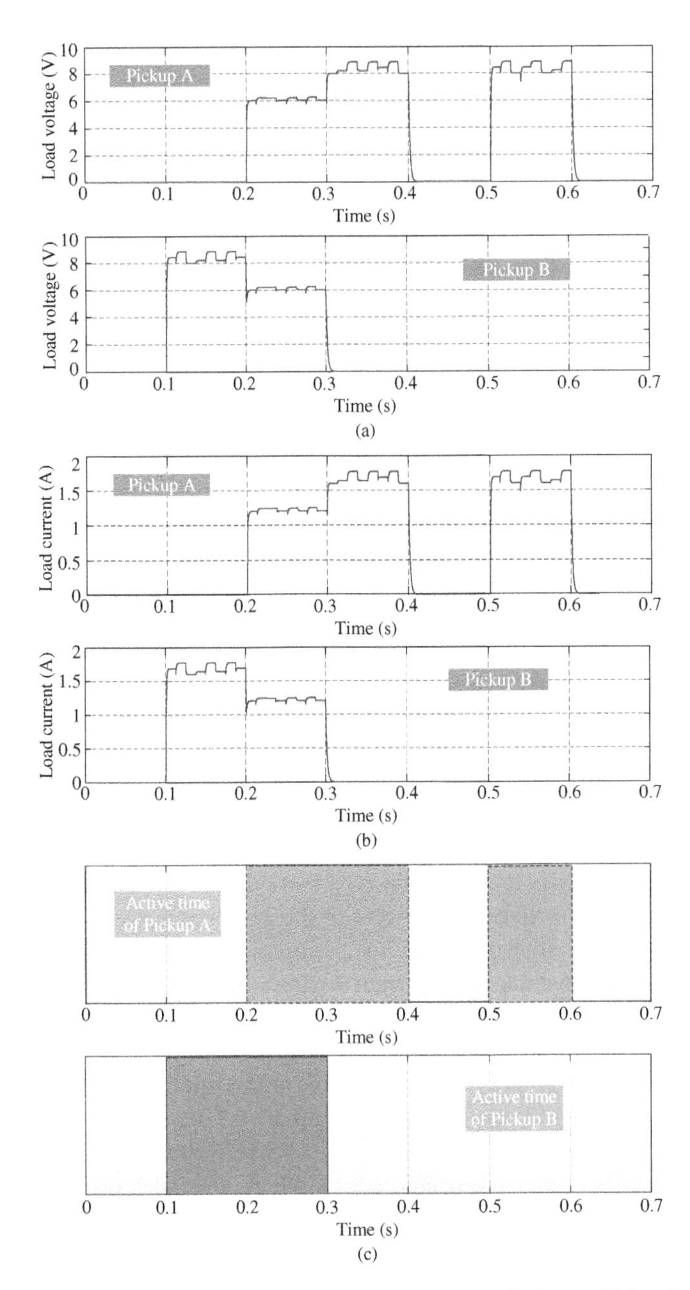

Figure 7.10 Simulation results of Case 1: (a) Load voltage; (b) Load current; (c) Active time.

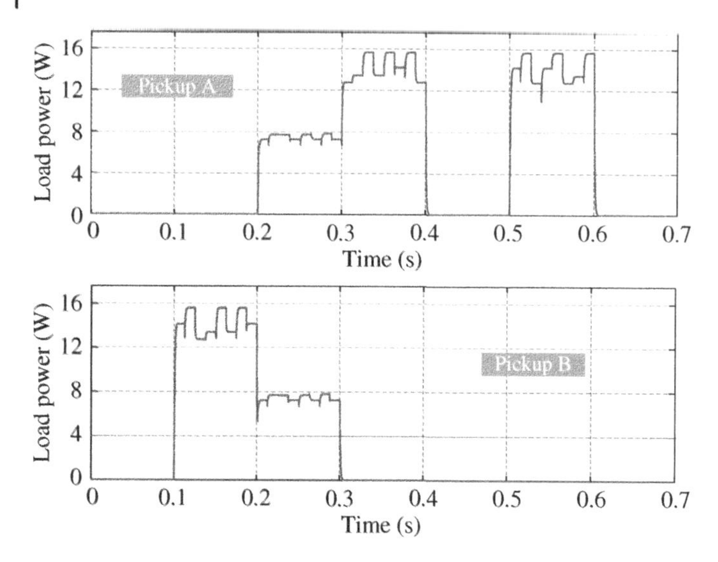

Figure 7.11 Simulation results of Case 1: Load power.

authorized pickup, unauthorized pickup, and unauthorized pickup that attempts to steal energy. Case 2 aims to verify that the energy is effectively transferred to the authorized pickup, while it will not be obtained by the unauthorized pickup.

Firstly, take the WPT system with one single transmitter for two pickups consisting of the authorized pickup and the unauthorized pickup as exemplification. Figure 7.12 depicts the simulation results of one authorized pickup and one unauthorized pickup, including the load current, the load voltage, and the active time. From Figure 7.12a,b, it can be observed that during 0.1–0.3 seconds and 0.4–0.6 seconds, the load voltage and current of the authorized pickup can reach about 8 V and 1.6 A as shown in blue line, respectively. As similar with Case 1, the waveforms of load voltage and current are fluctuating within a slight range during the active time. For the unauthorized pickup, the load voltage and current are suppressed at an extremely low level as shown in red line. In Figure 7.12c, the shaded area represents the active time of the authorized signal. It shows that the active time of the authorized pickup are 0.1–0.3 seconds and 0.4–0.6 seconds, which well corresponds to Figure 7.10a,b. Furthermore, the curve of the load power is also given in Figure 7.13, which illustrates that the load power of authorized pickup can reach 13 W, while it is suppressed at an extremely low level for the unauthorized pickup. The simulation result of the load power well agrees with that shown in Figure 7.12, which can verify that the energy encryption scheme can ensure the power transmitted to authorized pickup instead of unauthorized pickup. Taking the EV charging system as an

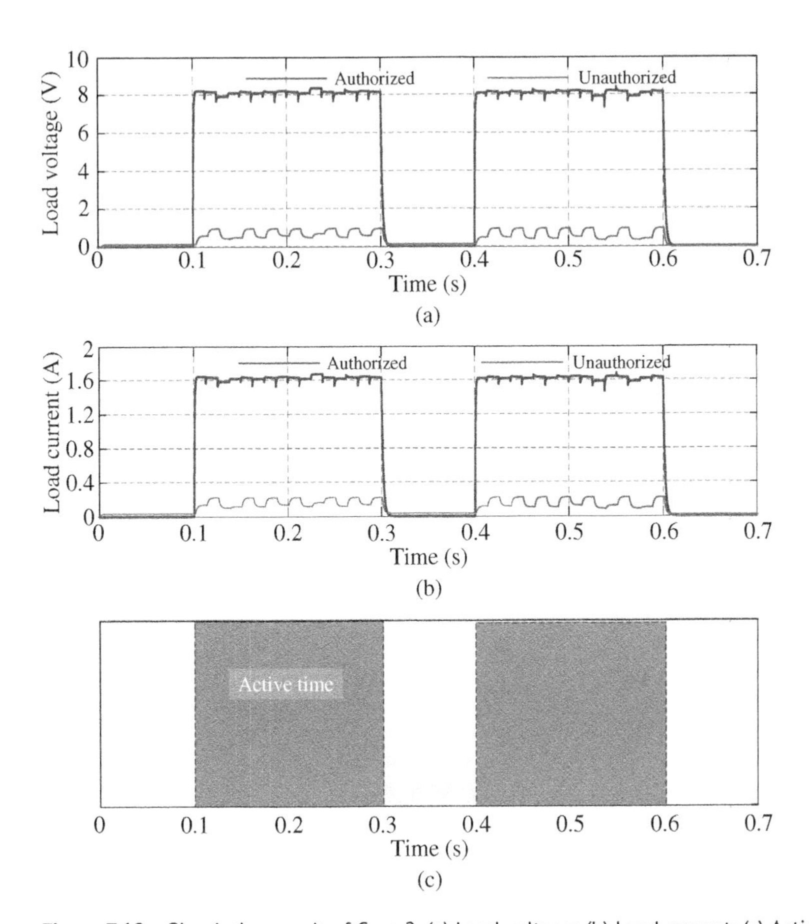

Figure 7.12 Simulation result of Case 2: (a) Load voltage; (b) Load current; (c) Active time.

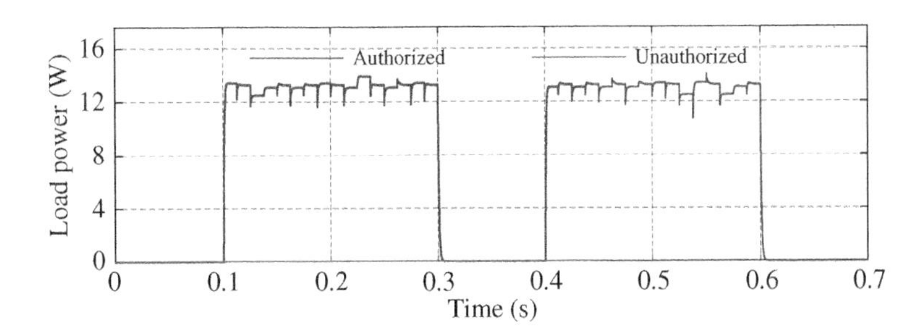

Figure 7.13 Simulation results of Case 1: Load power.

example, the unauthorized pickup may be unpaid or illegal, which means that the energy cannot be transferred to it.

The second test sets the WPT system with one single transmitter for three pickups consisting of one authorized pickup and two unauthorized pickups as an exemplification. One of the unauthorized pickups Pickup B changes the capacitance by capacitor array based on the chaotic sequence, which aims to track the frequency of the primary side to steal the power. Figure 7.14 shows the corresponding simulation results, including the load current, the load voltage, and the active time of the authorized signal, respectively. From Figure 7.14a,b, it can be observed that during 0.1–0.6 s, the load voltage and the load current of the authorized pickup can reach around 8 V and 1.6 A as shown in authorized

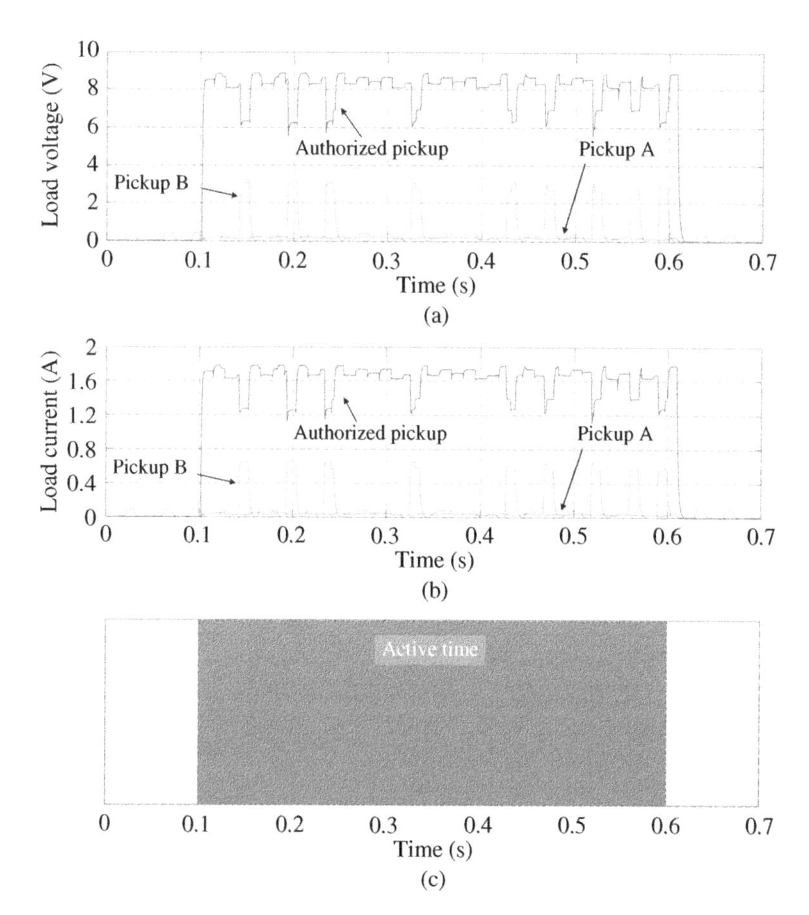

Figure 7.14 Simulation result of Case 2: (a) Load voltage; (b) Load current; (c) Active time.

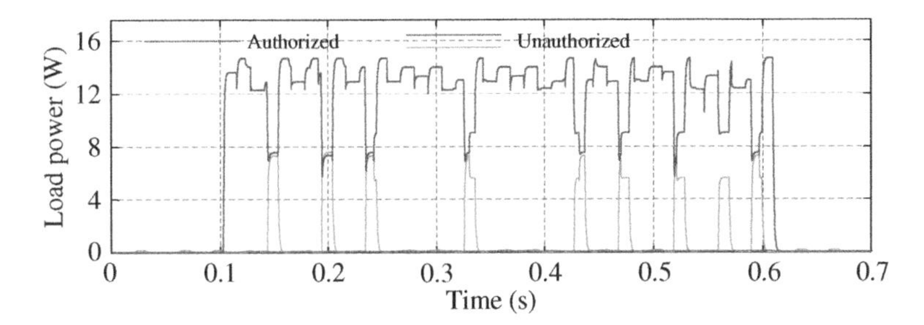

Figure 7.15 Simulation result of Case 2: Load power.

pickup, respectively. For unauthorized Pickup A, the load voltage and current are suppressed at an extremely low level as shown in Pickup A, which is similar to the first condition. For unauthorized Pickup B, the initial values of the logistic map and the capacitance are not understood by Pickup B. Due to the chaotic characteristic of the logistic map and frequency sensitivity, the load voltage and current are also suppressed at a relatively low level as shown in Pickup B. In addition, it can be observed that the energy is transferred to unauthorized Pickup B in some short time. However, in the whole transmission time, the energy that is transferred to it only occupies a small percentage of all the transferred energy, which has little influence on the energy security of WPT systems. In Figure 7.14c, the shaded area represents the active time of the authorized signal. It shows that the active time of the authorized pickup is 0.1–0.6 seconds, which well corresponds to Figure 7.14ab.

Furthermore, the curve of the load power is also given in Figure 7.15, which illustrates that the load power of the authorized pickup can reach 13 W, while it is suppressed at a low level for other two unauthorized pickups. The simulation result of the load power well agrees with that shown in Figure 7.14, which can verify that the energy encryption scheme can ensure that the power is effectively transferred to the authorized pickup as well as prevent the unauthorized pickup from obtaining and stealing the energy. In addition, Figure 7.16 shows the simulation results of the electromagnetic field for this condition. It can be observed that the magnetic flux line of the authorized pickup distributes more intensive and uniform than that of the unauthorized pickup, which agrees well with the circuit simulation result. Accordingly, the energy encryption scheme is further validated to improve the security performance of WPT systems.

7.4.2 Experimentation

To verify the energy encryption scheme, the experimentation is carried out by taking a WPT system as an exemplification. The experimental prototype is set

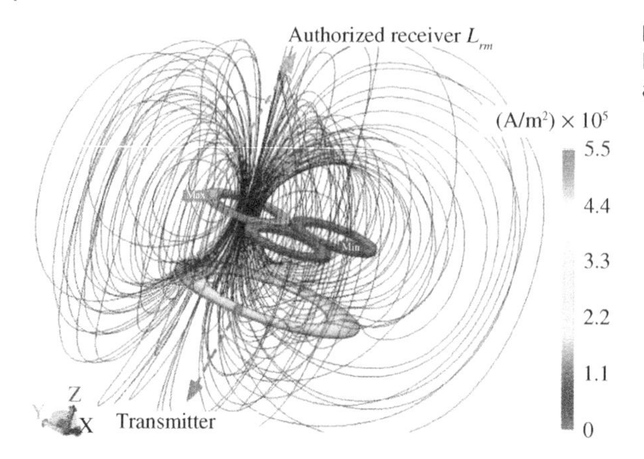

Figure 7.16 Electromagnetic field analysis of Case 2.

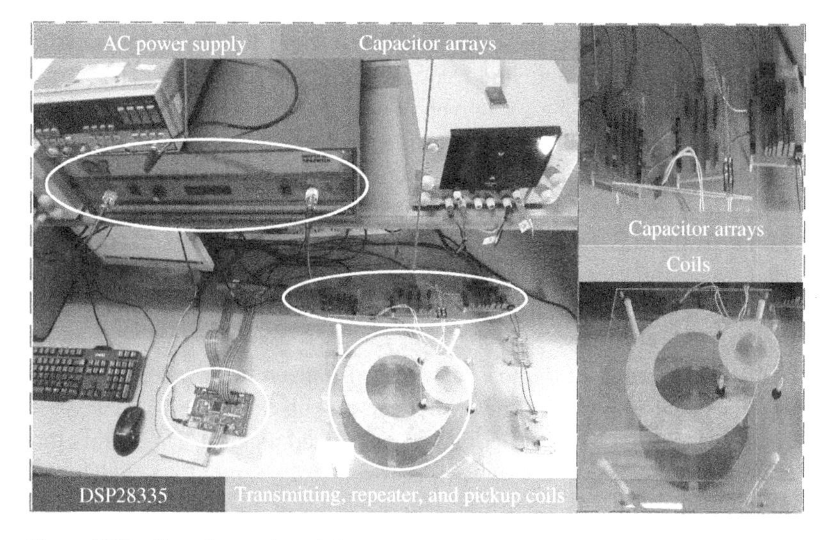

Figure 7.17 Experimental prototype.

up as shown in Figure 7.17, which consists of an AC power supply (Amplifier Research 75A250A), a current sensor (LA25-NP/AD637), a DSP28335 microcontroller to implement the energy encryption scheme, and a power analyzer (Tektronix TM502A) to analyze and display the measured waveforms. The key parameters of the coils are listed in Table 7.4, and the parameters of the capacitor array are the same as that listed in Table 7.3.

First, taking the authorized pickup as an example, it can work at resonant state using the security key and correspondingly adjusting the capacitor array. Accordingly, Figure 7.18a,b shows the experimental results of the authorized pickup, including the load current and the load voltage. It can be observed that the load

Table 7.4 Key parameters of the transmitting, resonant, and pickup coils.

Item	Value (mH)
Transmitting coil inductance (L_p)	0.09589
Repeater coil inductance (L_r)	0.09477
Pickup coil inductance (L_s)	0.009372
Mutual inductance (L_{pr})	0.005305
Mutual inductance (L_{rs})	0.007958

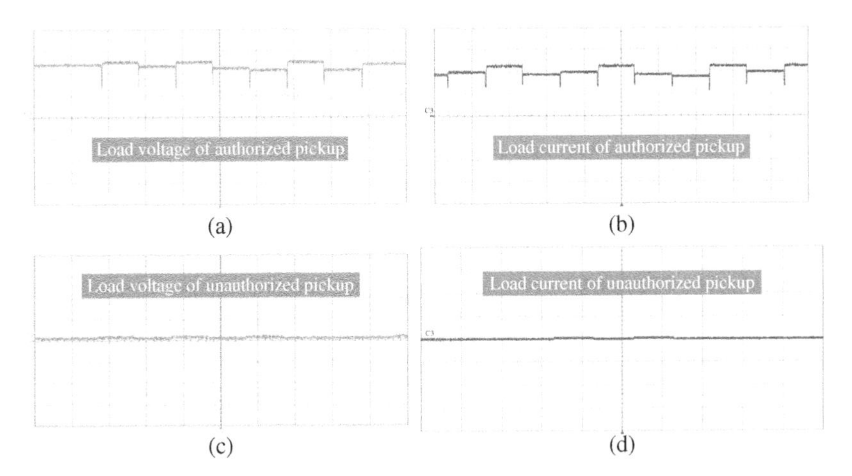

Figure 7.18 Experiment waveforms: (a) Load voltage of authorized pickup (X: 1s/div, Y: 4V/div); (b) Load current of authorized pickup (X: 1s/div, Y: 1A/div); (c) Load voltage of unauthorized pickup (X: 1s/div, Y: 4V/div); (d) Load current of unauthorized pickup (X: 1s/div, Y: 1A/div).

voltage and the load current can reach about 10 V and 2 A, respectively, namely the load power can reach 20 W. The waveforms of the authorized pickup have fluctuation in a slight range, which is mentioned in the simulation results. Then, for the experiment waveforms of the unauthorized pickup as shown in Figure 7.18c,d, the load voltage and the load current are suppressed at an extremely low level, which results from the nonresonant state of the circuit.

To sum up, the experimental results well validate the feasibility energy encryption scheme. It means that the energy encryption scheme can effectively transfer the energy to the authorized pickup and prevent the unauthorized pickup from obtaining or stealing the energy.

7.5 Opportunities

In this chapter, the energy encryption mechanism and scheme are introduced and verified. The security performance of WPT systems is significantly improved by the energy encryption scheme. However, there are still some opportunities and challenges that are worth to be considered.

Firstly, the energy encryption scheme can be further optimized. The logistic map is utilized to encrypt the energy, which can be further improved such as using other more complex chaotic systems or using multiple chaotic systems to encrypt the frequency. Additionally, the MRC-based WPT system has a higher quality factor, which can improve the sensitivity of frequency. Accordingly, the MRC-based WPT system can be a better choice for the encrypted WPT systems.

Secondly, the capacitor array method is utilized in the energy encryption scheme as an exemplification, which can be further improved such as increasing the number of capacitors [15, 16] and constructing the structure of the capacitor array to expand the available section range and enhancing the algorithm complexity. Meanwhile, the transmission performance should also be taken into consideration when improving the security performance in the future. The complexity of the circuit, the extra power loss, and the additional communication of the primary and secondary sides have a negative impact on the transmission performance, undoubtedly. Accordingly, the balance between the security performance and transmission performance in specific applications is worth to be considered. In addition, by continuously adjusting the capacitance instead of in a discrete way, the number of capacitances is significantly increased, which reduces the possibility of stealing the energy by unauthorized pickup. Accordingly, the security performance can be significantly improved in WPT systems.

Thirdly, in WPT systems, many difficulties can be effectively solved by utilizing frequency sensitivity. For example, a problem occurs in power distribution for STMP-WPT. Accordingly, a selective WPT methodology for a multiple-receiver system has been demonstrated, and the proposed WPT system is modeled with distinct resonant frequencies for the receivers that form passbands at separate center frequencies [17]. Using the selective WPT methodology, the problem of power distribution has been effectively solved. The frequency sensitivity is positively used in the proposed WPT system.

References

1 Cheon, S., Kim, Y., Kang, S. et al. (2011). Circuit-model-based analysis of a wireless energy transfer system via coupled magnetic resonances. *IEEE Transactions on Industrial Electronics* 58 (7): 2906–2914.

2 Kim, Y., Kang, S., Cheon, S. et al. (2010). Optimization of wireless power transmission through resonant coupling. In: *SPEEDAM 2010*, Pisa, 1069–1073.
3 Cannon, B.L., Hoburg, J.F., Stancil, D.D., and Goldstein, S.C. (2009). Magnetic resonant coupling as a potential means for wireless power transfer to multiple small receivers. *IEEE Transactions on Power Electronics* 24 (7): 1819–1825.
4 Huang, R., Zhang, B., Qiu, D., and Zhang, Y. (2014). Frequency splitting phenomena of magnetic resonant coupling wireless power transfer. *IEEE Transactions on Magnetics* 50 (11): 8600204: 1–8600204: 4.
5 Zhang, Y. and Zhao, Z. (2014). Frequency splitting analysis of two-coil resonant wireless power transfer. *IEEE Antennas and Wireless Propagation Letters* 13: 400–402.
6 Sample, A.P., Meyer, D.A., and Smith, J.R. (2011). Analysis, experimental results, and range adaptation of magnetically coupled resonators for wireless power transfer. *IEEE Transactions on Industrial Electronics* 58 (2): 544–554.
7 Zhang, Y., Zhao, Z., and Chen, K. (2014). Frequency-splitting analysis of four-coil resonant wireless power transfer. *IEEE Transactions on Industry Applications* 50 (4): 2436–2445.
8 Choi, S.Y., Gu, B.W., Lee, S.W. et al. (2014). Generalized active EMF cancel methods for wireless electric vehicles. *IEEE Transactions on Power Electronics* 29 (11): 5770–5783.
9 Lyu, Y., Meng, F., Yang, G. et al. (2015). A method of using non-identical resonant coils for frequency splitting elimination in wireless power transfer. *IEEE Transactions on Power Electronics* 30 (1): 6097–6107.
10 Huang, S., Li, Z., and Lu, K. (2016). Frequency splitting suppression method for four-coil wireless power transfer system. *IET Power Electronics* 9 (15): 2859–2864.
11 Zhang, Z., Chau, K.T., Liu, C. et al. An efficient wireless power transfer system with security considerations for electric vehicle applications. *Journal of Applied Physics* 115 (17): 17A328: 1–17A328: 3.
12 Zhang, Z., Chau, K.T., Qiu, C., and Liu, C. (2015). Energy encryption for wireless power transfer. *IEEE Transactions on Power Electronics* 30 (9): 5237–5246.
13 Zhang, Z., Liu, C., and Qiu, C. (2015). Energy-security-based contactless battery charging system for roadway-powered electric vehicles. In: *2015 IEEE PELS Workshop on Emerging Technologies: Wireless Power (2015 WoW)*, Daejeon (5, 6 June 2015), 1–6. https://ieeexplore.ieee.org/document/7132806.
14 Ahene, E., Ofori-Oduro, M., and Agyemang, B. (2017). Secure energy encryption for wireless power transfer. In: *2017 IEEE 7th International Advance Computing Conference (IACC)*, Hyderabad, 199–204.
15 Liu, W., Chau, K., Lee, C. et al. (2018). A switched-capacitorless energy-encrypted transmitter for roadway-charging electric vehicles. *IEEE Transactions on Magnetics* 54 (11): 8401006:1–8401006:5.

16 Lim, Y., Tang, H., Lim, S., and Park, J. (2014). An adaptive impedance-matching network based on a novel capacitor matrix for wireless power transfer. *IEEE Transactions on Power Electronics* 29 (8): 4403–4413.

17 Zhang, Y., Lu, T., Zhao, Z. et al. (2015). Selective wireless power transfer to multiple loads using receivers of different resonant frequencies. *IEEE Transactions on Power Electronics* 30 (11): 6001–6005.

8

Omnidirectional Wireless Power Transfer

In this chapter, omnidirectional wireless power transfer (WPT) technologies, including 2-dimensional (2D) WPT and 3-dimensional (3D) WPT with multiple pickups, are introduced. Then, the mathematical analyses of omnidirectional WPT, based on the basic theoretical knowledge, are described. In addition, the design of the transmitting coils for synthetic magnetic field is elaborated in detail. Furthermore, the design and control considerations for pickup coils to receive power at any position are discussed, respectively. Then, the technology of load detection is introduced for the application of the omnidirectional WPT. Finally, the omnidirectional WPT technologies are discussed and summarized.

8.1 Introduction

Omnidirectional WPT, which is capable of transmitting power efficiently to huge areas of 3D space with arbitrary directions, is a huge challenge. Accordingly, in recent years, omnidirectional WPT has been researched intensely between the academia and the industry. In many scenarios, the sufficient geometric freedom of the charged device is required for the WPT system, namely, the device can be placed anywhere in the 3D space to wirelessly receive power. This means that wireless power transmission is necessary regardless of any angle between the transmitting coil and the load. More importantly, with the addition of multiple pickups, the WPT system should be capable to seamlessly achieve full coverage and ensure high efficiency. Undoubtedly, omnidirectional WPT has salient significance and practical value to modern automation systems, medical appliances, and especially for the consumer electronics of intelligent houses. For example, in the application of omnidirectional WPT in smart houses, the electronic devices such as people's mobile phones and watches are automatically detected once people return home. After judging the remaining power of the

Wireless Power Transfer: Principles and Applications, First Edition. Zhen Zhang and Hongliang Pang.

device, the charging process will automatically start, where omnidirectional efficient wireless charging will guarantee the charging speed.

With omnidirectional WPT gradually occupying the mainstream position, the transmitters made of orthogonal coils are generally accepted by researchers because they regulate a full range of charging mode, perfectly suited for omnidirectional WPT systems. Besides, some different transmitting and pickup coils are also considered to be designed to achieve omnidirectional WPT [1–4]. In the design of the transmitting coil, a unique bowl-shaped structure with multitransmitter coils was proposed to charge mobile devices at any position and in any direction [2]. Besides, in order to dynamically charge the drone during the flight, a novel quasi-omnidirectional dynamic WPT system with double 3D coils was proposed [3]. With the proposed 3D intermediate coil, the specific quadrant of double 3D coils can be designed to artfully align to the center of the flying area, and the drone in the specific space can be flexibly and efficiently powered. In the design of pickup coils, a novel quadrature-shaped pickup coil topology was proposed to realize an angular-misalignment insensitive omnidirectional wireless charging system [4]. The simulated and experimental verifications prove that the proposed pickup coil topology can compensate the huge fluctuations of the transmitted power caused by the self-rotation of the pickup unit, resulting in the angular misalignment insensitive omnidirectional wireless charging system.

In addition to the design of the transmitting and pickup coils in the omnidirectional WPT system, the employment of phase and current control methods is also required to control the current of the transmitters in the system. As an important technology and on theoretical basis, the nonidentical current control technique of 2D and 3D omnidirectional WPT was introduced, which can make the magnetic field vector point uniformly in all directions and realize omnidirectional power transmission [5]. In addition, in order to achieve higher transmission efficiency, a novel cubic transmitter was provided in the omnidirectional WPT system [6], where a single power source is utilized to drive the current of the transmitter instead of other complicated technologies, such as the phase and current control methods.

With the deepening of the research, higher technical requirements have been put forward on the application of omnidirectional WPT, which not only needs to transmit energy to all areas of the 3D space but also needs to direct concentrated transmission power to the position where the load is located. Accordingly, a current amplitude control technique is proposed in [7], including omnidirectional scanning and directional power flow control. In the 3D space WPT system, the structure composed of three orthogonal transmitting coils and multiple pickups can generate magnetic field vectors uniformly on the sphere. In [8] and [9], based on the rotating magnetic field generated by the transmitting coil, the basic mathematical analysis including the expressions of transmitted power and energy efficiency is introduced for 2D and 3D WPT systems. In [10], a selective omnidirectional magnetic resonant coupling (MRC) WPT system for multiple pickups is proposed, which can extend the service life of the device. The omnidirectional transmitter consists of three orthogonal transmitting coils, and the pickups utilize the band-pass filter principle to selectively acquire power at specific resonant frequencies.

8.2 Mathematical Analysis

8.2.1 2-Dimensional WPT with Multiple Pickups

For the convenience of analysis, omnidirectional WPT system can be divided into the 2D WPT and the 3D WPT from a dimensional point of view. In the 2D WPT system, an appropriate technique, current vector control, is provided to generate the magnetic field vector at any direction. According to the mathematical analysis presented in this section, it is found that the total input power and the energy efficiency of the WPT system can be represented by the angle of the generated magnetic field vector. Moreover, the two corresponding functions follow the Lemniscates of Bernoulli where the important parameter, such as the directivity, is shown. In theory, with the property of direction of the function curves, wireless power can be efficiently transferred to the pickup coils. After the experimental verification, the 2D theory is strongly supported [8]. Additionally, by understanding the 2D theory, the 3D theory discussed in the following section will become relatively easy to comprehend.

In this section, the relationships of the input power, output power, and energy efficiency from different angular positions between the pickup and two orthogonal transmitting coils are analyzed. According to the Lemniscates of Bernoulli which the total input power, output power, and energy efficiency of the system follow, the energy efficiency can reach maximum point or minimum point along the angular position like the sine function. Thus, after measuring several parameters on the primary side, the optimum positions of pickups can be determined to receive wireless power from the transmitting coils.

Figure 8.1 illustrates a familiar 2D omnidirectional WPT system with two transmitting coils and one pickup coil. The first (gray) coil and the second (dark gray) coil are two orthogonal transmitting coils that have no mutual inductance theoretically, and the pickup coil is the green coil loaded with a series resonant capacitor and a resistive load R_L. Different forms of nonidentical currents can be chosen to generate the rotating magnetic field vector. Here, the two orthogonal transmitting coils are excited by two nonidentical AC current sources $\vec{I}_1$ and $\vec{I}_2$ which have the same frequency and phase and different amplitude modulation functions. Equation (8.1) shows the nonidentical AC current sources,

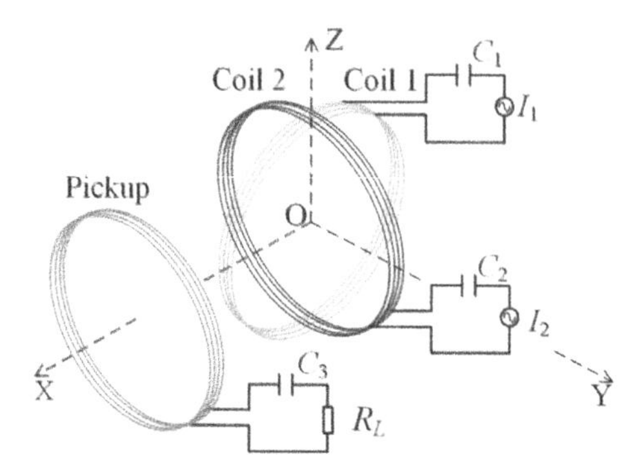

Figure 8.1 Two-dimensional omnidirectional WPT system with a loaded pickup coil resonator.

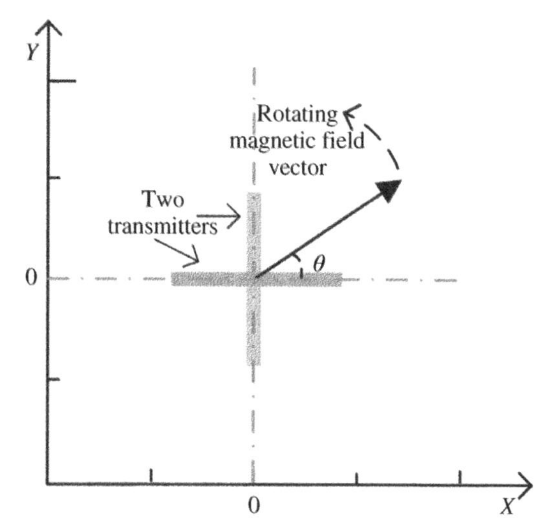

Figure 8.2 Schematic diagram of the two orthogonal transmitting coils and a typical rotating magnetic field vector.

$$\begin{bmatrix} \vec{I}_1 \\ \vec{I}_2 \end{bmatrix} = \begin{bmatrix} \cos\theta \\ \sin\theta \end{bmatrix} I \tag{8.1}$$

where $\cos\theta$ and $\sin\theta$ are the amplitude modulation functions, respectively, and I is a sinusoidal time function. θ represents the physical angle of the synthetic magnetic field vector on the 2D planes as shown in Figure 8.2, and $0° \leq \theta < 360°$. The synthetic magnetic field vector is the sum of two magnetic field vectors produced by two transmitting coils. In Figure 8.3, the plane of the pickup coil (green) is placed to face the center of the transmitting coils. Accordingly, the position of the pickup coil can be expressed in polar coordinates: $d\angle\alpha$ with the angle α as shown in Figure 8.3. Then, a 2D omnidirectional WPT system is formed with the power transmission from transmitters to pickups.

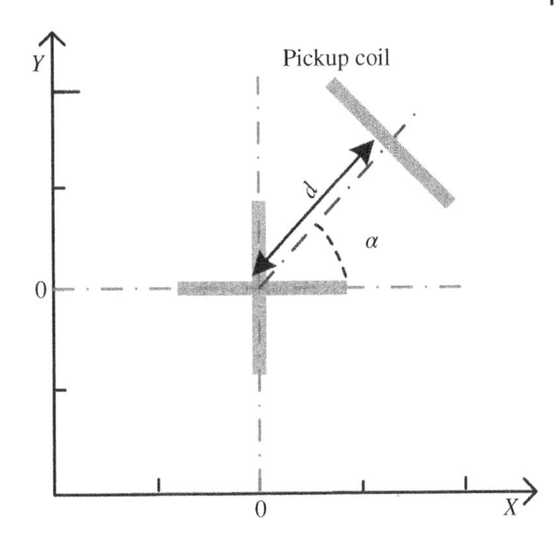

Figure 8.3 Geometrical relationship of a loaded pickup coil resonator in a 2D omnidirectional structure.

Equation (8.2) shows the coupled circuit equation of the three-coil WPT system, where $\vec{U}_1$ and $\vec{U}_2$ are the equivalent voltages of the primary sides, which provide the current sources $\vec{I}_1$ and $\vec{I}_2$ for the two transmitting coils. The transmission power and the overall efficiency in the system can be calculated on account of the known parameters in Eq. (8.2).

$$\begin{bmatrix} \vec{U}_1 \\ \vec{U}_2 \\ 0 \end{bmatrix} = \begin{bmatrix} R_1 + j\left(\omega L_1 - \frac{1}{\omega C_1}\right) & j\omega M_{12} & j\omega M_{13} \\ j\omega M_{12} & R_2 + j\left(\omega L_2 - \frac{1}{\omega C_2}\right) & j\omega M_{23} \\ j\omega M_{13} & j\omega M_{23} & R_3 + R_L + j\left(\omega L_3 - \frac{1}{\omega C_3}\right) \end{bmatrix} \begin{bmatrix} \vec{I}_1 \\ \vec{I}_2 \\ \vec{I}_3 \end{bmatrix} \tag{8.2}$$

The simulated energy efficiency result with respect to the angle α as the x-coordinate is shown in Figure 8.4 for a 2D omnidirectional WPT system. Here, two current sources that drive the transmitting coils have the same magnitude, and the phase difference is 90°. Figure 8.4 shows that the efficiency of the pickup load is between 0.7 and 0.8 at different angles. When d is equal to 0.3 m, and α is equal to 45°, the energy efficiency as the angle θ changes from 0° to 360° is as shown in Figure 8.5. By comparing the two results, it is seen that the maximum efficiency in Figure 8.5 is higher than the maximum efficiency in Figure 8.4. Accordingly, the omnidirectional detection of load positions is required, and then

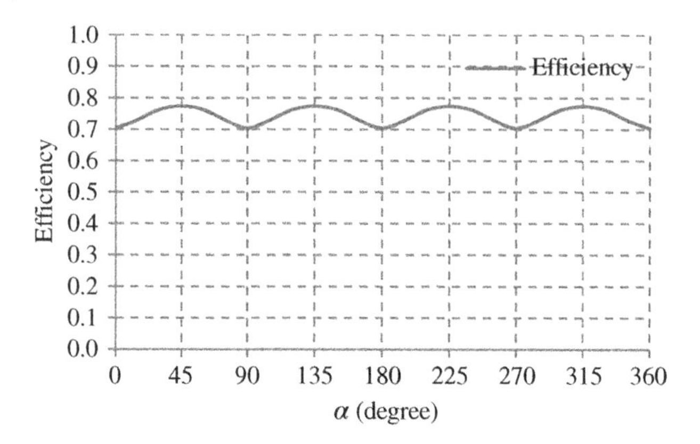

Figure 8.4 Energy efficiency of the 2D WPT system when the pickup load is placed around the origin of the two transmitting coils (which are excited with another nonidentical current control with the two currents of the same magnitude and a phase difference of 90°).

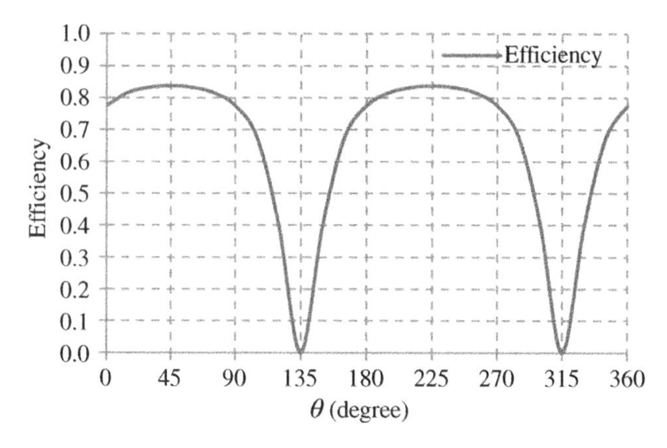

Figure 8.5 Efficiency when the magnetic vector is rotating around the origin and the load is placed at 45°.

the power can be focused toward the load directions. In addition, the calculations of the input power, output power, and the efficiency will be shown later.

In fact, M_{12} in Eq. (8.2) is equal to zero, since the two transmitting coils are orthogonal to each other. Then, substituting Eq. (8.1) into Eq. (8.2), we get Eq. (8.3).

$$\begin{bmatrix} \vec{U}_1 \\ \vec{U}_2 \\ 0 \end{bmatrix} = \begin{bmatrix} R_1 + jX_1 & 0 & j\omega M_{13} \\ 0 & R_2 + jX_2 & j\omega M_{23} \\ j\omega M_{13} & j\omega M_{23} & R_3 + R_L + jX_3 \end{bmatrix} \begin{bmatrix} I\cos\theta \\ I\sin\theta \\ \vec{I}_3 \end{bmatrix} \tag{8.3}$$

where

$$X_1 = \left(\omega L_1 - \frac{1}{\omega C_1}\right), X_2 = \left(\omega L_2 - \frac{1}{\omega C_2}\right), X_3 = \left(\omega L_3 - \frac{1}{\omega C_3}\right).$$

8.2.1.1 Load Current Calculation

According to Eq. (8.3), it can be shown that

$$j\omega M_{13} I \cos\theta + j\omega M_{23} I \sin\theta + (R_3 + R_L + jX_3)\vec{I}_3 = 0 \tag{8.4}$$

Reorganizing Eq. (8.4) to Eq. (8.5), the load current information is obtained:

$$\vec{I}_3 = -\frac{j\omega I\sqrt{M_{13}^2 + M_{23}^2}}{R_3 + R_L + jX_3} \sin\left(a\tan\frac{M_{13}}{M_{23}} + \theta\right) \tag{8.5}$$

where the amplitude of $\vec{I}_3$ is

$$I_3 = \frac{I\omega\sqrt{M_{13}^2 + M_{23}^2}}{\sqrt{(R_3 + R_L)^2 + X_3^2}} \left|\sin\left(a\tan\frac{M_{13}}{M_{23}} + \theta\right)\right| \tag{8.6}$$

8.2.1.2 Output Power Calculation

8.2.1.2.1 P_L: Load Power The load power can be calculated by Eq. (8.7)

$$P_L = I_3^2 R_L = I^2\left(\frac{R_L\omega^2\left(M_{13}^2 + M_{23}^2\right)}{(R_3 + R_L)^2 + X_3^2} \times \sin^2\left(a\tan\frac{M_{13}}{M_{23}} + \theta\right)\right) \tag{8.7}$$

Based on Eq. (8.7), Figure 8.6 can be plotted, which shows the Lemniscate of Bernoulli. The vector in Figure 8.6 is used to represent the load power, and its positive direction is determined by the right-hand rule of the current. Eq. (8.7) can be simplified as Eq. (8.8)

$$P_L = K_{2-D} I^2 R_L \sin^2(\gamma_{2-D} + \theta) \tag{8.8}$$

where

$$K_{2-D} = \frac{\omega^2\left(M_{13}^2 + M_{23}^2\right)}{(R_3 + R_L)^2 + X_3^2}, \gamma_{2-D} = a\tan\frac{M_{13}}{M_{23}}$$

According to Eq. (8.8), when $\sin^2(\gamma_{2-D} + \theta) = 1$, i.e. $\gamma_{2-D} + \theta = \pi/2$ or $3\pi/2$, it can be found that P_L can reach its maximum value

$$P_{L\_\max} = K_{2-D} I^2 R_L \tag{8.9}$$

And when $\sin^2(\gamma_{2-D} + \theta) = 0$, i.e. $\gamma_{2-D} + \theta = 0$ or π, the load power can reach its minimum value

$$P_{L\_\min} = 0 \tag{8.10}$$

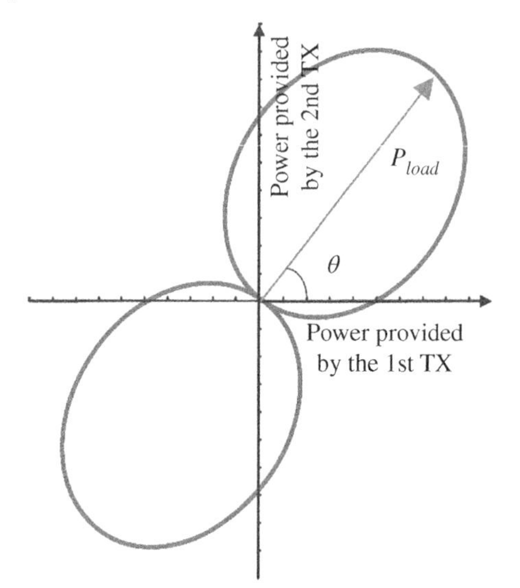

Figure 8.6 Schematic diagram of load power (in the form of a Lemniscate of Bernoulli).

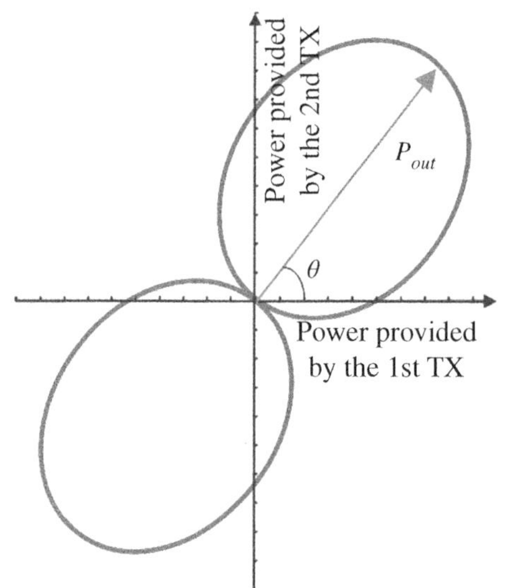

Figure 8.7 Schematic diagram of the power picked up by the pickup (in the form of the Lemniscate of Bernoulli).

8.2.1.2.2 P_{out}: Power Picked Up by the Pickup The acquired power of the pickups from the transmitters can be expressed as

$$P_{out} = I_3^2 R_3 + I_3^2 R_L = K_{2-D} I^2 (R_3 + R_L) \sin^2(\gamma_{2-D} + \theta) \tag{8.11}$$

Figure 8.7 shows the plot of the power picked up by the pickup. It is obvious from Figure 8.7 that the output power, which is expressed as a vector with an amplitude and a direction, belongs to a Lemniscate of Bernoulli. In addition, P_{out} reaches its maximum value, when $\sin^2(\gamma_{2-D}+\theta) = 1$, i.e. $\gamma_{2-D}+\theta = \pi/2$ or $3\pi/2$. When $\sin^2(\gamma_{2-D}+\theta) = 0$, i.e. $\gamma_{2-D}+\theta = 0$ or π, P_{out} reaches its minimum value:

$$P_{out\_max} = K_{2-D}I^2(R_3 + R_L) \tag{8.12}$$

$$P_{out\_min} = 0 \tag{8.13}$$

8.2.1.3 Input Power Calculation

The total input power includes the power losses in the two transmitting coils, the pickup coil and the load power. According to Eq. (8.6), the input power can be expressed as follows:

$$\begin{aligned} P_{in} &= P_{loss1} + P_{loss2} + P_{loss3} + P_L \\ &= I_1^2R_1 + I_2^2R_2 + I_3^2R_3 + I_3^2R_L \\ &= I^2R_1\cos^2\theta + I^2R_2\sin^2\theta + I_3^2(R_3 + R_L) \\ &= I^2(R_1 + (R_2 - R_1)\sin^2\theta + K_{2-D}(R_3 + R_L)\sin^2(\gamma_{2-D} + \theta)) \end{aligned} \tag{8.14}$$

If $R_1 = R_2 = R_3$, then

$$P_{in} = I^2R + K_{2-D}I^2(R_3 + R_L)\sin^2(\gamma_{2-D} + \theta) \tag{8.15}$$

Figure 8.8 shows a deformed Lemniscate of Bernoulli where the deformation occurs near the origin. By comparing Eqs. (8.11) and (8.15), it can be found that the difference is the term I^2R, which is the reason for the deformation of the Lemniscate near the origin. P_{in} has a maximum value, when $\sin^2(\gamma_{2-D}+\theta) = 1$, i.e.

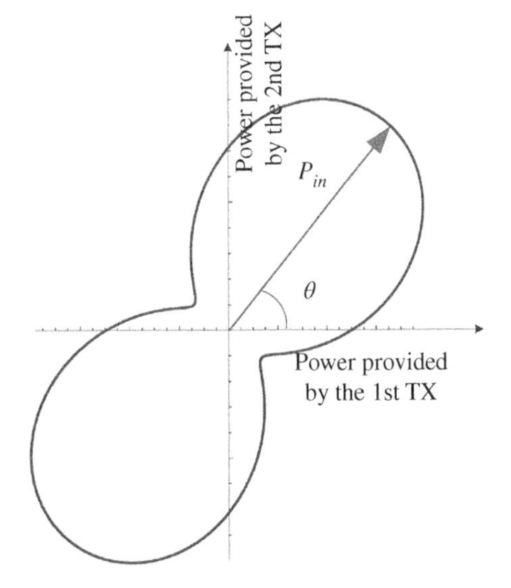

Figure 8.8 Schematic diagram of the input power (in the form of a deformed Lemniscate of Bernoulli).

$\gamma_{2-D}+\theta=\pi/2$ or $3\pi/2$. When $\sin^2(\gamma_{2-D}+\theta)=0$, i.e. $\gamma_{2-D}+\theta=0$ or π, P_{in} has a minimum value:

$$P_{in\_max} = I^2(R + K_{2-D}(R_3 + R_L)) \tag{8.16}$$

$$P_{in\_min} = I^2R \tag{8.17}$$

8.2.1.4 Efficiency Calculation

The energy efficiency is the ratio of load power to input power:

$$\begin{aligned}\eta = \frac{P_L}{P_{in}} &= \frac{K_{2-D}I^2R_L\sin^2(\gamma_{2-D}+\theta)}{I^2R + K_{2-D}I^2(R_3+R_L)\sin^2(\gamma_{2-D}+\theta)}\\ &= \frac{R_L}{\frac{R}{K_{2-D}\sin^2(\gamma_{2-D}+\theta)} + (R_3+R_L)}\end{aligned} \tag{8.18}$$

Accordingly, the energy efficiency η can reach its maximum value, when $\sin^2(\gamma_{2-D}+\theta) = 1$, i.e. $\gamma_{2-D}+\theta = \pi/2$ or $3\pi/2$. When $\sin^2(\gamma_{2-D}+\theta) = 0$, i.e. $\gamma_{2-D}+\theta=0$ or π, η has a minimum value:

$$\eta_{max} = \frac{R_L}{\frac{R}{K_{2-D}} + (R_3+R_L)} \tag{8.19}$$

$$\eta_{min} = 0 \tag{8.20}$$

Figure 8.9 shows the energy efficiency of the 2D omnidirectional WPT system with respect to the physical angle θ of the synthetic magnetic field vector. Here, the fixed distance between the transmitters and the pickups is 0.3 m, and the angular positions α are 0°, 45°, 90°, and 135°, respectively.

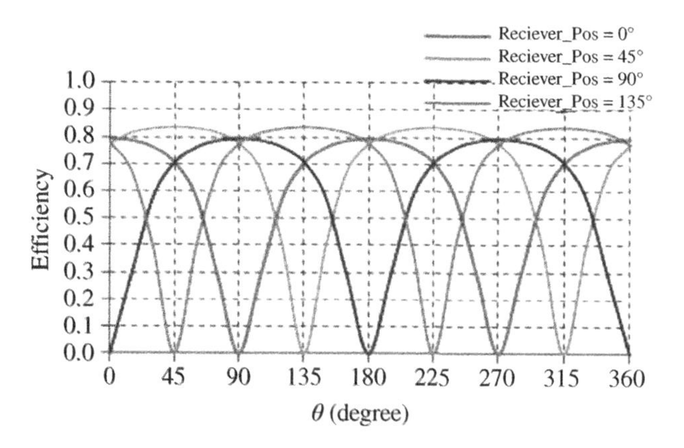

Figure 8.9 Energy efficiency plots when the pickup is placed at different angle α and the pickup's center $d = 0.3$ m, and the two transmitting coils produce a rotating magnetic field.

8.2.1.5 Physical Implications of the Input Power in the Form of the Lemniscate of Bernoulli

The Lemniscate of Bernoulli of input power indicates a lot of important information, which is in respect to the direction of the pickup coil, the load resistance, and the load power. In Eq. (8.15), the input power can be calculated by the five known parameters (ω, I, R, R_3, and X_3) and three unknowns (M_{13}, M_{23}, and R_L), which can be theoretically calculated at different angles (θ_1, θ_2, θ_3) based on Eq. (8.21). The angular position of the load can be determined after we calculate the unknown R_L with the parameters that have been obtained. Meanwhile, the input power on the primary side can be measured, which is rewritten as Eq. (8.22).

$$\begin{cases} P_{in1} = I^2R + K_{2-D}I^2(R_3 + R_L)\sin^2(\gamma_{2-D} + \theta_1) \\ P_{in2} = I^2R + K_{2-D}I^2(R_3 + R_L)\sin^2(\gamma_{2-D} + \theta_2) \\ P_{in3} = I^2R + K_{2-D}I^2(R_3 + R_L)\sin^2(\gamma_{2-D} + \theta_3) \end{cases} \tag{8.21}$$

$$P_{in} = I^2R + I^2A_{2-D}\sin^2(\gamma_{2-D} + \theta) \tag{8.22}$$

where $A_{2-D} = K_{2-D}\ (R_3 + R_L)$, which is a constant. Firstly, with the load resistance R_L determined, the parameters A_{2-D} and γ_{2-D} can be obtained by two measurements of the total input power at different current vector angles (θ_1, θ_2) in Eq. (8.21). Then, the angle θ_m can be predicted where the load power, the output power, and the energy efficiency simultaneously reach their maximum, thus acquiring the accurate angular position of the pickup load. Moreover, the locations of the pickups can also be determined by the power vectors, which is explained in [7]. It is worth noting that the above analysis is based on the assumption that the center of the pickup coil faces the transmitting coils. When the pickup coil is parallel to the system and crosses the origin of the coordinates of the transmitting coils instead of facing the transmitter system, the pickup coil cannot receive any power in theory, which is a terrible circumstance for power transmission. To avoid this situation, the pickup coil can be designed with two orthogonal coils.

8.2.1.6 Electromagnetic Position

When the maximum power flow is reached, the angle of the magnetic field vector θ may be different from the actual physical angle α of the pickup coil. The reason is that the magnetic flux in the magnetic vector analysis is assumed to flow infinitely along a straight line. However, it will deviate from the straight trajectory to form closed loop paths around the transmitting coils. In addition, the influence of the spatial distribution on the magnetic field should be considered in the calculation of the mutual inductance [11]. Figure 8.10 shows the predicted angular positions (the maximum power flows) relative to the physical angular position of the pickup coil based on the mathematical model of the 2D WPT system. Here, the pickup coil

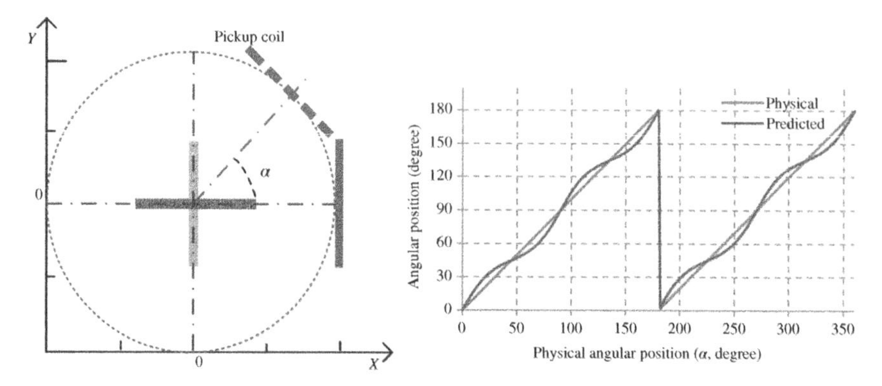

Figure 8.10 (a) Relative location of the pickup coil with the two orthogonal transmitting coils (with the pickup coil facing the central of the transmitting coil system and moved along the dashed circle). (b) Predicted angle of the maximum power flow against the physical angle of the pickup coil (corresponding to various positions in Figure 8.10a).

plane faces the center of the 2D transmitting coil system. It can be found that only at 45°, 90°, 135°, and 180° the angular angle of the maximum power flow is the same as the physical angular position of the pickup coil. It is because the symmetry of the magnetic flux, which is produced by two transmitters, at these four angles cancels out the effects caused by the spatial distribution of the magnetic field.

Although there are slight differences between the maximum power flow angle and the actual physical angle, a general control method has been developed in [7] for omnidirectional WPT systems. It contains a scanning process with the magnetic vector toward the circular space for a 2D system. Then the power related to the magnetic vector can be obtained in the system, which has guiding significance to transfer the energy to the really needed pickups.

In summary, the relationships among the total input power, load power, output power, and efficiency are analyzed theoretically for 2D omnidirectional WPT systems. As described earlier, the total input power is the sine function of the angle of the rotating current vector on a 2D plane. The functions including the total input power, load power, output power, and energy efficiency follow the Lemniscates of Bernoulli, and the energy efficiency curve of the system approximately follows the curve of the total input power. Based on the theoretical knowledge, the direction of maximum efficiency can be obtained with only a few measurements on the transmitting side, so that the pickups can be placed in this position to achieve maximum power transmission. More importantly, this analysis method can be extended to 3D omnidirectional WPT systems, which will be introduced below.

8.2.2 3-Dimensional WPT with Multiple Pickups

The mathematical analysis of the 3D omnidirectional WPT is an extended knowledge of the 2D omnidirectional WPT. In the 3D omnidirectional WPT system, the pickup coil can be freely located in the position of the 3D space; however, it is inevitable that this method was previously considered energy inefficient. That is because a lot of energy is transferred to the unloaded end, and a lot of waste is caused. For instance, as shown in Figure 8.11, the rotating magnetic field covers the entire sphere to make sure that the pickup coil can get energy anywhere from the transmitter. However, it is obvious that the distribution of magnetic field vectors is not uniform; in other words, the pickups in different regions get different amounts of power. The magnetic field vectors in the poles of the sphere are more than the other space.

Figure 8.12 describes the application of the 3D omnidirectional WPT system with the proposed method of discrete magnetic field vector control [5]. In the technique, first the positions of the pickups are detected, which is called "tracking," and then the magnetic flux vector is focused toward the targeted pickups, which is called "firing." Indeed, the system achieves efficient energy transfer in 3D space. The theoretical analysis of the 3D WPT system to explain the physical phenomenon is provided in [9]. The mathematical expressions of the input power vector and the output power vector are provided. And it is proved that the geometry of this 3D distribution space is the revolution of the Lemniscate of Bernoulli along its vertical axis. Furthermore, the proposed 3D omnidirectional WPT theory is verified by experiments in [9].

Figure 8.13 shows a 3D WPT system with three orthogonal transmitting coils and one pickup coil. The AC current sources $\vec{I}_1$, $\vec{I}_2$, and $\vec{I}_3$ connected in series to the transmitting coils have the same frequency and phase which are set in Eq. (8.23). Accordingly, the three current vectors can produce a synthetic

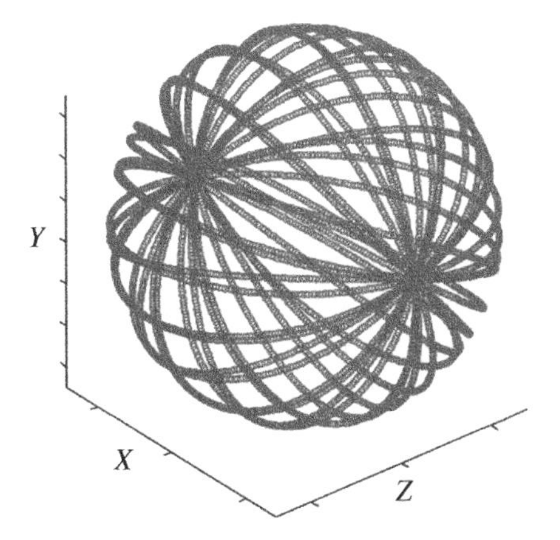

Figure 8.11 Trajectory of the magnetic field vector at the center under controlling scheme proposed in [5].

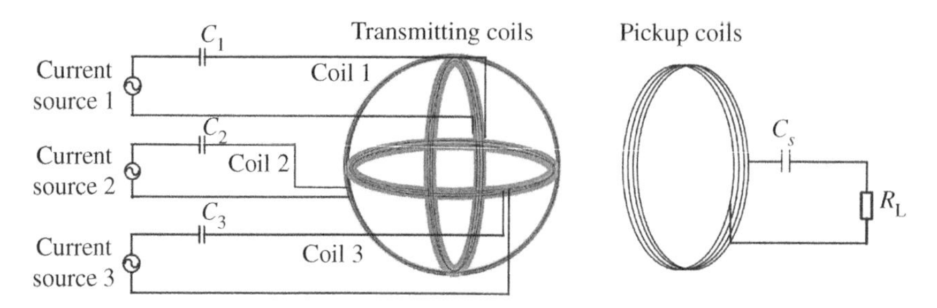

Figure 8.12 Three-orthogonal-coil structure for the proposed omnidirectional WPT control method.

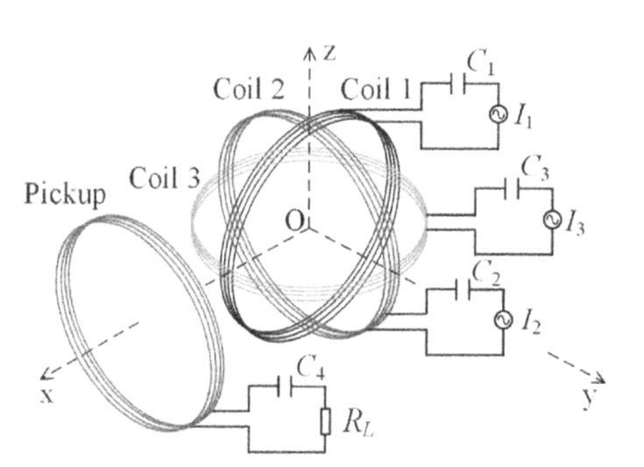

Figure 8.13 Three-dimensional omnidirectional WPT system comprising three orthogonal transmitting coils and a pickup coil.

current vector. As shown in Figure 8.14, the synthetic current vector has the same direction as the resultant magnetic field vector in the 3D system.

$$\begin{bmatrix} \vec{I}_1 \\ \vec{I}_2 \\ \vec{I}_3 \end{bmatrix} = \begin{bmatrix} \sin\theta\cos\varphi \\ \sin\theta\sin\varphi \\ \cos\theta \end{bmatrix} I \tag{8.23}$$

In Eq. (8.23), θ and φ are the angles in the spherical coordinates, which is shown in Figure 8.14, and I is a sinusoidal function of time. In the nonidentical current control, θ and φ are the control variables to generate the synthetic magnetic vector with different directions. Then, the rotating magnetic field in the omnidirectional WPT system can be generated, which helps to simplify the analysis. Actually, the resultant unity current vector will generate the resultant magnetic field vector whose directions are the same as shown in Figure 8.14.

Figure 8.15 shows the synthetic input current vector in the 3D space, where the resultant magnetic vector can be pointed to the surface of the sphere. With the nonidentical current control, the synthetic current vector by three transmitting coil currents is produced to cover all N points of the surface of the sphere which realizes the omnidirectional WPT in the 3D space. If N is 200, the input current vector of the ith node is as represented in Figure 8.15b, its magnitude is I, and its

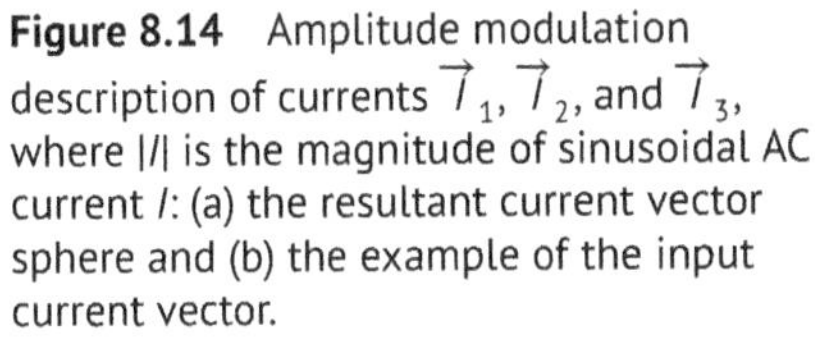

Figure 8.14 Amplitude modulation description of currents $\vec{I}_1$, $\vec{I}_2$, and $\vec{I}_3$, where $|I|$ is the magnitude of sinusoidal AC current I: (a) the resultant current vector sphere and (b) the example of the input current vector.

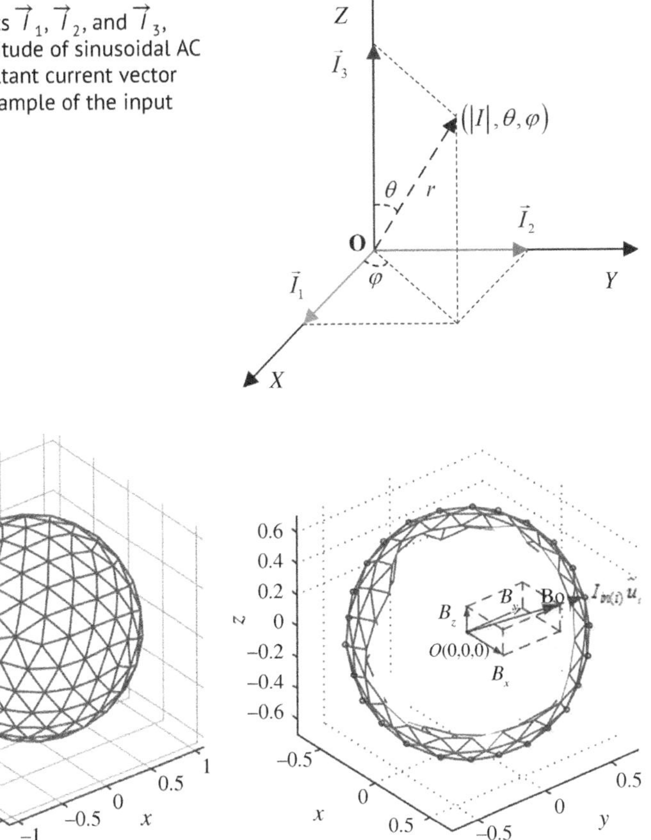

Figure 8.15 Current vector plotted in space [9]: (a) the resultant current vector sphere and (b) the example of the input current vector.

direction is the same as the unit vector u_i.

$$\hat{u}_i = \left(\frac{I_{1i}}{I}, \frac{I_{2i}}{I}, \frac{I_{3i}}{I} \right) \tag{8.24}$$

$$\vec{I}_{in(i)} = I_{in(i)} \hat{u}_i \tag{8.25}$$

Figure 8.16 shows the amplitude modulation functions of I_1, I_2, and I_3, which can generate the magnetic vector to scan all the points of the sphere in turn. There are N points on the sphere. It is supposed that the scanning process of the magnetic vector is from top to bottom. As the scanning moves from the top of the sphere to the center, the number of discrete points scanned will increase. And then the

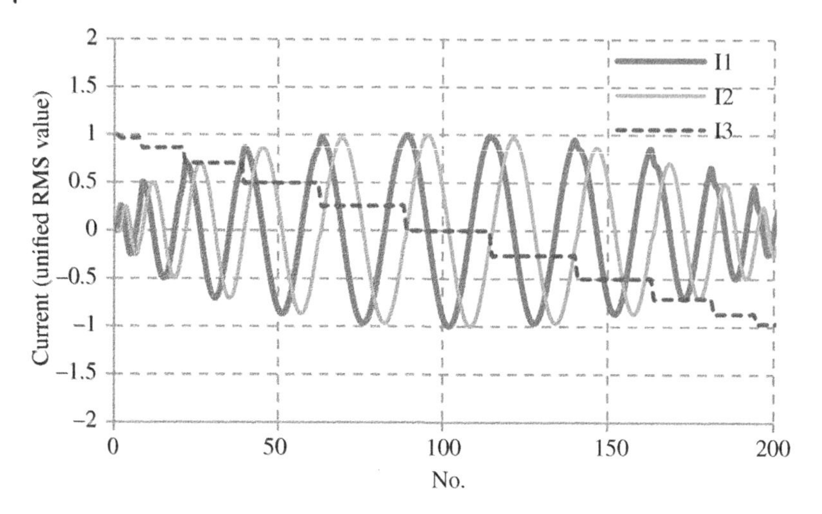

Figure 8.16 Variations of the amplitude modulation functions of Eq. (8.23) as the magnetic field vector moves from the top of the spherical surface progressively in discrete steps to the bottom of the spherical surface.

number of discrete points will decrease as the scanning continues to descend to the bottom. Therefore, according to Eq. (8.23), the amplitude modulation function of I_3 steps down with a ladderlike shape, while the amplitude modulation function of I_1 and I_2 with a constant θ is sinusoidal, which is in line with the facts.

It is set that R_i represents the coil resistance of the ith coil, L_i represents the self-inductance of the ith coil, and M_{ij} represents the mutual inductance between the ith coil and the jth coil. Meanwhile, $\vec{U}_1$, $\vec{U}_2$, and $\vec{U}_3$ represent the voltages of the current sources $\vec{I}_1$, $\vec{I}_2$, and $\vec{I}_3$. Then, the circuit equation which the 3D WPT system satisfies is shown below:

$$\begin{bmatrix} \vec{U}_1 \\ \vec{U}_2 \\ \vec{U}_3 \\ 0 \end{bmatrix} = \begin{bmatrix} R_1 + jX_1 & j\omega M_{12} & j\omega M_{13} & j\omega M_{14} \\ j\omega M_{12} & R_2 + jX_2 & j\omega M_{23} & j\omega M_{24} \\ j\omega M_{13} & j\omega M_{23} & R_3 + jX_3 & j\omega M_{34} \\ j\omega M_{14} & j\omega M_{24} & j\omega M_{34} & R_4 + R_L + jX_4 \end{bmatrix} \begin{bmatrix} \vec{I}_1 \\ \vec{I}_2 \\ \vec{I}_3 \\ \vec{I}_4 \end{bmatrix} \tag{8.26}$$

where

$$X_1 = \left(\omega L_1 - \frac{1}{\omega C_1}\right), X_2 = \left(\omega L_2 - \frac{1}{\omega C_2}\right), X_3 = \left(\omega L_3 - \frac{1}{\omega C_3}\right),$$

$$X_4 = \left(\omega L_4 - \frac{1}{\omega C_4}\right).$$

In Eq. (8.26), $M_{12} = M_{13} = M_{23} = 0$. That is because the three transmitting coils are orthogonal to each other. In addition, all the parameters can be obtained or calculated. Substitution of Eq. (8.23) into Eq. (8.26) leads to Eq. (8.27).

$$\begin{bmatrix} \vec{U}_1 \\ \vec{U}_2 \\ \vec{U}_3 \\ 0 \end{bmatrix} = \begin{bmatrix} R_1 + jX_1 & 0 & 0 & j\omega M_{14} \\ 0 & R_2 + jX_2 & 0 & j\omega M_{24} \\ 0 & 0 & R_3 + jX_3 & j\omega M_{34} \\ j\omega M_{14} & j\omega M_{24} & j\omega M_{34} & R_4 + R_L + jX_4 \end{bmatrix} \begin{bmatrix} I\sin\theta\cos\varphi \\ I\sin\theta\sin\varphi \\ I\cos\theta \\ \vec{I}_4 \end{bmatrix} \tag{8.27}$$

Accordingly, the vectors $\vec{I}_4$, the power from the transmitters to the pickup, and the total efficiency of the system can be calculated, which will be described below.

8.2.2.1 Load Current Calculation

According to the last row of Eq. (8.27), load current $\vec{I}_4$ can be calculated with its amplitude.

$$\vec{I}_4 = \frac{-j\omega I}{R_4 + R_L + jX_4} \times (M_{14}\sin\theta\cos\varphi + M_{24}\sin\theta\sin\varphi + M_{34}\cos\theta) \tag{8.28}$$

$$I_4 = \frac{\omega I}{\sqrt{(R_4 + R_L)^2 + X_4^2}} \times |M_{14}\sin\theta\cos\varphi + M_{24}\sin\theta\sin\varphi + M_{34}\cos\theta| \tag{8.29}$$

8.2.2.2 Output Power Calculation

8.2.2.2.1 P_L: Load Power After determining the load current, the output power can be acquired as

$$P_L = I_4^2 R_L = \frac{\omega^2 I^2 R_L}{(R_4 + R_L)^2 + X_4^2} \times (M_{14}\sin\theta\cos\varphi + M_{24}\sin\theta\sin\varphi + M_{34}\cos\theta)^2 \tag{8.30}$$

Furthermore, the equation of the output power can be simplified as

$$\begin{aligned} P_L &= \frac{\omega^2 I^2 R_L}{(R_4 + R_L)^2 + X_4^2} \times (M_{14}\sin\theta\cos\varphi + M_{24}\sin\theta\sin\varphi + M_{34}\cos\theta)^2 \\ &= \frac{\omega^2 I^2 R_L}{(R_4 + R_L)^2 + X_4^2} \left(\sqrt{M_{14}^2 + M_{24}^2} \times \sin\left(\arctan\frac{M_{14}}{M_{24}} + \varphi\right)\sin\theta + M_{34}\cos\theta\right)^2 \\ &= \frac{\omega^2 I^2 R_L}{(R_4 + R_L)^2 + X_4^2} \left(\left(M_{14}^2 + M_{24}^2\right) \times \sin^2\left(\arctan\frac{M_{14}}{M_{24}} + \varphi\right) + M_{34}^2\right) \\ &\quad \times \sin^2(\gamma_{3-D} + \theta) \\ &= I^2 R_L K_{3-D(\varphi)} \sin^2(\gamma_{3-D} + \theta) \end{aligned} \tag{8.31}$$

where

$$\gamma_{3-D} = \arctan \frac{M_{34}}{\sqrt{M_{14}^2 + M_{24}^2} \sin\left(\arctan \frac{M_{14}}{M_{24}} + \varphi\right)} \tag{8.32}$$

$$K_{3-D(\varphi)} = \frac{\omega^2 \left(\left(M_{14}^2 + M_{24}^2\right) \sin^2\left(\arctan \frac{M_{14}}{M_{24}} + \varphi\right) + M_{34}^2 \right)}{(R_4 + R_L)^2 + X_4^2} \tag{8.33}$$

Then, the load power can be drawn as the surface of revolution of Lemniscate of Bernoulli along the X-axis, which is dumbbell shaped, as shown in Figure 8.17.

The load power P_L takes its maximum value, when $\sin^2(\arctan(M_{14}/M_{24})+\varphi) = 1$ and $\sin^2(\gamma_{3-D} + \theta) = 1$, i.e. $\arctan(M_{14}/M_{24})+\varphi = \pi/2$ or $3\pi/2$ and $\gamma_{3-D} + \theta = \pi/2$ or $3\pi/2$. When $\sin^2(\gamma_{3-D} + \theta) = 0$, i.e. $\gamma_{3-D} + \theta = 0$ or π, P_L has a minimum value.

$$P_{L\_max} = I^2 K_{3-D} R_L \tag{8.34}$$

where

$$K_{3-D} = \frac{\omega^2 \left(M_{14}^2 + M_{24}^2 + M_{34}^2\right)}{(R_4 + R_L)^2 + X_4^2} \tag{8.35}$$

$$P_{L\_min} = 0 \tag{8.36}$$

8.2.2.2.2 P_{out}: Power Picked Up by the Pickup The output power consists of the load power and the conduction power loss in the pickup resonant coil.

$$\begin{aligned} P_{out} &= I_4^2 R_4 + I_4^2 R_L \\ &= \frac{\omega^2 I^2 (R_4 + R_L)}{(R_4 + R_L)^2 + X_4^2} \times (M_{14} \sin\theta \cos\varphi + M_{24} \sin\theta \sin\varphi + M_{34} \cos\theta)^2 \\ &= I^2 (R_4 + R_L) K_{3-D(\varphi)} \sin^2(\gamma_{3-D} + \theta) \end{aligned} \tag{8.37}$$

According to Eq. (8.37), the output power can also be drawn as the surface of revolution of Lemniscate of Bernoulli along the X-axis, as shown in Figure 8.18.

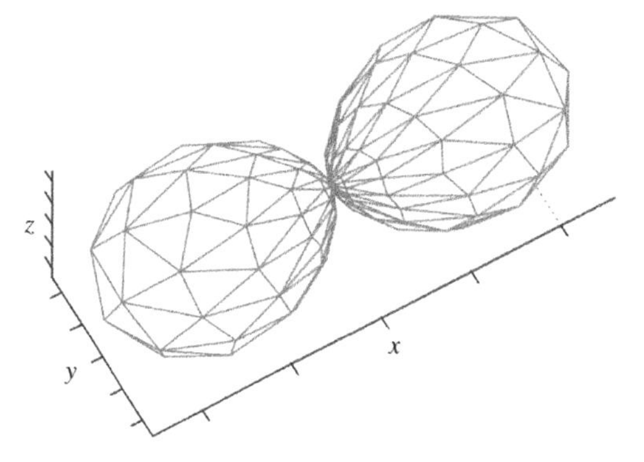

Figure 8.17 Load power: the surface of revolution of Lemniscate of Bernoulli along the longitudinal axis [9].

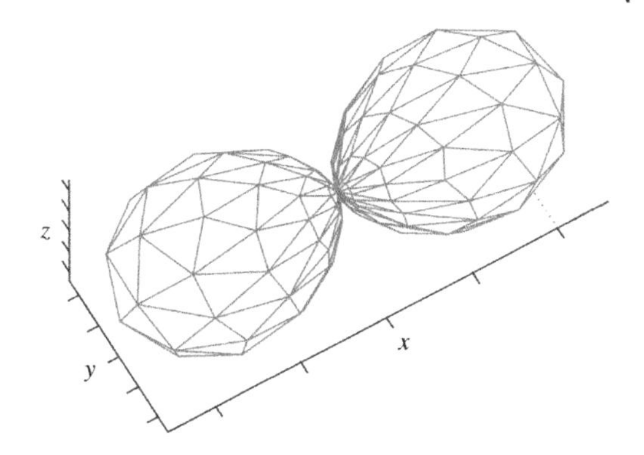

Figure 8.18 Power picked up by the pickup: the surface of revolution of Lemniscate of Bernoulli curve along the longitudinal axis [9].

Meanwhile, the output power includes the conduction power loss compared with the load power; thus, the dumbbell figure in Figure 8.18 has a larger volume than that in Figure 8.17.

The output power P_{out} will maximize when $\sin^2(\arctan(M_{14}/M_{24})+\varphi) = 1$ and $\sin^2(\gamma_{3-D}+\theta) = 1$, i.e. $\arctan(M_{14}/M_{24})+\varphi = \pi/2$ or $3\pi/2$ and $\gamma_{3-D}+\theta = \pi/2$ or $3\pi/2$. P_{out} will reach its minimum value when $\sin^2(\gamma_{3-D}+\theta) = 0$, i.e. $\gamma_{3-D}+\theta = 0$ or π.

$$P_{out\_max} = I^2 K_{3-D}(R_4 + R_L) \tag{8.38}$$

$$P_{out\_min} = 0 \tag{8.39}$$

8.2.2.3 Input Power Calculation

The input power includes the conduction power losses in the transmitter and the pickup coils and the load power:

$$\begin{aligned} P_{in} &= P_{loss1} + P_{loss2} + P_{loss3} + P_{loss4} + P_L \\ &= I_1^2 R_1 + I_2^2 R_2 + I_3^2 R_3 + I_4^2 R_4 + I_4^2 R_L \\ &= I^2 R_1 \sin^2\theta \cos^2\varphi + I^2 R_2 \sin^2\theta \sin^2\varphi + I^2 R_3 \cos^2\theta + I_4^2 R_4 + I_4^2 R_L \end{aligned} \tag{8.40}$$

If $R_1 = R_2 = R_3 = R$, substituting Eq. (8.29) into Eq. (8.40) results as follows:

$$\begin{aligned} P_{in} &= I^2 R + \frac{\omega^2 I^2 (R_4 + R_L)}{(R_4 + R_L)^2 + X_4^2} \times (M_{14} \sin\theta \cos\varphi + M_{24} \sin\theta \sin\varphi + M_{34} \cos\theta)^2 \\ &= I^2 R + I^2 (R_4 + R_L) K_{3-D(\varphi)} \sin^2(\gamma_{3-D} + \theta) \end{aligned} \tag{8.41}$$

The input power P_{in} will reach the maximum value when $\sin^2(\arctan(M_{14}/M_{24})+\varphi) = 1$ and $\sin^2(\gamma_{3-D}+\theta) = 1$, i.e. $\arctan(M_{14}/M_{24})+\varphi = \pi/2$ or $3\pi/2$ and $\gamma_{3-D}+\theta = \pi/2$ or $3\pi/2$. P_{in} will reach its minimum value when $\sin^2(\gamma_{3-D}+\theta) = 0$, i.e. $\gamma_{3-D}+\theta = 0$ or π.

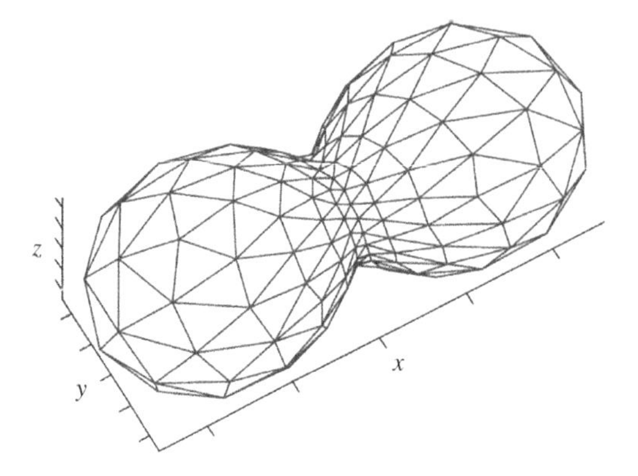

Figure 8.19 Total input power of the three transmitters: the surface of revolution of deformed Lemniscate of Bernoulli along the longitudinal axis [9].

$$P_{in\_max} = I^2R + I^2K_{3-D}(R_4 + R_L) \tag{8.42}$$

$$P_{in\_min} = I^2R \tag{8.43}$$

Figure 8.19 shows that the input power also follows the surface of revolution of Lemniscate of Bernoulli along the X-axis, which is dumbbell shaped.

8.2.2.4 Efficiency Calculation

The energy efficiency is mathematically expressed as

$$\begin{aligned} \eta = \frac{P_L}{P_{in}} &= \frac{I^2R_LK_{3-D(\varphi)}\sin^2(\gamma_{3-D}+\theta)}{I^2R + I^2(R_4+R_L)K_{3-D(\varphi)}\sin^2(\gamma_{3-D}+\theta)} \\ &= \frac{R_LK_{3-D(\varphi)}\sin^2(\gamma_{3-D}+\theta)}{R + (R_4+R_L)K_{3-D(\varphi)}\sin^2(\gamma_{3-D}+\theta)} \\ &= \frac{R_L}{\frac{R}{K_{3-D(\varphi)}\sin^2(\gamma_{3-D}+\theta)} + (R_4+R_L)} \end{aligned} \tag{8.44}$$

The energy efficiency η has a maximum value when $\sin^2(\arctan(M_{14}/M_{24})+\varphi) = 1$ and $\sin^2(\gamma_{3-D}+\theta) = 1$, i.e. $\arctan(M_{14}/M_{24})+\varphi = \pi/2$ or $3\pi/2$ and $\gamma_{3-D}+\theta = \pi/2$ or $3\pi/2$, and a minimum value when $\sin^2(\gamma_{3-D}+\theta) = 0$, i.e. $\gamma_{3-D}+\theta = 0$ or π.

$$\eta_{max} = \frac{R_L}{\frac{R}{K_{3-D}} + (R_4+R_L)} \tag{8.45}$$

$$\eta_{min} = 0 \tag{8.46}$$

According to Eqs. (8.34), (8.42), and (8.45), these three indexes including load power, input power, and energy efficiency have the same maximization conditions, which have great significance for energy efficiency and load power direction. The two conditions are as follows:

(1) $\sin^2\left(\arctan\dfrac{M_{14}}{M_{24}}+\varphi\right)=1\Rightarrow\arctan\dfrac{M_{14}}{M_{24}}+\varphi=\dfrac{\pi}{2}$ or $\dfrac{3\pi}{2}$.

(2) $\sin^2(\gamma_{3-D}+\theta)=1\Rightarrow\gamma_{3-D}+\theta=\dfrac{\pi}{2}$ or $\dfrac{3\pi}{2}$.

Accordingly, under the abovementioned two conditions, the direction of the pickup coil can be chosen with the maximum of the energy efficiency, the position of the pickup coil can be detected, and the magnetic field can be controlled to maximize the power obtained by the pickup. In practice, after controlling the input power on the primary side, the corresponding load power can be controlled to the maximum point, which reduces the wireless feedback from the pickup. Based on the theoretical analysis above, the approaches (mathematical method and two-plane method) are both proposed to find the maximum point for the input power in [9].

In conclusion, the mathematical formulas for the total input power, output power, and energy efficiency in 3D omnidirectional WPT system are introduced above. The theoretical result demonstrates that the geometry of the distribution of the total input power, load power, and efficiency follows the revolution of Lemniscate of Bernoulli. In addition, the direction of the maximum input power vector is proved to be identical with the direction of maximum energy efficiency, which is critical for the load detection. These characteristics derived from the theory have been identified in the experimental facility.

8.3 Design of Transmitting Coils for Synthetic Magnetic Field

In omnidirectional WPT, the design of the transmitting coils is a key problem to optimize and control the shape of the magnetic field, which has attained more attention from the scholars. By controlling the shape of the synthetic magnetic field, the magnetic field vector can be pointed in the desired direction, which can greatly improve the energy transmission efficiency and reduce the flux leakage. In general, the technology of the design of the magnetic field consists of two types: the active and passive methods. In the passive method, the magnetic core can be used in the transmitting coils to reduce the reluctance in the magnetic circuit, which in turn helps to improve the coupling between the transmitting coil and the pickup coil. In the active mode, most of the methods generate the synthetic magnetic field by controlling the amplitude and phase of the current vector in series with the transmitting coil. To achieve the free adjustment of the magnetic field strength and the direction in the 3D space, many design schemes of different transmitting coils have been proposed, such as the structure of three orthogonal transmitting coils [12], the reticulated planar transmitter with four interleaved and overlapped meander coils [13], and crossed dipole coils with an orthogonal phase difference [14].

Designed to address a large leakage of magnetic flux between the transmitting coil and the pickup coil, an active flux orientation approach has been proposed, which elaborates the design of the structure of three orthogonal transmitting coils and the magnetic field orientation based on current control [12]. With the technology, the 3D omnidirectional magnetic orientation can be formed in the WPT system. In consideration of the different positions of multiple pickup coils, the magnitude and direction of the magnetic field can be adjusted accordingly to realize the high utilization of the magnetic field vector with very little magnetic flux leakage.

Figure 8.20 demonstrates the diagram of the proposed WPT system, which adopts the 3D magnetic orientation technology based on the amplitude-phase control of three independent currents. In terms of the structural design of the transmitting coil, the theoretical analysis has been done of four different methods of placing two wires, which can cover all the cases [12]. The conclusion is extended to the three coils, that is when the three coils are perpendicular to each other and have the same length, the mutual inductance between the coils can be greatly reduced. Accordingly, the performance of the system can be improved, and it is beneficial to the control of the synthetic magnetic field. If the structure changes slightly due to external factors, the synthetic magnetic field can still be synthesized by three nonorthogonal B-field vectors, which is computationally complicated.

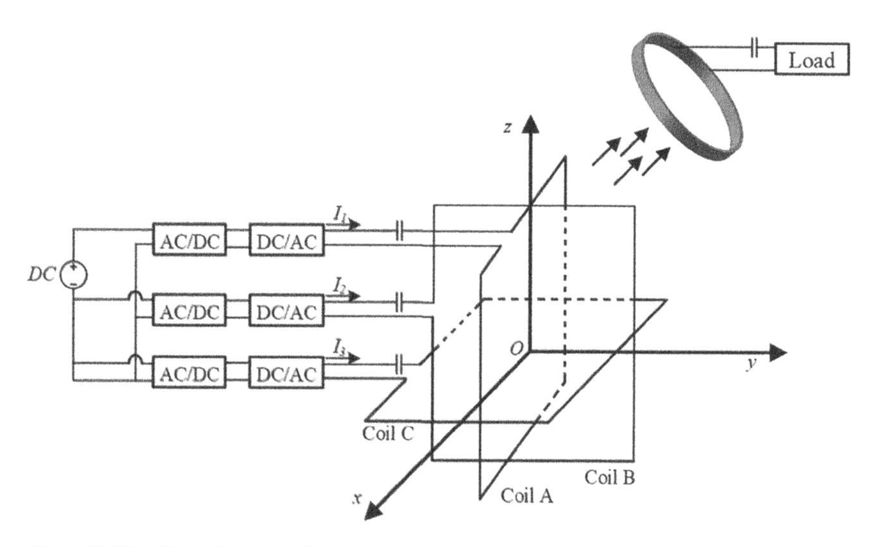

Figure 8.20 Block diagram of the proposed WPT system.

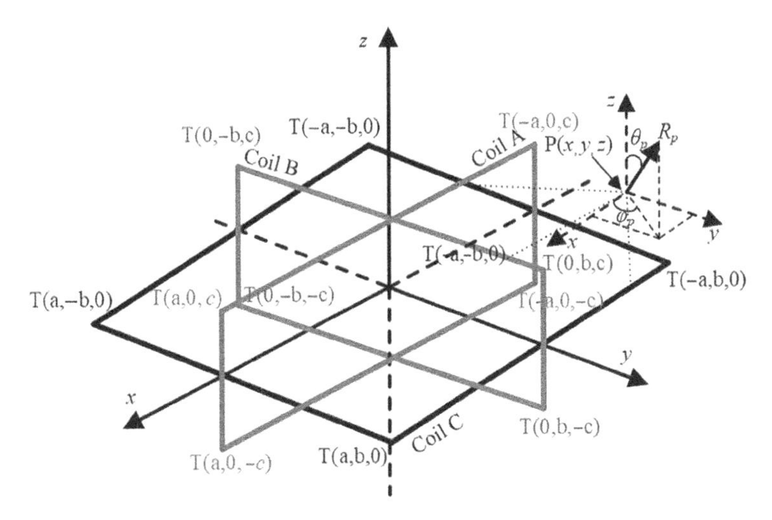

Figure 8.21 Detailed structure of the proposed coils.

Figure 8.21 shows the structure of three transmitter coils that are perpendicular to each other, ensuring that the coupling between the coils is as low as possible. Here, suppose that the given currents in each of the three coils are $I_A\cos(\omega t+\theta_A)$, $I_B\cos(\omega t+\theta_B)$, and $I_C\cos(\omega t+\theta_C)$, respectively, where ω is the angular frequency of the three currents, θ_A, θ_B, and θ_C are the corresponding phases, and I_A, I_B, and I_C are the corresponding amplitudes. To simplify the analysis, the sides of the three square-shaped coils are equal to $2l$, and the number of turns of every coil is N. Then, according to the Biot–Savart law, the component of the magnetic flux density of the three coils at the origin (0,0,0) can be expressed by Eq. (8.47):

$$\begin{cases} B_{x_0} = -\dfrac{\sqrt{2}\mu_0 N I_B \cos(\omega t+\theta_B)}{\pi l} \\ B_{y_0} = -\dfrac{\sqrt{2}\mu_0 N I_A \cos(\omega t+\theta_A)}{\pi l} \\ B_{z_0} = -\dfrac{\sqrt{2}\mu_0 N I_C \cos(\omega t+\theta_C)}{\pi l} \end{cases} \tag{8.47}$$

In the spherical coordinate system, the magnetic flux density can be expressed as

$$\begin{cases} R_p = \sqrt{B_x^2+B_y^2+B_z^2} \\ \theta_p = \tan^{-1}\left(\sqrt{B_x^2+B_y^2}/B_z\right) \\ \varphi_p = \tan^{-1}(B_y/B_x) \end{cases} \tag{8.48}$$

where R_p represents the magnitude of the magnetic field at point p, θ_p represents the angle between the positive Z-axis and the magnetic field, and φ_p represents the angle of rotation from the positive X-axis counterclockwise to the projection of the magnetic field onto the X–Y-plane ($0 \leq \theta_p \leq \pi$, $0 \leq \varphi_p \leq 2\pi$). According to Eqs. (8.47) and (8.48), the synthetic magnetic field at the origin (0,0,0) is obtained as

$$\begin{cases} R_p = \sqrt{2}\mu_0 N \sqrt{I_A^2\cos^2(\omega t + \theta_A) + I_B^2\cos^2(\omega t + \theta_B) + I_C^2\cos^2(\omega t + \theta_C)}/\pi l \\ \theta_p = \tan^{-1}\left(-\sqrt{I_A^2\cos^2(\omega t + \theta_A) + I_B^2\cos^2(\omega t + \theta_B)}/I_C\cos(\omega t + \theta_C)\right) \\ \varphi_p = \tan^{-1}(I_A\cos(\omega t + \theta_A)/I_B\cos(\omega t + \theta_B)) \end{cases} \tag{8.49}$$

It can be seen from Eq. (8.49) that an arbitrary magnetic field is realized by controlling the amplitude and phase of the current; in other words, for the magnitude and direction of the given magnetic field at a certain point, the corresponding set of currents can be calculated. In the application, the phase angle of the current can be determined in accordance with the pickup's position in the 3D space, and then the three currents I_A, I_B, and I_C can be calculated in terms of the desired magnetic field amplitude. Based on the theoretical analysis, both the simulation and experiment in [12] verify the feasibility of the general algorithm of field orientation, which can realize the desired magnetic field vector at any position in the 3D space.

In addition, for the charging technology of the low-power devices such as radio frequency identification devices (RFID) and sensors, a nonidentical current control approach is proposed and applied on the 2D and 3D WPT systems [5]. It has been proved that the method of exciting the orthogonal coils with the identical current cannot guarantee the magnetic field vector with a full range in the 3D space. Figure 8.22 illustrates the coil structure of a 3D omnidirectional transmitter composed of three orthogonal coils. The coil is connected in series with a capacitor to form a coil resonator and is driven by an AC power supply. The three currents can be expressed in Eq. (8.50):

$$\begin{cases} \vec{I}_1 = I_{m1}\sin(\omega t) \\ \vec{I}_2 = I_{m2}\sin(\omega t + \alpha) \\ \vec{I}_3 = I_{m3}\sin(\omega t + \beta) \end{cases} \tag{8.50}$$

where ω is the angular frequency of the currents, α and β are the current phases of the last two coils, and I_{m1}, I_{m2}, and I_{m3} are the current amplitudes of the three coils.

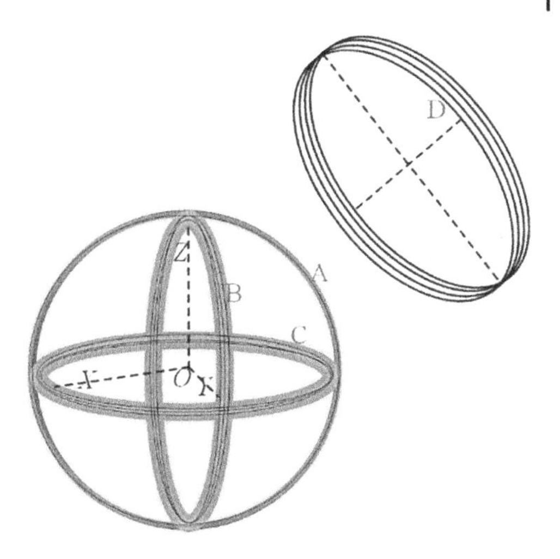

Figure 8.22 Use of three separate orthogonal coils A, B, and C connected in series as a transmitter structure for wireless power transfer.

In order to realize the omnidirectional transmission of power, the rotating magnetic field vector can be generated in these three ways: current amplitude modulation, current frequency modulation, and current phase angle control.

First, for the amplitude modulation method, we set the amplitude $I_{m1} = I_m$, $I_{m2} = I_m \sin(\omega_2 t)$, and $I_{m3} = I_m \sin(\omega_2 t + \pi/2)$, where ω_2 is the angular frequency different from ω, and the phase angle $\alpha = \pi/2$ and $\beta = \pi/2$. Then, Eq. (8.50) can be changed to Eq. (8.51). In this case, the amplitude of the current of the first coil is a constant value, and the current amplitudes of the last two coils are sinusoidal functions with the phase angle difference $\pi/2$. Through the couple circuit model and magnetic field solver, the current value and phase angle can be calculated and sampled in the excitation period of the angular frequency ω. Then, the sampled data can be used to plot the magnetic field vector. On the basis of the trajectory of the peaks of the magnetic field vectors, Figure 8.11 can be plotted, which shows that the magnetic field vector is pointed to every place in the 3D space through this mode of amplitude modulation,

$$\begin{cases} \vec{I}_1 = I_m \sin(\omega t) \\ \vec{I}_2 = I_m \sin(\omega_2 t) \sin(\omega t + \pi/2) \\ \vec{I}_3 = I_m \sin(\omega_2 t + \pi/2) \sin(\omega t + \pi/2) \end{cases} \tag{8.51}$$

Second, the method of current frequency modulation can be realized with Eq. (8.52),

$$\begin{cases} \vec{I}_1 = I_{m1} \sin(\omega t) \\ \vec{I}_2 = I_{m2} \sin[\omega t + |\alpha_m| \sin(\omega_2 t)] \\ \vec{I}_3 = I_{m3} \sin[\omega t + |\alpha_m| \sin(\omega_2 t)/k] \end{cases} \tag{8.52}$$

where ω_2 is the angular frequency at which the phase angle changes, α_m is a constant value, and k is also a constant real number. Third, the current phase angle control is an approach with the shift of phase angle. One example is shown in Eq. (8.53).

$$\begin{cases} \vec{I}_1 = I_m \sin(\omega t) \\ \vec{I}_2 = I_m \sin(\omega t + \alpha) \\ \vec{I}_3 = I_m \sin(\omega t + k\alpha) \end{cases} \tag{8.53}$$

In addition to the three circular or square transmitting coils that are perpendicular to each other, there are the reticulated planar transmitter and the crossed dipole coils to realize the omnidirectional WPT. In [13], the reticulated planar transmitting coil is proposed to achieve the free-positioning omnidirectional WPT system, where the multiple pickups can be powered simultaneously in any position and directions. As shown in Figure 8.23, the magnetic coupler consists of a reticulated planar transmitter and multiple pickups. The reticulated planar transmitter is equipped with four interleaved and overlapped meander coils to generate the 3D rotating magnetic field. Accordingly, with the proposed reticulated planar transmitter and excitation current modulation strategy, the free-positioning omnidirectional WPT can be realized with only planar single-coil pickups.

As shown in Figure 8.24a, in order to address the problems of free positioning and omnidirectional power supply, two crossed dipole transmitting coils with orthogonal phase difference have been designed to generate a direct and quadrature (DQ) rotating magnetic field with six degrees of freedom for the 3D WPT

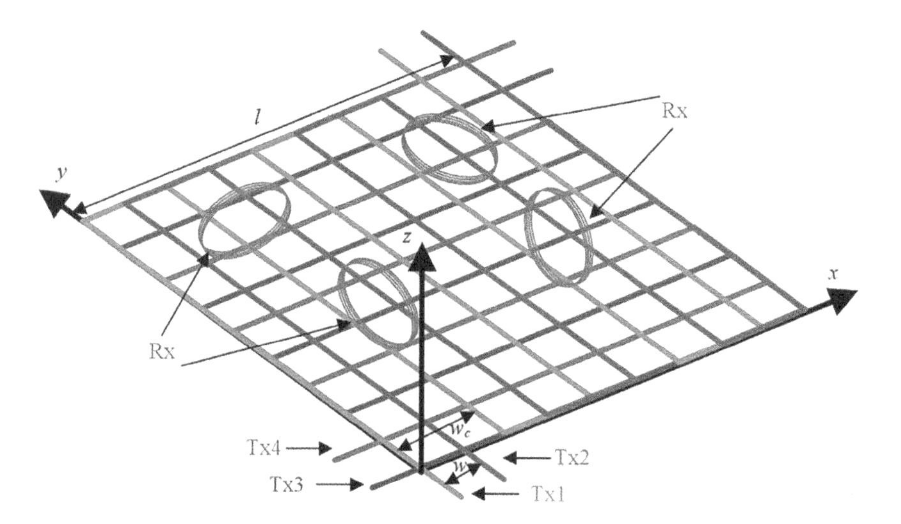

Figure 8.23 The structure of the magnetic coupler (3D view of the magnetic coupler).

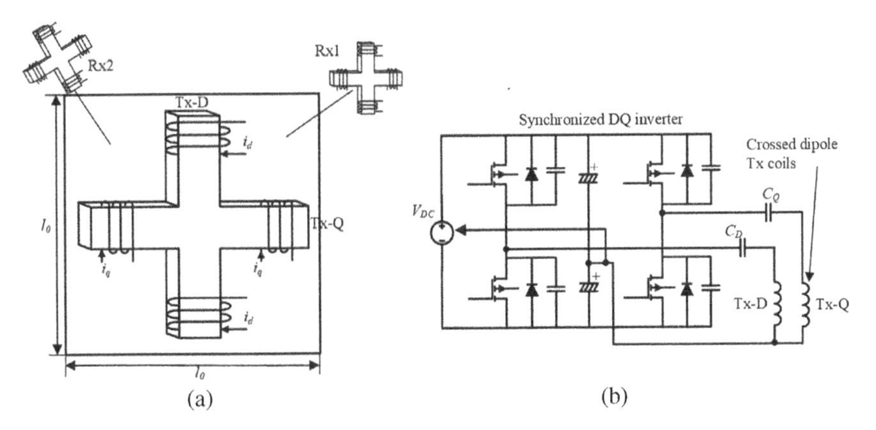

Figure 8.24 Proposed crossed dipole coils for six degrees of freedom.

system [14]. Actually, the planar coil structure of the transmitting and pickup coils is more conducive to the practical application of wireless charging. Besides, the six degrees of freedom in the 3D space refer to the position vector and the rotation vector of the load. The position vector $\vec{P}\ (x, y, z)$ consists of the coordinates x, y, and z on three axes, and the rotation vector $\vec{R}\ (\theta_x, \theta_y, \theta_z)$ embraces three angles, namely the pitch angle θ_x, the roll angle θ_y, and the yaw angle θ_z.

Figure 8.24b is the circuit diagram of the transmitting coil, where the synchronous DQ inverter is used to provide two currents with phase difference of 90° for the transmitting coil. The switching angular frequency of the DQ inverter is set a little higher than the resonant frequency of the transmitting coil and the compensating capacitor; thus, the inverter switch works under zero-voltage switching condition. The two orthogonal dipole coils can generate an appropriate rotating magnetic field under the excitation of the current. The amplitude of the current is the same, which can be expressed as

$$\begin{cases} \vec{I}_d \equiv I_d \angle 0 \\ \vec{I}_q \equiv I_d \angle \pi/2 \end{cases} \tag{8.54}$$

The corresponding generated magnetic flux density vector can be expressed as

$$\begin{cases} \vec{B}_d \equiv (\vec{B}_{dx}, \vec{B}_{dy}, \vec{B}_{dz}) = (B_{dx}\angle 0, B_{dy}\angle 0, B_{dz}\angle 0) \\ \vec{B}_q \equiv (\vec{B}_{qx}, \vec{B}_{qy}, \vec{B}_{qz}) = (B_{qx}\angle \pi/2, B_{qy}\angle \pi/2, B_{qz}\angle \pi/2) \end{cases} \tag{8.55}$$

It is worth noting that the phase of $\vec{B}_q$ is 90° ahead of $\vec{B}_d$, and the magnitude of $\vec{B}_q$ is independent of the magnitude of $\vec{B}_d$. The magnetic field produced by the transmitting coil will allow pickup coils of the plane structure to receive power vertically or horizontally, which is suitable for the charging scenarios in practical life. Furthermore, the simulated and experimental verifications illustrate that the generated DQ rotating magnetic field is omnidirectional in the 3D space [14].

In conclusion, various methods to design the transmitting coils are aimed to generate omnidirectional magnetic field vectors that can be transmitted wirelessly to the pickup coils in the 3D space. Furthermore, the design of the pickup coils, which is equally important for the process of the omnidirectional wireless power transmission, will be introduced in the following section.

8.4 Design and Control Considerations for Pickup Coils

The pickup coil is an important part of the omnidirectional WPT system, which is placed in a suitable place to pick up the power quickly, especially for the consideration of practical applications. As mentioned in the previous section, power can be transmitted to any region of the 3D space by generating an omnidirectional synthetic magnetic field with the design of the transmitting coil. Nevertheless, it is often the case that at certain positions, such as when the pickup coil is perpendicular to the transmitting coil, the transmitting power becomes so low that the load cannot receive energy. Meanwhile, there are power requirements for special positions in the omnidirectional WPT system; accordingly, it is necessary to design and optimize the pickup coil to satisfy the power and efficiency requirements of different positions.

In [15], a quadrature-shaped pickup coil assisted with magnetic core is developed to solve the problem of low power at certain positions, and the influence of key dimensions on transmission performance is analyzed by simulation, which provides an effective design reference for various requirements of practical applications.

In the omnidirectional WPT system, the transmitting coils are three orthogonal coils with their independent current sources and with three independent compensated capacitors that ensure the resonant state. The designed quadrature-shaped pickup coil consists of three groups of coils of planar structure in series to the compensation capacitor and the load, as depicted in Figure 8.25. Two sets of cross-coils are wound around a cross-shaped magnetic core, namely PC40, and a third set is wound in a circular fashion around the center of the cross-coil. The length l and the width w of the crossed coils are the key dimensional factors affecting the performance, which can be designed for different power requirements. With the 3D finite element analysis software, namely JMAG, the effects of different lengths l and width w at specific locations on transmission power and transmission efficiency in omnidirectional WPT can be verified.

On the one hand, the length l of the central crossed structure, which determines the section area of the cross-coil, influences the coil self-induction and the

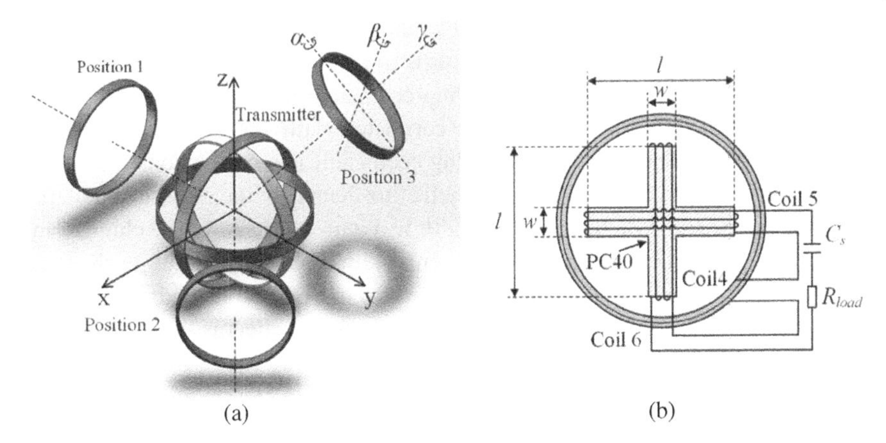

Figure 8.25 Schematics of (a) omnidirectional WPT system and (b) quadrature-shaped pickup.

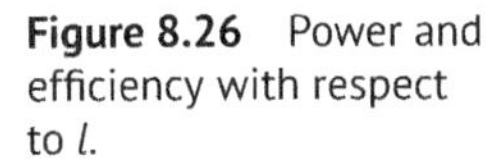
Figure 8.26 Power and efficiency with respect to l.

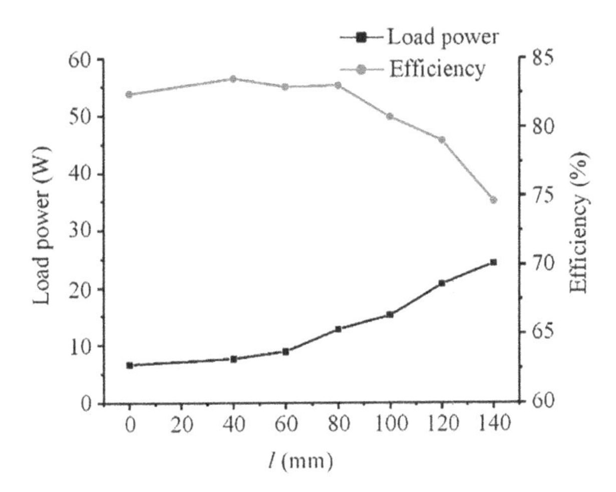

coupling coefficient between the transmitting coils and the pickup coils. With the increase in the coupling coefficient, the magnetic flux density of the pickup coil plane improves correspondingly, which also means the increase in the transmitting power at the special position. As shown in Figure 8.26, when the width w is 30 mm and the length l changes from 0 to 140 mm, the load power of the proposed pickup coil at a specific location increases from 6.6 to 24.2 W, while the transmission efficiency decreases from 83% to 75%. The simulation results show that the length l of the crossed-coil structure can be set according to different application requirements to balance the load power and transmission efficiency.

On the other hand, another key variable, the width w, is related to the number of turns of the crossed coil. When the turns of coils increase, the self-induction coefficient increases. In fact, the coupling coefficient between the transmitting coil and the pickup coil is positively correlated with the width w, and the enhancement effect of w on the coupling coefficient is more obvious than that of l. With the length l 120 mm, the magnetic flux density of the crossed coil plane is enhanced with the increase in the width w. Figure 8.27 shows the changes in the load power and efficiency with different widths. When the width w changes from 0 to 60 mm, the load power increases from 6 to 54 W, while the transmission efficiency decreases from 83% to 64%. The performance effect is the same as the length effect but more significant. Accordingly, the value of the width w can be selected for different performance requirements.

In the traditional omnidirectional WPT system, the resulting angle misalignment will degrade the power performance when the pickup coil is not aligned with the transmitting coil. Accordingly, based on the above theoretical analysis of the critical dimensions, this novel pickup coil topology is introduced to avoid the problem of angle misalignment and realize high power transmission for an angular-misalignment-insensitive omnidirectional WPT system [4]. The structure of the transmitting and pickup coils has been described above as depicted in Figure 8.25. The three current sources in series with the three transmitting coils, namely I_1, I_2, and I_3, have the same frequency and different amplitudes, respectively. By adjusting the amplitudes, the omnidirectional magnetic field vector can be generated, so that the energy can be received as long as the pickup coil is facing the transmitting coil at any position in the 3D space. If the pickup coil rotates along the roll (α) and yaw (β) axis as in Figure 8.25a, it will cause an angular misalignment between the transmitting coil and the pickup coil, while

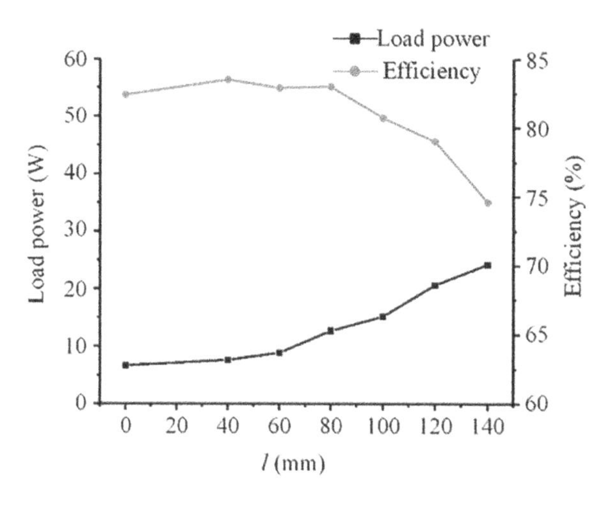

Figure 8.27 Power and efficiency with respect to w.

Figure 8.28 Schematic of magnetic field strength.

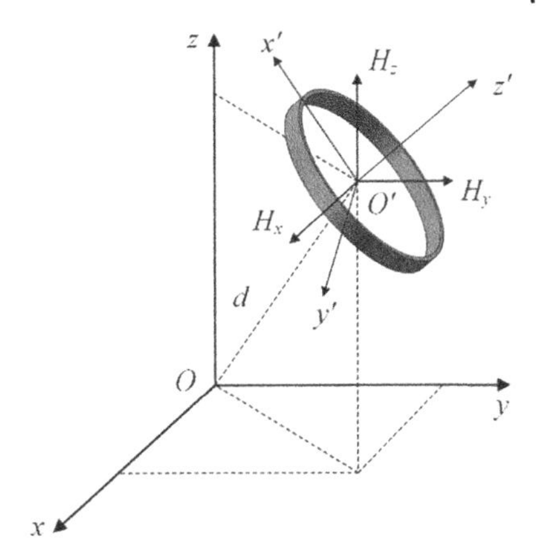

when the pickup coil rotates around the pitch (γ) axis, its plane center is always facing the transmitting coil.

The strength of the magnetic field $\vec{H}$ generated by the current at a particular location can be calculated by the Biot–Savart Law. To facilitate the analysis, the magnetic field strength is broken down into three axes as shown in Figure 8.28. These three components H_x, H_y, and H_z can be expressed as

$$\begin{cases} H_x = \dfrac{I}{4\pi}\left(\displaystyle\int_{y_1}^{y_2} \frac{z_0 - z}{r^3}dy - \int_{z_1}^{z_2} \frac{y_0 - y}{r^3}dz\right) \\ H_y = \dfrac{I}{4\pi}\left(\displaystyle\int_{z_1}^{z_2} \frac{x_0 - x}{r^3}dz - \int_{x_1}^{x_2} \frac{z_0 - z}{r^3}dx\right) \\ H_z = \dfrac{I}{4\pi}\left(\displaystyle\int_{x_1}^{x_2} \frac{y_0 - y}{r^3}dx - \int_{y_1}^{y_2} \frac{x_0 - x}{r^3}dy\right) \end{cases} \tag{8.56}$$

where r represents the distance between the current unit (x, y, z) and the position (x_0, y_0, z_0) of the magnetic field, namely $r = (x_0 - x, y_0 - y, z_0 - z)$. The magnetic flux density $\vec{B}$ can be obtained by $\vec{B} = (B_x, B_y, B_z) = \mu(H_x, H_y, H_z)$. Owing to the magnetic core PC40, the magnetic permeability μ is greater than that of air; hence, the magnetic flux through the coil is strengthened.

As shown in Figure 8.28, a new rectangular coordinate system (x', y', z') is established with the center of the pickup coil as the origin and the roll (α), the yaw (β), and the pitch (γ) axis as the new coordinate axis. Then, the induced magnetic flux density can be divided into three components along the new coordinate axis, namely (B_x', B_y', B_z'). When the pickup coil is placed near the transmitting coil and facing its center, a magnetic field perpendicular to the pickup coil can be generated by adjusting the current amplitude. It indicates that the whole induced magnetic flux density through the pickup coil is synthesized along the z'-axis, and the components on the x'- and y'-axis are both 0. According to Faraday's law of electromagnetic induction, the induced voltages of the three pickup coils can be obtained as given in Eq. (8.57).

$$\begin{cases} U_1 = jU_{m1}(\cos\alpha\cos\beta)B'_{z'} \\ U_2 = jU_{m2}(\sin\alpha\cos\gamma + \cos\alpha\sin\beta\sin\gamma)B'_{z'} \\ U_3 = jU_{m3}(\sin\alpha\sin\gamma - \cos\alpha\cos\beta\cos\gamma)B'_{z'} \end{cases} \tag{8.57}$$

where $U_{mi} = \omega N_i S_i$, ω represents the operating angular frequency, N_i represents the turns of the ith pickup coil, and S_i represents the cross-sectional area of the ith pickup coil. In addition, α, β, and γ represent the angular displacement of rotation along the x'-, y'-, and z'-axis, respectively, and $B'_{z'}$ represents the component of the mean flux density decomposed in the z'-axis. It can be found that the induced voltages of the three pickup coils are affected by the angular misalignment. Then, the sum of the induced voltages generated by the structure of the proposed pickup coil under different angle misalignments is shown in Figure 8.29. Compared with the conventional coil, namely U_1, the induced voltage of the proposed pickup coil does not drop to 0 at some special angle, such as at $\pi/2$ of α or at $\pi/2$ of β, and is in any case higher than that of the conventional coil. Obviously, as the pickup coil rotates along the γ-axis, its plane is always aligned with the transmitting coil; therefore, the induced voltage remains constant with the change in angular displacement γ. According to the above theoretical analysis, the proposed quadrature-shaped pickup coil can effectively solve the problem of angular misalignment in the omnidirectional WPT system.

In addition, Figure 8.30 shows the simulated result of the power transmitted when two different pickup coils rotate at several angles with the α-, β-, and γ-axis, respectively. It can be found that the power drops to zero as the conventional pickup coil rotates along the roll and the yaw axis. With the proposed quadrature-shaped pickup coil, the power will not be greatly affected, even if there is a large angular displacement, which indicates that there is always magnetic flux through the coil, and the power remains at 45 W with little change, as the coil rotates along the γ-axis. The simulation results indicate that the designed quadrature-shaped pickup coil is capable of greatly improving the angular misalignment insensitivity and enhancing the transmission performance in the omnidirectional WPT system.

In addition, aiming to deal with the low degree of freedom for wireless charging mobile devices, a novel type of nonorthogonal pickup coil is proposed, and it realizes omnidirectional wireless charging with the minimum number of pickup coils [16]. In the omnidirectional WPT system, the conventional pickup coils are generally the structure of the orthogonal coils, which leads to the number of the transmitting coils and pickup coils of at least four and is not conducive to practical application. The proposed plane-type pickup coil is identical in material selection and coil thickness as the traditional one, whose design is not very complicated.

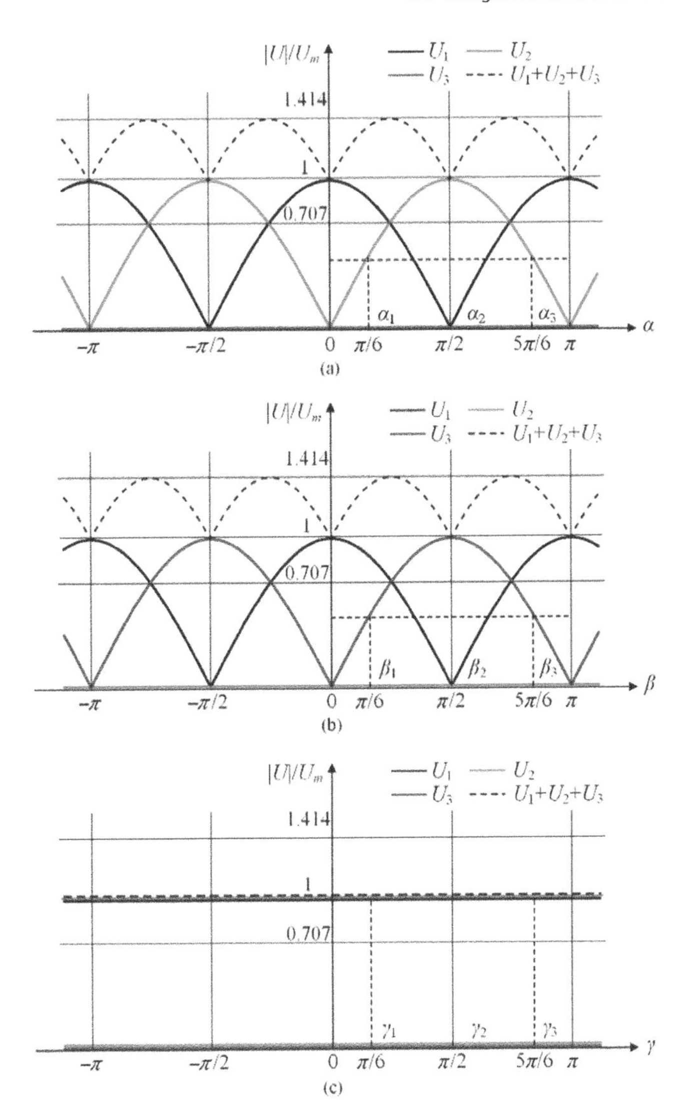

Figure 8.29 Induced pickup coil voltages with angular misalignment around: (a) α-axis, (b) β-axis, and (c) γ-axis.

Besides, by reducing the number of coils, it strongly demonstrates the high economy in the application of wireless charger. As depicted in Figure 8.31, the proposed pickup coil structure is planar and consists of an inner coil, an outer coil, and an intermediate orthogonal magnetic core. The two square coils are embedded in a cross-shaped ferrite core, and the sides of the two coils are placed across the legs of the core.

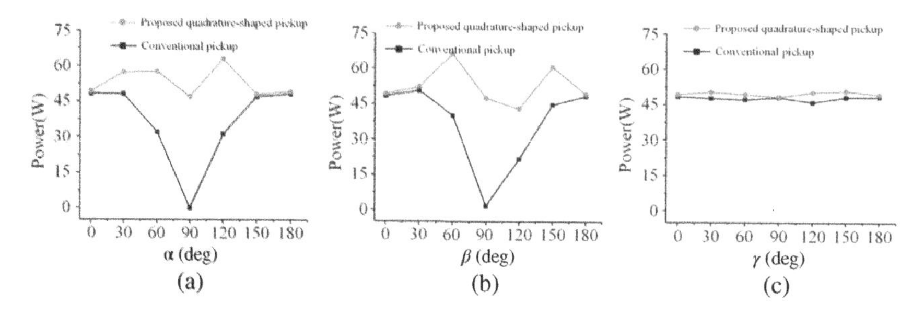

Figure 8.30 Comparative analysis: (a) roll, (b) yaw, and (c) pitch.

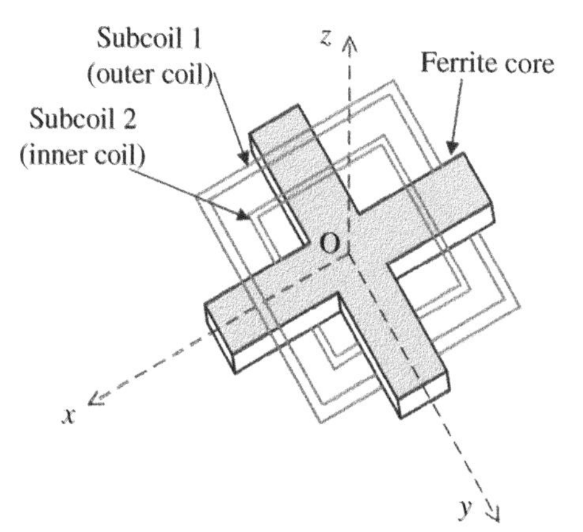

Figure 8.31 Configuration of the proposed pickup coil with minimum number of coils for omnidirectional wireless power transfer.

To illustrate the ability of the proposed pickup coil to receive power in an arbitrary omnidirectional magnetic field, it is assumed that there is only an outer coil as shown in Figure 8.31. It can be found that one side of the ferrite core is above the coil and the other side is below the coil. If the magnetic field is going in the positive direction of the z-axis, the flux is going through the ferrite core from beside the ferrite core toward the negative direction of the x-axis and the positive direction of the y-axis. A similar conclusion is reached when the magnetic field points in the negative direction of the z-axis. In addition, the flux can flow through the core wound by the coil when the magnetic field is parallel to the x- or the y-axis. To illustrate the situation further, an example of a magnetic field in the y-axis direction is shown in Figure 8.32a.

However, in the case of a single coil with only the outer coil, there will still be some "dead zones" that cannot accept the power, similar as the magnetic flux through the coil makes the effective magnetic flux zero. Figure 8.32b shows one example of the "dead zone" when the direction of the magnetic field is parallel to $y = x$. Under the circumstances, the flux B_x passing through the top to the bottom

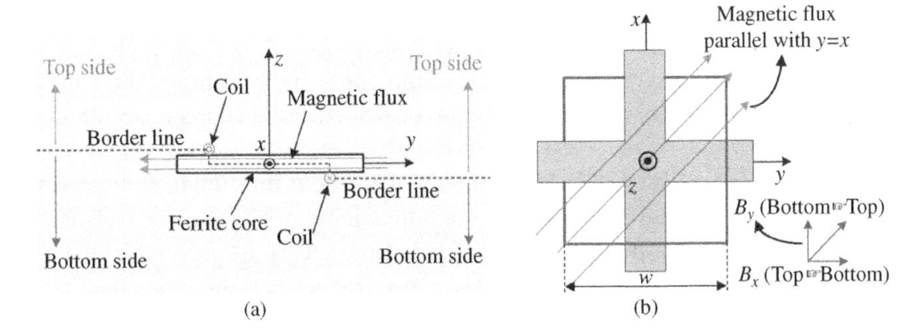

Figure 8.32 Diagrams indicating how induced voltage is generated or canceled with various types of magnetic flux: (a) magnetic flux parallel with y-axis intersecting the outer coil and (b) example of the first dead zone canceling the induced voltage.

of the core is identical with the flux B_y passing through the bottom to the top of the core. Accordingly, the induced voltage produced by the coil will be canceled out. To eliminate the "dead zone," another conventional coil is wound around the ferrite core, namely the inner coil of the proposed structure, as shown in Figure 8.31. It can be seen that the inner coil is rotated 90° clockwise relative to the outer coil, and the inner coil is obviously smaller than the outer coil. With this structure, the magnetic flux can pass through the inner coil when the magnetic field is parallel to $y = x$. Besides, since the areas of the inner and the outer coils are different, and the induced voltage generated is not equal, the power can be accepted with the pickup coil in any direction. In addition, the induced voltage is amplified by means of the compensating network around the pickup coil, and the omnidirectional WPT system is guaranteed by improving the coil quality factor.

8.5 Load Detection

Load detection in the WPT system is critical to establishing a stable and efficient wireless power source of high quality for WPT applications, such as the portable home electronics. An effective load detection method, namely the transient load detection model, is proposed, which can detect the load state with the energy injection and free resonance modes [17]. The transient load detection method rather than the steady-state method is needed because the load information must be obtained by detecting the load conditions before different rated power transmission modes. During the load detection, the energy injection mode is in the process of providing energy, which prepares the free resonance mode to detect the load condition. Besides, a smart detection scheme to identify the position and direction of the pickup coil based on the input power information is proposed to focus power transfer toward targeted loads [18]. Meanwhile, the total efficiency of the WPT system is further improved with the proposed maximum efficiency point-tracking scheme. The experimental results confirm the accuracy of the load detection scheme where multiple loads can be detected and charged simultaneously.

In addition, in order to avoid a large amount of magnetic flux flowing into places without load, a control technique of omnidirectional WPT, namely weighted time-sharing power control, is proposed, which first detects the power demand in the 3D space and then directs the wireless power to focus on the target load [7]. It is worth noting that this process is applicable to any WPT system with three orthogonal transmitter coils and multiple pickups with coil resonators.

The weighted time sharing of power flow control consists of the following steps:

(1) In the first step, the resultant magnetic field vector can be scanned across the entire sphere by controlling the circuit current of the transmitting coil as shown in Figure 8.15a. It can be assumed that the magnetic field vectors at N points on the surface of the sphere represent the entire sphere. After scanning, the amplitude and phase of the drive current and voltage of each vector are recorded, and the input power vector can also be determined. According to the information contained in the power vector, the power demand at each point can also be known.
(2) The second step is to concentrate on the power flow to the place where the power demand is higher, namely the position of the load, on the basis of the power demand obtained in the last step. It is the core content of weighted time-sharing power transmission, which does not require power to be transferred to every location, because there is virtually no power demand at many locations.
(3) The third step is to carry out a local scan of the 3D space to monitor whether there is a big change in the power transmission and the receiving at the load end. It is worth noting that in order to allocate more time to the system for power transmission, the local scan only takes less time. In other words, the local scan only needs to scan a part of the surface of the sphere, and multiple local scans can cover the whole sphere.
(4) In addition, if it is detected that the load demand is close to zero at N positions in the whole space during the working process, then the circuit of the transmitting coil will go into a dormant state. Furthermore, in order to detect the addition of load, the first step of omnidirectional scanning can be activated on time. If a load is found, the power flow control process will go from a dormant state to work.

This load detection and control approach based on weighted time-sharing scheme improves the efficiency of power distribution by means of directing power to the appropriate location, and it can be applied not only to single load but also to multiple loads. With further research, the ideal distribution of the induced magnetic field in an omnidirectional WPT system is not a full coverage of the 3D space but should be purposefully concentrated on the load, no matter where the load is in the space.

8.6 Discussion

This chapter introduces the realization of omnidirectional WPT system, which is of great significance to the application of WPT. Firstly, in view of the common 2D and 3D omnidirectional WPT systems, how to realize the energy transmission with orthogonal transmitting coil and orthogonal pickup coil is theoretically analyzed. The omnidirectional magnetic field vector is generated by current control, and the load current, power, and efficiency can be calculated. It is found that the total input power and output power follow the Lemniscates of Bernoulli. Then, the design structure of various transmitting coils is introduced to realize the omnidirectional magnetic field distribution, such as the structure of three mutually perpendicular transmitting coils based on current control and the cross-dipole coil which can generate a DQ rotating magnetic field with six degrees of freedom. The simulated and experimental verifications are both given to illustrate the capability of omnidirectional power transmission. In addition, in the structural design of the pickup coil, a quadrature-shaped pickup coil is proposed to solve the problem of angle misalignment in 3D wireless charging technology, which is verified to be feasible by simulation and experimental results. Another plane-type pickup coil with minimum number of coils is proposed to guarantee the power reception at any angle, and the superiority of the structure is verified by simulation model and experimental equipment. Furthermore, in order to realize the directional transmission of power flow, which can improve the efficiency of the system, a method of load detection is introduced.

In addition, for the omnidirectional WPT systems, space vector control and optimal power control need to be discussed for future development. As aforementioned, the space vector control is challengeable, which needs to adjust the magnetic field direction dynamically with the load moving in the process of charging. An efficient omnidirectional WPT system should allocate space vectors according to different positions of multiple loads. How to design the control strategy of space vectors will be the key point of future research. Besides, optimal power control is another key technical challenge for the future development of omnidirectional WPT. In the face of various problems in omnidirectional WPT, such as angle misalignment and position misalignment, it is crucial to realize optimal power control at the arbitrary positions. Although there have been many studies that have addressed different problems by designing the structure of the transmitting coil and the pickup coil to improve the power transmission efficiency, the power transmission is still influenced by circuit control and unknown perturbation. Then, how to control the optimal power transmission is a hot research topic.

References

1 Chabalko, M.J. and Sample, A.P. (2015). Three-dimensional charging via multimode resonant cavity enabled wireless power transfer. *IEEE Transactions on Power Electronics* 30 (11): 6163–6173.

2 Feng, J., Li, Q., Lee, F.C., and Fu, M. (2019). Transmitter coils design for free-positioning omnidirectional wireless power transfer system. *IEEE Transactions on Industrial Informatics* 15 (8): 4656–4664.

3 Han, W., Chau, K.T., Jiang, C., and Lam, W.H. (2019). Design and analysis of quasi-omnidirectional dynamic wireless power transfer for fly-and-charge. *IEEE Transactions on Magnetics* 55 (7): 1–9.

4 Zhang, Z. and Zhang, B. (2020). Angular-misalignment insensitive omnidirectional wireless power transfer. *IEEE Transactions on Industrial Electronics* 64 (4): 2755–2764.

5 Ng, W.M., Zhang, C., Lin, D., and Hui, S.Y.R. (2014). Two-and three-dimensional omnidirectional wireless power transfer. *IEEE Transactions on Power Electronics* 29 (9): 4470–4474.

6 Ha-Van, N. and Seo, C. (2018). Analytical and experimental investigations of omnidirectional wireless power transfer using a cubic transmitter. *IEEE Transactions on Industrial Electronics* 65 (2): 1358–1366.

7 Zhang, C., Lin, D., and Hui, S.Y. (2016). Basic control principles of omnidirectional wireless power transfer. *IEEE Transactions on Power Electronics* 31 (7): 5215–5227.

8 Lin, D., Zhang, C., and Hui, S.Y.R. (2017). Mathematic analysis of omnidirectional wireless power transfer—part-I: two-dimensional systems. *IEEE Transactions on Power Electronics* 32 (1): 625–633.

9 Lin, D., Zhang, C., and Hui, S.Y.R. (2017). Mathematic analysis of omnidirectional wireless power transfer—part-II three-dimensional systems. *IEEE Transactions on Power Electronics* 32 (1): 613–624.

10 Dai, Z., Fang, Z., Huang, H., and Wang, J. (2018). Selective omnidirectional magnetic resonant coupling wireless power transfer with multiple-receiver system. *IEEE Access* 6: 19287–19294.

11 Babic, S. and Akyel, C. (2000). Improvement in calculation of the self-and mutual inductance of thin-wall solenoids and disk coils. *IEEE Transactions on Magnetics* 36 (4): 1970–1975.

12 Zhu, Q., Su, M., Sun, Y. et al. (2018). Field orientation based on current amplitude and phase angle control for wireless power transfer. *IEEE Transactions on Industrial Electronics* 65 (6): 4758–4770.

13 Feng, T., Zuo, Z., Sun, Y. et al. (2022). A reticulated planar transmitter using a 3-D rotating magnetic field for free-positioning omnidirectional wireless power transfer. *IEEE Transactions on Power Electronics* https://doi.org/10.1109/TPEL.2022.3155251.

14 Choi, B.H., Lee, E.S., Sohn, Y.H., and Jang, G.C. (2016). Six degrees of freedom mobile inductive power transfer by crossed dipole Tx and Rx coils. *IEEE Transactions on Power Electronics* 31 (4): 3252–3272.

15 Zhang, Z., Zhang, B., and Wang, J. (2018). Optimal design of quadrature-shaped pickup for omnidirectional wireless power transfer. *IEEE Transactions on Magnetics* 54 (11): 1–5.

16 Kim, J.H., Choi, B.G., Jeong, S.Y. et al. (2020). Plane-type receiving coil with minimum number of coils for omnidirectional wireless power transfer. *IEEE Transactions on Power Electronics* 35 (6): 6165–6174.

17 Wang, Z., Li, Y., Sun, Y. et al. (2013). Load detection model of voltage-fed inductive power transfer system. *IEEE Transactions on Power Electronics* 28 (11): 5233–5243.

18 Feng, J., Li, Q., and Lee, F.C. (2022). Load detection and power flow control algorithm for an omnidirectional wireless power transfer system. *IEEE Transactions on Industrial Electronics* 69 (2): 1422–1431.

Part IV

Application

9

WPT for High-power Application – Electric Vehicles

In this chapter, origination of WPT for electric vehicle (EV), static and dynamic wireless charging, and relevant regulations, are first introduced. Then, the working principle, typical prototypes and demonstration project, will be elaborated for the static and the dynamic wireless charging. After that, technical specifications and elimination methods for electromagnetic field reduction was presented. Finally, some key technologies will be addressed to further elaborate the EV wireless charging systems.

9.1 Introduction

9.1.1 Origination of WPT for EVs

The EV was invented dating back to the 1830s and commercialized before 1839 [1, 2]. In the early 1900s, however, large reserves of petroleum were discovered in Texas, California, and Oklahoma [3]. Particularly, Ford Motor Company developed the first assembly line in the automotive industry in 1913, which significantly reduces the cost of gasoline cars. As a consequence, EVs disappeared from the automotive stage gradually, while the internal combustion engine (ICE) vehicles dominated the market. Along with the increasing requirements and the development of transportation industry, the large consumption of gasoline leads to the energy crisis around the world. Furthermore, the global warming caused by the emission of the greenhouse gases as well as the air pollution has become increasingly concerned, especially in recent years. As a result, EVs are regaining the attentions from the industry and the academia. From 1999 to 2015, the total number of sold EVs is approximately 2.1 million [4]. Now, the EV has come back to the center of the stage.

The EVs can be classified into several categories such as the pure electric vehicles (PEVs) [5], the hybrid electric vehicles (HEVs) [6], and the plug-in hybrid electric vehicles (PHEVs) [7]. Each kind of EV has its own technical advantages,

Wireless Power Transfer: Principles and Applications, First Edition. Zhen Zhang and Hongliang Pang.

while there are still critical technical challenges ahead, especially for the energization. Specifically, the mainstream technology adopts the cable to connect the power supply and EVs. In such a way, the corresponding disadvantages are the manual handling, the cable loss, and the flexibility. More importantly, the manual plugging and unplugging of cables may result in electrical sparks. Besides, the weather with high humidity may cause short circuits, thus leading to safety hazards. In contrast to the conventional charging technology with cables, the wireless charging can energize EVs in a cordless way, which shows considerable convenience and safety. Technically, the EV wireless charging techniques can be achieved through two mainstream modes: the static charging and the dynamic charging. The static charging has advantages of its convenience, flexibility, and reliable security compared with the conventional cable charging, although the disadvantages of batteries still impede the development of EVs significantly, such as the bulky volume, the low energy capacity, the high cost, the long charging time, and the short life time [8]. In particular, the long charging time of EVs deteriorates the experience of end users compared with the IEC vehicles. In order to address this kind of problem, various quick charging techniques were proposed to decrease the charging time to 20 minutes [9, 10]. Nevertheless, it is still too long compared with the refueling process. Then, the concept of roadway-powered EV appears in our vision, which can realize the energization for EVs in a cordless way even though running on the road. Such a dynamic wireless charging system has been taken as an effective solution to address the above issues, where EVs can wirelessly acquire the energy from the power track mounted beneath the ground even in motion [11, 12]. It means that EVs are no long required to stay at the charging station. Hence, the WPT shows salient research significance and practical value for EVs by reducing the dependence on batteries and improving the user experiences. Meanwhile, we also realize that there remains a long way ahead to achieve the real large-scale commercialization for wireless charging of EVs.

9.1.2 Development of WPT for EVs

9.1.2.1 Static Wireless Charging

Literally, the static wireless charging means that the vehicle can be charged in a cordless way when it is parked at a specified position, so it is also called *park-and-charge*. It is similar to the gas station for ICE vehicles, but the most key difference is that the static wireless charging is more flexible, which means that the charging pad can be mounted in the parking lot itself rather than a special area such as the gas station. In other words, the park-and-charge can be offered in the supermarket, the office building, the underground of apartments, etc. Hence, the salient advantage has attracted increasing attentions in recent years. In the 1990s, PATH performed an experiment prototype of EV static wireless charging systems at a frequency of 400 Hz for 7.6-cm air gap with the efficiency of 60% [13]. In 1997–1998, the IPT Technology demonstrated an EV with wireless charging at Rotorua Geothermal Park in New Zealand [4]. In 2009, Showa Aircraft Company realized a 30-kW power and 14-cm air gap wireless charging system for EVs, which is operated at a frequency of 22 kHz with an efficiency of 92% [14]. In 2015, Qualcomm released the Qualcomm Halo Wireless Electric Vehicle Charging (WEVC) technology, which has been tested rigorously in the FIA Formula E race. Meanwhile, the official BMW i8 can be fully charged within one hour using the 7.2-kW Qualcomm Halo WEVC system. The corresponding charging efficiency reaches 90%, which is almost the same as that of cable charging [15]. Additionally, WiTricity also released wireless charging technologies at various power levels from 3.6 to 11 kW, which aims to meet the wide-range requirement of PHEVs and EVs [16]. Regarding higher power requirement of public transits and delivery vans, Momentum Dynamics presented wireless charging system with power from 50 to 450 kW [17]. Hence, the EV static wireless charging has obtained a significant progress in recent more than 30 years, whether in the area of transmitting power or the system efficiency. It has already become a strong competitor against traditional cable charging technologies, although there are still some issues ahead to be addressed, for example the misalignment, the detection of foreign objectives, and electromagnetic interference.

9.1.2.2 Dynamic Wireless Charging

The history of dynamic wireless charging for EVs began in 1976, when the 8-kW prototype was demonstrated by the Lawrence Berkley National Laboratory [18]. In 1979, the Santa Barbara Electric Bus project was set up by adopting a 4.3-m length of track and a 1-m width pickup as well as utilizing the switched capacitors for the regulation of the transmitted power [19]. The feasibility of EV dynamic wireless charging systems was demonstrated, but the shortcomings of high cost, large size, and low efficiency were also exposed. Accordingly, the prospective of commercialization was still unpromising if critical technical issues cannot be settled. In recent years, a great number of efforts have been made for improving the performance of the dynamic charging. As one of the most world-renowned research teams in this field, Korea Advanced Institute of Science and Technology (KAIST) worked on the optimizing design of the inverter, the transmitting/pickup coil, the power converter, and the core structure for EV dynamic wireless charging systems. They set up a demonstration project which realized the 100-kW transmitting power with an efficiency of 80% under 26-cm air gap [20]. Additionally, the Oak Ridge National Laboratory (ORNL) has been investigating the dynamic wireless charging for EVs since 2011 [21]. In 2019, ORNL implemented and tested the wireless charging for Toyota RAV4 vehicle, which realized the 14-kW transmitting power over a 16-cm magnetic air gap. The corresponding dc-to-dc efficiency can achieve 95.16% [22]. Besides, Bombardier PRIMOVE developed a 250-kW EV dynamic charging system in Augsburg, Germany. The move-and-charge, namely the dynamic wireless charging, is anticipated.

9.1.3 Regulations

Along with the increasing technical maturity and application promotion, the focus of attentions on EV wireless charging systems gradually expands to the standards or regulations, which is the right way to the technical development and the wide-range commercialization. This section summarizes the mainstream international organizations as well as the related works on the wireless charging of EVs, which aims to unify the technical specifications as well as the mitigation of electromagnetic radiation [23].

9.1.3.1 IEC

As the earliest international electrotechnical standardization organization in the world, the International Electrotechnical Commission (IEC) was founded in 1906. Its responsibility is to deal with the international standardization in the field of electrical and electronic engineering. IEC Technical Committee devotes to the publication on electrical power transfer systems for electrically propelled road vehicles and rechargeable energy storage system for industrial trucks. The business involves the conductive power transfer, inductive power transfer, and battery swap [24]. The Technical Committee issued the standard IEC 61980-1 *Electric Vehicle Wireless Power Transfer Systems – Part 1*, which specifies the general requirement, classification, installation, interoperability, orientation, measurement convention of the parking space, offset, the air gap, etc. Regarding specific requirements of EV wireless charging systems, the standard covers the leakage-touch current, insulation resistance, dielectric withstand voltage, overload protection, and the short circuit withstand. From the perspective of constructional requirements, this standard gives detailed regulations for the power cable, breaking capacity of switching devices, and the strength of materials and parts. Lastly, the electromagnetic compatibility, ambient air temperature, ambient humidity, and the air pressure are required additionally. Besides, the publication of IEC TS 61980-2 *Electric Vehicle Wireless Power Transfer Systems – Part 2* is now in progress, which specifies the requirements of the communication between EVs and infrastructure as well as the communication between EVs and WPT systems [25]. In addition, the specification of the magnetic field is also considered for EV wireless charging systems by IEC TS 61980-3 *Electric Vehicle Wireless Power Transfer Systems – Part 3* [26].

9.1.3.2 SAE

Society of Automotive Engineers (SAE) is a technical society which falls into the category of materials and manufacturing of cars, trucks and engineering vehicles, aircrafts, and engines. The standards formulated by SAE are authoritative and widely used not only in automobile industry but also in other industries. As the first SAE standard for EV wireless charging applications, SAE TIR J2954 was drafted in 2011 and eventually revised in 2017 and 2019 [27]. Companies involved in J2954 consist of BMW, Chrysler, Ford, GM, Honda, Mitsubishi, Honda, Toyota, etc. The main features of J2954 are listed as follows:

- WPT power-level classifications and minimum efficiency at nominal alignment or at offset position on each level
- Positioning tolerance for the EV and the charging unit
- Frequency (nominal frequency is 85 kHz, and the frequency range is 79–90 kHz)
- Testing requirements (test stand configuration, environment, efficiency, safety, foreign object detection (FOD), and live object protection)
- Wireless communication and software
- Interoperability
- Electromagnetic compatibility
- Electromagnetic fields (EMFs) exposure to humans and implanted medical devices
- Thermal considerations.

In addition to the aforementioned aspects, SAE also issued other standards to specify specific requirements for EV wireless charging systems, for example J2847-6 *Communication Between Wireless Charged Vehicles and Wireless EV Chargers* [28] and *J2931-6 Signaling Communication for Wirelessly Charged Electric Vehicles* [29].

9.1.3.3 Other Works

In Japan, the Broad Wireless Forum (BWF) was established in 2009. The Wireless Power Transmission Working Group (WPT–WG) was formed within BWF to discuss about the frequency band of WPT, facilitate the standardization publication, establish guidelines for protection of human immunity, and develop other matters with regard to WPT technologies. There is a technical team in the Japan Automobile Research Institute (JARI), namely Inductive Wireless Charging Subworking Group (SWG), which also devotes to the EV wireless charging technologies [30].

In China, the launching ceremony for the establishment of EV wireless charging technology standards was held in 2016, which was hosted by the China Electric Power Research Institute. Participants included scientific research institutions, charging facility manufacturers, automobile companies, and charging service operation companies. Since then, China has been working on the standards for EV wireless charging technologies. These standards involve various areas, including the system architecture, the device (technical requirements and device requirements), the interface (communication and interoperability), the testing (safety test and interoperability), and the installation (charging station) [31].

9.2 EV Wireless Charging

9.2.1 Introduction

Applying WPT technologies into EVs, namely EV wireless charging, is an ideal technical solution to address the problems aforementioned (e.g. cable loss, manual handling, electrical sparks, and short circuits). From the perspective of implementation, the EV wireless charging system can be classified into static wireless charging and dynamic wireless charging, which can energize the vehicle either parking or moving, respectively. In this section, the working principle, architecture, demonstration project, and the product will be introduced for the static and the dynamic wireless charging for EVs.

9.2.2 Static Wireless Charging

9.2.2.1 Introduction

As depicted in Figure 9.1, the EV static wireless charging system consists of the primary and the secondary sides. The primary side is composed of the rectifier, the inverter, the primary compensation network, and the transmitting coils, while the secondary side owns the pickup coils, the secondary compensation network, the rectifier, and the DC–DC converter. The rectifier converts the utility power source of 50/60 Hz to DC power, which is then modulated to the high-frequency AC power. The compensation network is used to match with the transmitting coil for ensuring the resonant state of the primary side at the frequency of the AC power. Then, the AC current of the transmission coil induces

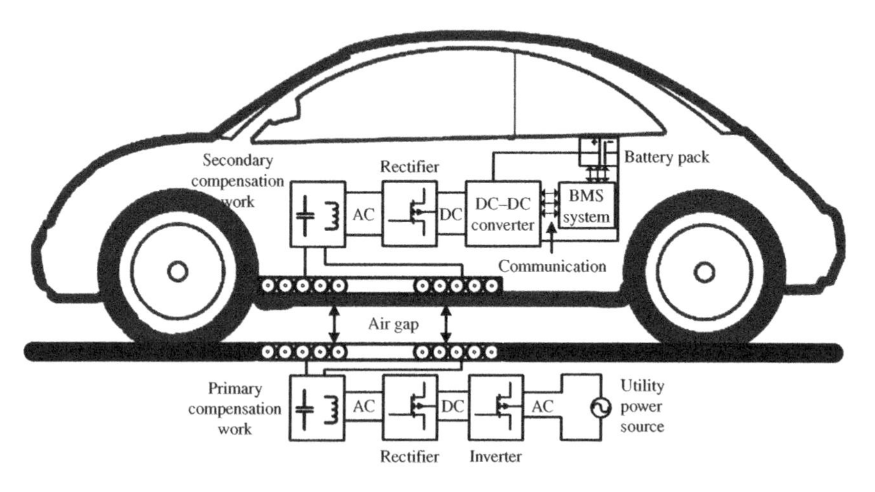

Figure 9.1 EV static wireless charging system.

a time-varying EMF. According to the Faraday's law of electromagnetic induction, the change in magnetic flux produces the induced electromotive voltage. It means that through the magnetic coupling between the transmitting and the pickup coils, the electric energy can be transmitted wirelessly from the primary side to the secondary side. In such a way, the battery pack of EVs can be charged by the power wirelessly transmitted from the charging pad. Practically, there is a battery management system (BMS) existing between the converted DC power and the battery pack, which is used to monitor the state of batteries and control the process of charging/discharging.

As another technical solution to realize the wireless charging for EVs, the capacitive WPT adopts a high-frequency AC electric field to transfer the power without direct electrical connection. By comparing with inductive WPT as aforementioned, the capacitive WPT system shows salient and unique advantages. For example, the transmission performance of capacitive WPT systems is not affected by metal objects, while the efficiency of inductive WPT systems is deteriorated due to the eddy current loss caused by the foreign metal involved in the induced magnetic field. Similar to the inductive WPT, the resonance also becomes the most concerned issue for capacitive WPT systems. Specifically, the nonresonant capacitive WPT system requires the high capacitance at tens or hundreds of nF. Since the coupling capacitors work as a power storage component, it can only realize the transmission distance of less than 1 mm at the nonresonant state [3]. When the capacitive WPT system works at the resonant state, the situation is different, that is the value of coupling capacitors is significantly reduced as long as the operating frequency is high enough.

Accordingly, there are increasing research institutions to work on capacitive WPT technologies for EV wireless charging, aiming to improving the transmission performance and the misalignment tolerance. In [32], for example a capacitive WPT system is proposed to reduce the voltage stress on the coupling capacitors by adopting a double-sided *LCLC* compensation network. In order to verify the feasibility, a 2.4-kW capacitive WPT system is designed using four $610 \times 610\text{-mm}^2$ copper plates. The experimental prototype can realize 150-mm transmission distance with a dc–dc efficiency of 90.8% at the output power of 2.4 kW. To improve the misalignment tolerance of capacitive WPT systems, an *LC–CLC* compensation topology is presented in [33], where the characteristic of the output power is given by taking into account the variation of the coupler capacitance caused by the misalignment of the coupler. Meanwhile a 1-kW capacitive WPT system with an air gap of 150 mm is set up to validate the proposed compensation topology. The measured results show that it can effectively stabilize the power output within the fluctuation range of 15%. More interestingly, a foam-based charging pad is present in capacitive WPT technologies for EV static wireless charging [34]. When an EV is parked within the charging area, the weight of the vehicle presses on the surface of the foam-based charging pad, which results in decreasing the air gap, and thus strengthening the capacitive coupling effect. An experimental parking station was built to charge the 156-V battery pack of a Corbin Sparrow EV, where the efficiency reached 90% at 1-kW output power.

9.2.2.2 Typical Prototypes and Demonstration Projects

In recent years, a great number of efforts have been made to verify the feasibility of the static wireless charging for EVs. Accordingly, this section discusses about the typical prototypes and demonstration projects of EV static wireless charging systems, with emphasis on the system architecture, the power conversion technologies, and the control strategies.

9.2.2.2.1 Wireless Charging of Toyota RAV4 The ORNL is one of the most world-renowned research institutes that devotes to the inductive WPT for EVs. In 2016, the ORNL implemented a static wireless charging system for a Toyota RAV4, which can realize the maximum output power of 12 kW, the air gap of 16 cm, and the dc–dc efficiency of 95.14% [35].

In the primary side, the system adopted a single-phase H-bridge active boost rectifier to convert the 50/60-Hz AC mains voltage to the DC and then fed to the inverter. The most important difference from conventional designs is to utilize a high-frequency transformer for the isolation between the inverter and the primary circuit, as shown in Figure 9.2. What's more, ORNL adjusts the turns ratio of the transformer to regulate the maximum rate power. The step-down characteristic of the transformer allows the inverter to operate at higher voltages and

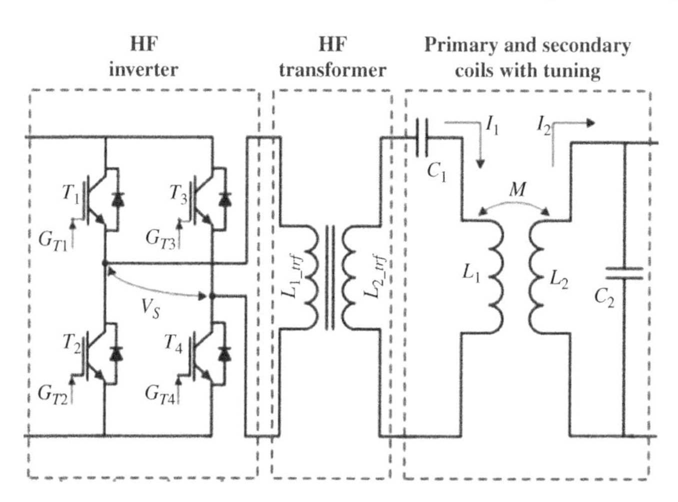

Figure 9.2 Transformer between the inverter and the primary circuit. Source: Onar et al. [35].

Figure 9.3 Experimental prototype. Source: Adapted from Onar et al. [35].

lower currents. In such a way, the conduction loss of the inverter can be reduced, and thus the transmitted power can be increased. In other words, the battery pack receives the power at higher level, while keeping the input voltage of the inverter under the limit.

In the secondary side, the state of charge (SOC) of batteries is nonconstant. Since the battery is charged in the constant current (CC) mode when the SOC is in the range of 10–90%, a CC control scheme was proposed and implemented, where the measured battery current is wirelessly fed back to the primary side via the communication. Then, the output voltage of the rectifier is regulated to ensure the CC mode for the battery charging. Using the proposed scheme, a 20-kW platform with dc–dc efficiency of 95% was set up and applied for Toyota RAV4, as shown in Figure 9.3 [36].

9.2.2.2.2 Qualcomm Halo Wireless Charging In 2011, Qualcomm acquired Halo IPT as well as its WPT technologies and relevant patents and then devoted to wireless charging technologies for EVs. In 2012, Delta Motorsport signed an agreement with Qualcomm, planning to integrate the Qualcomm Halo wireless charging system into its Delta E-4 coupe EVs. Then, Qualcomm became an official technology

Figure 9.4 Qualcomm Halo wireless charging system.

partner of the FIA Formula E Championship, which is an international championship featuring racing cars powered by electrical energy. In the second season of 2014/2015 FIA Formula E Championship, the Qualcomm Halo wireless charging technology was adopted into safety cars, which is depicted in Figure 9.4. In 2015, the wireless charging technique reached an output power of 7.2 kW, which can fully charge the BMW i8 safety cars within one hour [37]. However, it has not been commercialized on a large scale till now. In 2019, an acquisition of Qualcomm's WPT technologies was announced by WiTricity [38], which resulted in not only the end of the WPT story in Qualcomm but also the birth of a well-deserved overlord in the field of WPT technologies around the world, that is WiTricity.

9.2.2.2.3 WiTricity's Wireless Charging WiTricity was founded in 2007 and devoted to commercializing the WPT technologies based on the magnetic resonance theory, which was proposed by a Massachusetts Institute of Technology (MIT) research team. In [39], the demonstration can illuminate a 60-W light bulb from the power source over 2-m away. Since then, WiTricity is developing core WPT technologies for the EV wireless charging as well as other applications. As of 2019, WiTricity has possessed over 900 granted patents, which involves the architecture, design, performance enhancement, control, and typical applications of WPT technologies. In the aspect of standards and specifications, WiTricity is an active leader of SAE J2954 and involved as a member of the International Organization for Standardization (ISO), IEC, AirFuel Alliance, and China Automotive Technology and Research Center. Undoubtedly, WiTricity is one of the WPT leaders around the world.

Figure 9.5 WiTricity wireless charging system.

As one of the flag products developed by WiTricity, the DRIVE 11 evaluation system is an end-to-end reference design for static EV/HEV wireless charging systems, which can deliver the power of 11 kW with an efficiency of 94%. Figure 9.5 depicts the DRIVE 11 system. The primary side of DRIVE 11 is capable of charging vehicles based on the industry power standards of 3.6, 7.7, and 11 kW. The secondary side can energize vehicles involving passenger vehicles, sports utility vehicles (SUVs), and light trucks over the low (10–15 cm), mid (14–21 cm), and high (17–25 cm) air gap. In addition, DRIVE 11 adopts various auxiliary techniques to deal with abnormal situations caused by foreign objectives, including the FOD and the live object detection (LOD). In particular, the position detection sensor is utilized for DRIVE 11, which can be tolerant to parking misalignment of 10 cm (side to side) and 7.5 cm (front to back) [16].

9.2.2.2.4 Plugless Wireless Charging In 2012, Plugless announced the Apollo program to test wireless charging systems for EVs, with executive partners including Google, Hertz Rent-a-Car, Duke Energy, Los Angeles Department of Water and Power, and US National Laboratories. By 2016, the Plugless charging system was successively sold for public use with the Nissan LEAF, Chevrolet Volt, BMW i3, and Tesla Model S, which are all the milestones of the Plugless commercialization. As of February 2017, customers from 46 states in the United States and four provinces in Canada have completed 900,000 hours tests of wireless charging by adopting Plugless wireless charging systems, which undoubtedly raised the upsurge of wireless charging for EVs.

Figure 9.6 Plugless wireless charging system.

As depicted in Figure 9.6, the Plugless wireless charging system consists of three components, including the vehicle adapter, the parking pad, and the control panel. The vehicle adapter weighting approximate 12 kg is mounted underneath the vehicle, while the parking pad is mounted on the garage floor. The control panel supplies power to the parking pad and provides the alignment guidance as well as the charging status. The whole system is capable of achieving continuous output power at 3.3–7.2 kW within 10-cm air gap [40]. Besides, the FOD is also involved in Plugless EV wireless charging systems.

9.2.2.2.5 Wireless Charging at Momentum Dynamics As aforementioned, there are a great number of enterprises devoted to the research and development of static wireless charging technologies for EVs, where most of their products work at the level of several kilowatts. However, the concept of EV involves a wide range of electric-driven vehicles, including not only small-size passenger cars but also electric buses, vans, or trucks. It means that the wirelessly transmitted power of several kilowatts is not enough for the public transit and the cargo freight. As a consequence, it is vital to expand the output power level of EV wireless charging systems to tens of kilowatts or even several hundred kilowatts. Momentum Dynamics is a technology provider and a market leader of high-power wireless charging systems

Figure 9.7 Momentum dynamics wireless charging system.

for the automotive and transportation industries. Its product can realize output power up to 450 kW for charging large-size EVs. For buses, Momentum Dynamics' products can output power within 50–450 kW to realize the quick energization at the bus stops. In such a way, the mileage of vehicles can be extended without extra stop. For vans, the wireless charging system can offer the charging power level of 50–150 kW, which can ensure the charging process conducted during vehicles parked in the loading dock. To sum up, the high charging power charges vehicles at a short parking duration to guarantee vehicles more mileage, as shown in Figure 9.7.

In 2018, Link Transit, a public transportation provider in central Washington, commissioned a 200-kW wireless charging system for a BYD K9S electric bus from Momentum Dynamics. This marked the first wireless charging system in North America to operate at the 200-kW power level, which is a breakthrough in high-power wireless charging history. In the same year, Momentum Dynamics commissioned another 200-kW wireless charging system to power electric transit buses in Chattanooga, Tennessee, which is mounted inside the Chattanooga Area Regional Transportation Authority (CARTA) Shuttle Park. It was the second implementation at the 200-kW power level in North America. Soon after, the Vineyard Transit Authority (VTA) also purchased the high-power wireless charging system from Momentum Dynamics, which is composed of three 200-kW subsystems mounted at the Church Street Visitors Center in Edgartown, Massachusetts. Recently, Momentum Dynamics announced that it plans to equip three 300-kW wireless charging systems in Indianapolis [41]. Apparently, Momentum Dynamics is dominating the market of high-power wireless charging for EVs around the world.

9.2.3 Dynamic Wireless Charging

9.2.3.1 Introduction

The static wireless charging addresses the technical issues of EVs, such as the cable loss, manual handling, and electric sparks. However, such a charging mechanism determines that only the battery can offer the power after EVs are fully charged, which means the mileage of EVs is still limited by the energy capacity of the batter pack. In order to fundamentally reduce the dependence on the battery, a novel charging scheme, namely the dynamic wireless charging, is proposed for EVs, by which the vehicle can be energized when moving on the road. The reduced requirement of batteries cannot only save the space and the weight of vehicles but also reduce the cost of end users, thus increasing the market share and the penetration of EVs. Besides, the dynamic wireless charging can really take the advantage of wireless, that is, charging while driving, which is the ultimate goal of the combining EVs and WPT technologies.

Figures 9.8 and 9.9 show the basic structure of EV dynamic wireless charging systems with different track designs, including the lumped track and the stretched track. The overall system architecture seems similar as that of static wireless charging systems, whereas it is remarkably different from the point of view of the component level, such as the design of the transmitter, the operation and control modes, and the external disturbances.

9.2.3.1.1 Primary Side In order to transfer the power along the route of the vehicles, the primary side beneath the ground should be formed as a power track rather than the conventional single-transmitting-coil design used in EV static wireless

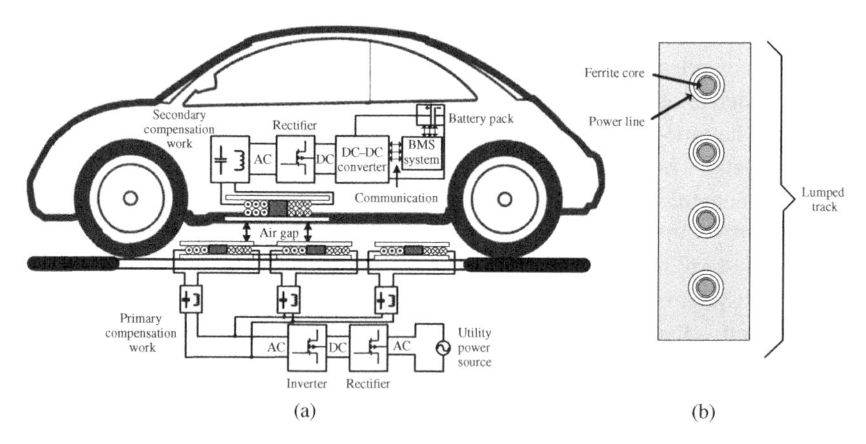

Figure 9.8 EV dynamic wireless charging system based on lumped track: (a) overall structure and (b) top view.

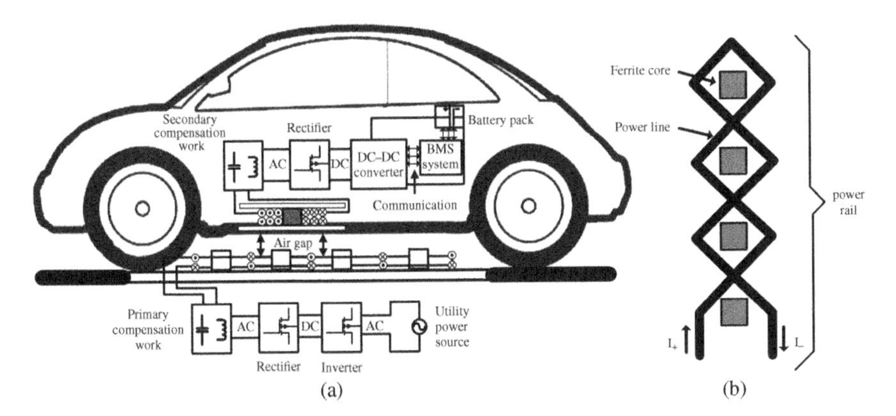

Figure 9.9 EV dynamic wireless charging system based on stretched track: (a) overall structure and (b) top view.

charging systems. There are two mainstream types of power tracks developed for EV dynamic wireless charging systems: lumped tracks and stretched tracks.

A lumped track is composed of a group of transmitting coils with approximately the same size as the pickup coil. This type is initiated by a research group at The University of Auckland. In such a structure, each transmitting coil can be separately excited based on the vehicle position. The lumped track requires a large number of compensation components and power electronic converters, which inevitably causes greater investment. Hence, several transmitter coils can be connected in series or parallel and then fed by the same power electronic converter so as to reduce the system cost [42]. On the other hand, the magnetic field between the two adjacent transmitting coils is weak and thus results in a power valley, where the performance of the EV dynamic wireless charging systems is deteriorated significantly. As a result, it should particularly deal with the deployment of the group of transmitting coils [43, 44].

As another technical solution, the stretched track is constituted by long-distance power supply rails, which are much longer than the pickup coil. Hence, it can power multiple vehicles simultaneously [45]. The corresponding research is led by the research team at KAIST [20]. Besides, in another study [46], each charging area is formed by nine power supply rails with a total length of 90 m. Such nine charging areas are deployed along a 3460-m track, which can output the maximum power up to 100 kW. However, the drawback of the stretched track is the lower efficiency since the huge size causes more magnetic flux leakage. Consequently, the corresponding optimization focuses on the optimal length, namely the segmentation, and the shape design of the power supply rails [47, 48].

9.2.3.1.2 Dynamic Variation of Parameters In EV dynamic wireless charging systems, the lateral and vertical misalignment might be considerable, which emerges more significant than the static wireless charging system. Besides, the difference in vehicle situations, such as different vehicle weights or loads, results in the changing of the air gap and thus affects the mutual inductance between the transmitting and the pickup coils. Then, the dynamic wireless charging system should possess an enhanced misalignment tolerance and an enhanced ability to deal with the non-resonance caused by the variation of parameters [49]. As a consequence, both the magnetic coupler and the compensation network require improvements for EV dynamic wireless charging systems.

For example, the optimal design of the magnetic core can reduce the impact of a slight misalignment on the coupling coefficient [50]. In [51], an I-type core is designed for dynamic wireless charging systems, which can effectively enhance the lateral misalignment tolerance. Besides, the asymmetric coil set which consists of a small pickup coil and a large transmitting coil can significantly increase the lateral misalignment tolerance [52]. More details about the power track design are discussed in Section 9.2.3.2.

9.2.3.1.3 Charging Object Detection In the static wireless charging system, the EV charging can be controlled easily by drivers or sensors. However, the dynamic wireless charging system with segmental power supply rails requires an additional process, that is when a vehicle moves from one power rail to another, the corresponding power source should be switched on promptly to ensure a continuous energy supply for the moving EVs [53, 54]. Simultaneously, the previous power supply should be switched off, waiting for another charging object. In order to achieve switching from one supply rail to another, the switching time of different power supplies can be determined through detecting the change of phase angle in the primary circuit [55]. Another way to address this problem is to add three detection coils, one lumped on the EVs and the other two installed beneath the ground. By detecting the variation in voltages in two coils underground, the primary power supply is able to energize the transmitting coil on time with the arrival of EVs [56].

9.2.3.2 Power Track

In order to achieve continuous power supply, the stretched power track is implemented and improved by various research teams around the world such as KAIST. Figure 9.10 shows the typical designs of power supply rails, where the research focus falls into the structure of ferrite cores to increase the coupling effect and the power transfer distance of the EV dynamic wireless charging systems.

The E-type power track is depicted in Figure 9.10a. The intersecting surface of the ferrite core is like the shape of a letter "E." Respectively, the corresponding

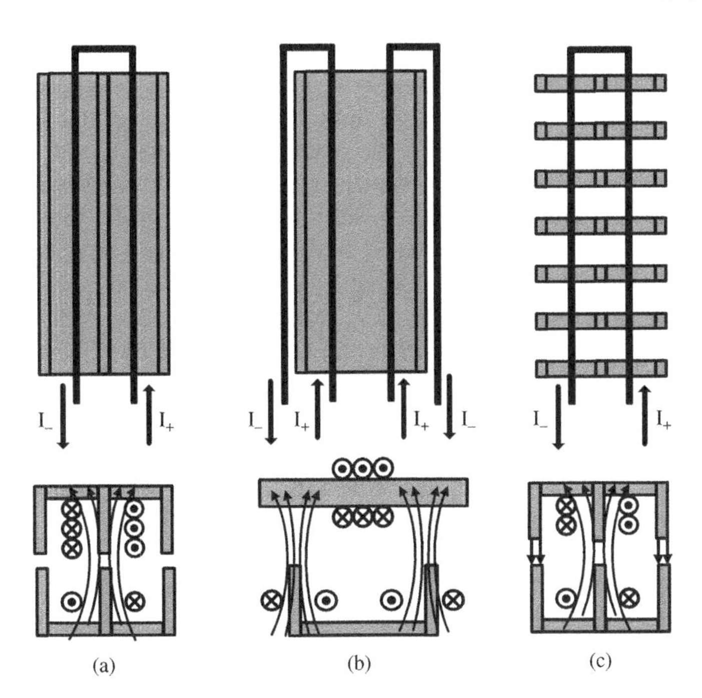

Figure 9.10 Top view and cross section of typical power track designs: (a) E-type; (b) U-type; and (c) W-type.

pickup coil is also designed as E-type. There are two power lines with opposite current directions underneath the road surface, which form a current loop. The current flowing in such power lines induces the magnetic fields, respectively, which are superimposed over the clearance between these two lines and thus forms a synthetic magnetic field in the air gap. Thus, the pickup coil can receive the power from the magnetic field induced by the two-line power rail. The prototype based on the E-type is demonstrated to reach the output power of 3 kW for each pickup coil with 1-cm air gap at an efficiency of 80%. However, the E-type structure is sensitive to lateral misalignment between the primary and the secondary sides [11].

In the case of small air gap and low misalignment tolerance, the KAIST developed the U-type structure as shown in Figure 9.10b. Differing from the E-type, the U-type structure adopts four power lines to form two current loops. The included magnetic flux is horizontal, so the pickup coils around ferrite core can clasp the magnetic flux which is parallel to the ground. It can effectively increase the air gap and desensitize the misalignment between the primary and the secondary sides. The U-type power track can energize each pickup coil at an output power of 6 kW. The whole system is designed for an electric bus mounted on 10 pickups, which can realize a total output power of 52 kW under 17-cm air gap. As a result of the parallel pickup design, the pole width of the primary ferrite core is smaller than that of the pickup coil, which aims to clasp more magnetic flux as the air gap increases. However, it should be noted that the upward magnetic leakage flux is more than that of the E-type, which results in the system efficiency of only 72%.

The U-type power track adopts four power lines to induce the horizontal magnetic flux, which inevitably increases the width of the power track; thus, the cost and size of the power track is increased. Additionally, the long-distance seamless ferrite core causes an increased magnetic resistance. To address these problems, the W-type power track is presented as shown in Figure 9.10c, which possesses W-shaped ferrite cores arranged in sets [57]. Compared with the U-type, it reduces the magnetic resistance by around three times [58]. Besides, using the narrower power track and the wider pickup ferrite core, it can realize transferring the high power with the large air gap. The measured output power is significantly increased to 15 kW for each pickup coil. Meanwhile, the power efficiency of the W-type-based WPT system is maintained at 71% with a 17-cm air gap.

9.2.3.3 Typical Demonstration Projects

The feasibility of the EV dynamic wireless charging systems has been verified by not only the theoretical analysis but also various demonstration projects, whereas there is still a long distance ahead regarding the commercialization due to its complexity, high cost, and lack of unified standards. Accordingly, this section introduces the typical demonstration projects of the EV dynamic wireless charging systems around the world.

9.2.3.3.1 The Great Effort of KAIST In 2009, the KAIST team started to work on the on-line electric vehicles (OLEVs), with emphasis on the design of the power track, the roadway construction methods, and the system optimization and control [59]. By now, KAIST has launched five generations of the OLEV. The first three generations were all developed in 2009, where the E-, the U-, and the W-type power tracks were designed and tested successively.

The first-generation OLEV is a small golf cart. In the case that the E-type structure is sensitive to lateral misalignment between the primary and the secondary sides, the golf cart is equipped with a controlled pickup coil which can align to the power supply rail automatically. This prototype was demonstrated to reach an output power of 3 kW for each pickup coil with 1-cm air gap at an efficiency of 80%. It verified the feasibility of the OLEV project and laid the first stone for further studies although it was just an experimental prototype operating on the accurately designed track within 3-mm lateral displacement.

A few months later, the second-generation OLEV was announced, which adopted the U-type power track to improve the air gap from 1 to 17 cm. As a consequence, the second-generation technology can be applied on an electric bus to meet the requirement of the practical applications. In view of the large battery capacity for electric bus, the secondary side was composed of 10 pickup coils, which can theoretically increase the total output power up to 60 kW. Since the width of the U-type power track with four power lines was 1.4 m, the pole width of the pickup coil was increased to 1.6 m to catch more magnetic flux, accordingly. In terms of the overall system, the second-generation OLEV operated on a 240-m route consisting of four 60-m long U-type power tracks with an efficiency of 72%. In contrast to the first-generation OLEV, it can maintain 50% of the maximum output in the case of a 23-cm lateral misalignment.

Aiming to reduce the magnetic resistance of the power track and meanwhile improve the maximum output power, the third-generation OLEV was implemented for a SUV, which adopted the W-type structure as aforementioned. This double-power-line structure can narrow down the width of the power track. In addition, the bone-like structure reduces the magnetic resistance three quarters smaller than the U-type. The overall efficiency can be maintained at 71% as the same level of the second-generation OLEV, whereas each pickup coil can increase the output power up to 15 kW, which is the most significant technical contribution of the third-generation products. In 2010, the third-generation OLEV was upgraded to promote the efficiency up to 83% under a 20-cm air gap by optimizing the rectifier, inverter, and on-board regulator. Two upgraded OLEV buses were deployed at the 2012 Yeosu EXPO, South Korea, and another two have been in full operation at the main campus of KAIST since 2012.

In the previous generations of OLEVs, the width of the power tracks is at tens of centimeter approximately, which causes not only the high construction cost but also the difficulty of deployment. Accordingly, the fourth-generation OLEV focuses on the practical application and adopts a 10-cm-width I-type power track. The optimized power track can not only reduce the width of the power track significantly but also increase the misalignment tolerance up to a lateral displacement of 24 cm. The maximum output power can achieve 25 kW under a 20-cm air gap. Another significant advantage of the fourth-generation OLEV is to reduce the EMF for the radiation safety. The corresponding discussion will be given in Section 9.3. As of 2015, the development of the fifth-generation OLEV is still in progress. The KAIST team has made a great number of efforts and contributions on the dynamic wireless charging for EVs.

9.2.3.3.2 The Prototype of ORNL The ORNL has been investigating the dynamic wireless charging system since 2011. Figure. 9.11 shows the development of the demonstration project funded by ORNL, which applies the WPT technology for EV dynamic charging.

Different from KAIST, ORNL focuses on the design of the lumped track consisting of multiple transmitting coils. The magnetic coupling coefficient fluctuates

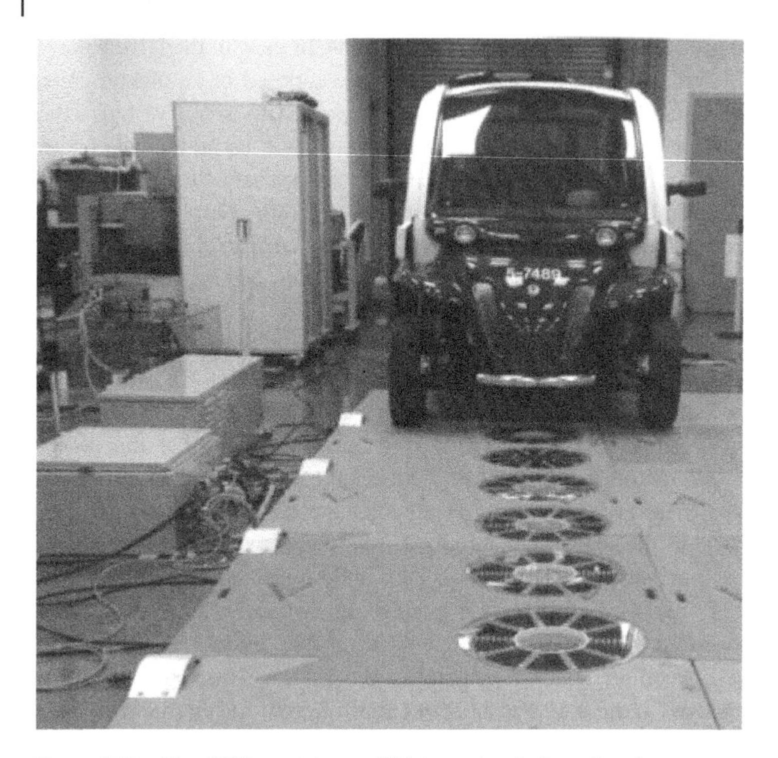

Figure 9.11 The ORNL prototype of EV dynamic wireless charging.

along the power track and reduces to an extremely low level around the adjacent transmitting coils, thus resulting in a power valley where the vehicle cannot acquire enough power. It means that the deployment of the group of transmitting coils should be particularly paid attention to. As shown in Figure 9.11, this prototype consists of six roadway embedded coils, where each two coils form a group. Three groups are energized in sequence, which is controlled by the trackside photocell interruption. The ultracapacitors from Maxwell Technologies are adopted to smooth the charging current for the EV battery pack. Moreover, the lithium-capacitor unit is parallelly connected to the grid, which effectively reduce the power pulsation by 84% [60].

In 2013, ORNL presented a novel dynamic wireless charging system based on a GEM EV, which operates at a frequency of 22 kHz and achieves a maximum power transfer of 2.2 kW with an efficiency over 70% [21]. It achieved in-vehicle power smoothing using ultracapacitors as well as grid side power smoothing using lithium capacitors as aforementioned. This prototype gives an ideal technical solution for lumped track.

9.2.3.3.3 Research at The University of Auckland In the 1990s, a research team at The University of Auckland proposed various wireless charging solutions. Meanwhile, this team is also a supporter of the lumped power track design. They proposed a solution for EV dynamic charging, that is as the EV travels along the road, only the transmitting coil which has the strongest coupling with the pickup coil can be energized. Evolved from the theory of single-coupled lumped track, the concept of double-coupled EV highway, which includes two stages of magnetic coupling, was developed in 2013 [61], where the buried primary side is composed of one power supply coil and several intermediary couplers. Each intermediary coupler is a two-port network, where the input and output ports are both coils. The bottom of each intermediary coupler couples with the power supply coil, which is the first stage of the magnetic coupling. The second stage of coupling happens on another side of each intermediary coupler and pickup coil mounted on the EVs. Thus, the power can be transferred from the power source to the EVs through several intermediary couplers. This system satisfies the regulations that specify the minimum depth of the mains conductors below the road surface. Besides, due to the existing first coupling stage, the buried intermediary couplers can be isolated from the mains supply and thus minimize the safety concerns. This prototype is capable of transferring 500 W within 2.5-ms reaction time under the air gap up to 250 mm. In the follow-up research, a 900-W average input and 5-kW peak output prototype of intermediary couplers have been constructed, which can achieve the maximum efficiency of the intermediary coupler at 92.5% [62].

In the field of the conventional lumped track adopting only transmitting coils without intermediary couplers, the team implemented a three-coil detection system to detect the approach of EVs, which is used to energize the corresponding transmitting coils in turn, where one of the three coils is mounted on the EV chassis around the pickup coil and is energized at a frequency of 432 kHz [63]. As EV approached, the detection voltage is induced in the other two detection coils mounted on the primary side. In the heading direction of the EVs, the two induced voltages undergo phase reversal. Then, the power supply can be energized and deenergized based on the position of the EVs.

Besides, this research team has made significant contributions to the design of coil pads including the square (or rectangular) quadrupole pad, the triangular DQ pad, and the bipolar pad [64–66]. The discussion about coil pads has been given in Chapter 3. In addition, multiple double-D pads are utilized to form the double coupled primary side mentioned above, while the bipolar pad is adopted for the secondary side. The demonstration project is capable of continuously delivering 15-kW power and allowing a lateral misalignment of ±200 mm [67].

9.2.3.3.4 Bombardier PRIMOVE Bombardier has been developing WPT technologies for the transportation since around 2010. Initially, the main applications

Figure 9.12 Bombardier prototype.

fall into the static wireless charging for public transportations such as trams and busses. Then, the dynamic wireless charging has also drawn attention by Bombardier; for example an 800-m power track is built for a tram in Augsburg, Germany, as shown in Figure 9.12 [68].

As a part of the "slide-in electric road project," Bombardier constructed the PRIMOVE highway track to test the functionality of the dynamic wireless charging systems, which is applicable for buses, trams, and trucks. The test track facility is 300 m in length, where four segments are built for the power supply. Meanwhile, the pickup coil is capable of receiving 200-kW output power. Besides, there exists a lifting device to raise the pickup at the locked position when the EV need not acquire the power from the PRIMOVE highway [69].

9.2.3.3.5 The FABRIC Project The project FABRIC is a large-scale integrated project, cofunded by the European Union's 7th Framework Programme implemented by 25 partner organizations from nine European countries. From 2014 to 2017, this project was conducted to address the technological feasibility, economic viability, and the socioenvironmental sustainability of dynamic on-road charging for EVs. It was implemented by two demonstration projects of dynamic EV wireless charging, one in France, and another in Italy. The test site in France, which is shown in Figure 9.13, is located in Satory, where the power level of 20–40 kW is tested on a 100-m track. The target of this project to achieve an

Figure 9.13 The FABRIC test site in Satory.

Figure 9.14 The FABRIC test site in Torino.

efficiency of 80% under the air gap within 12.5–17.5 cm, while the tolerance of misalignment is 20 cm. The Italian test site, which is shown in Figure 9.14, is located close to Torino, where the demonstration system is designed to realize the 20–100 kW power transfer with an efficiency of 70–80% under 25-cm air gap [70].

Figure 9.15 The VICTORIA dynamic wireless charging system.

9.2.3.3.6 The VICTORIA Project Endesa, the leading company in the Spanish electricity sector and the second-biggest operator in the electricity market in Portugal, devotes to developing the static and dynamic charging of EVs, which is also included as a part of the regional VICTORIA project. As shown in Figure 9.15, Endesa implements a 50-kW dynamic wireless charging system in a 100-m route section, which consists of eight transmitting coils with a 12.5-m interval between each coil. This demonstration project is a part of "Smart City Malaga" led by Endesa [71].

9.2.4 Market

Since 1990, General Motors took the lead in developing EVs, including PEVs, HEVs, and PHEVs, followed by Ford, BMW, Nissan, Honda, and Toyota. In 2015, the world's best-selling PEV was Nissan Leaf with the global sales of 200,000. The second and third were Chevy Volt with the sales of 104,000 and Tesla Model S with the sales of 100,000 [72], respectively. From the perspective of HEVs, more than 1.5 million HEVs were sold around the world in 2012. In particular, the sales of HEVs increased from 3% to 10% of the total Japanese automobile market, where Toyota occupied the majority of the EV market by selling 1.2 million HEVs, while other companies held the other market shares, such as 50,000 HEVs sold by Hyundai and 30,000 HEVs sold by Ford. From the perspective of EVs (including

PEVs, HEVs, and PHEVs), the corresponding global automobile market share is still a little slice in 2016, whereas its growth rate is about 100% each year, especially after entering the year of 2000 [3]. In addition, the International Energy Agency proposed a positive and ambitious roadmap to achieve widespread adoption and utilization of EVs and PHEVs worldwide by 2050. The large-scale application of EVs means not only the penetration of renewable energies into our daily life but also the reduction in the global CO_2 emission.

Accordingly, the EV wireless charging technology possesses a promising future with the rapid development of the EV market. In [73], an engineering system model is proposed to evaluate the economic feasibility and environmental impact of the transportation system adopting wireless charging technology, which reveals that a 20% penetration of the wireless charging technology in EVs can result in a 20% reduction in air pollution and 10% reduction in energy use. In order to further promote the development of dynamic wireless charging technologies, some researchers presented a large-scale study to thoroughly address the critical issues of dynamic wireless charging implementations. In [58], the research shows that the dynamic wireless charging can increase the probability of completing the whole trip for EVs from 75.3% to 97.7%. Besides, it can also reduce the total emissions of CO_2 from light-duty vehicles and trucks by 29.3 trillion kilograms in the first 50 years [74]. In South Korea, the analysis of the total cost is performed in Seoul by comparing the economic feasibility of the OLEV with other kinds of vehicles. The conclusion is that the OLEV is the cheapest solution among all types of vehicles. What's more, the total cost of the commercialization for OLEV, including the power track construction, the research investment, and the maintenance of power electronic devices, is up to $29,275 million, whereas the overall benefit can reach $393,900 million [58]. Recently, research institutions and organizations, especially for enterprises, have realized the market and investment value during the commercialization of EV wireless charging. The market prospect of the wireless charging techniques is obviously promising, though there are still implementation and construction issues to be addressed ahead. The EV wireless charging technologies including the static and the dynamic will realize the large-scale commercialization around the world along with gradually completing regulations of market admission and further reducing the investment cost.

9.2.5 Patent

As the core competitiveness of enterprises, patent is like a mirror to reflect the research and development capabilities of EV wireless charging technologies. This section discusses about the current enterprise-level research status via this mirror.

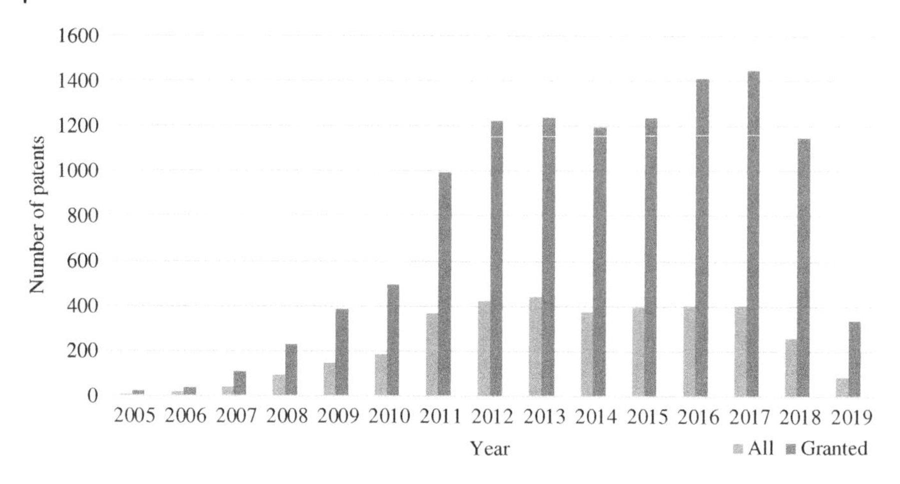

Figure 9.16 Statistics of patents related to wireless charging for EVs.

9.2.5.1 Previous Development

Searching on Google Patent, the summary of the global patents related to EV wireless charging techniques from 2005 to 2019 is shown in Figure 9.16. It can be found that the number of applications has increased gradually year by year since 2005, though the number of patent applications fluctuated during 2012–2015. The remarkable leap upward of the patent number happens in 2011; afterward, the application number per year stays in a relatively stable level of 1,000–1,500. Meanwhile, the number of granted patents is about 400 each year. In 2019, the number of applications and the granted patents both declined significantly, which may be caused by the delay in the information updation during preparing the manuscript in the early 2020. From the perspective of the general trend, obviously, the EV wireless charging technique has been obtaining attention in the area of intellectual property for 10 years since its explosive growth in 2010.

9.2.5.2 Patents from Enterprises

Figure 9.17 shows the share of patents with respect to different companies from 2005 to 2019. This section only counts granted patents instead of the total number of applications. It can be seen that the majority of patents related to EV wireless charging technologies are owned by automobile enterprises. In terms of the patent number, Qualcomm, WiTricity, and Toyota are the top three companies, which own about 75.3% share of the patent in total among the 11 companies listed in the figure. From the location of these companies, it shows that the EV wireless charging technologies are concentrated in a few counties, such as the United States, Japan, China, and South Korea.

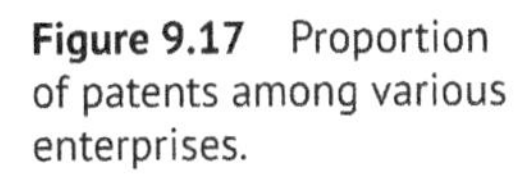

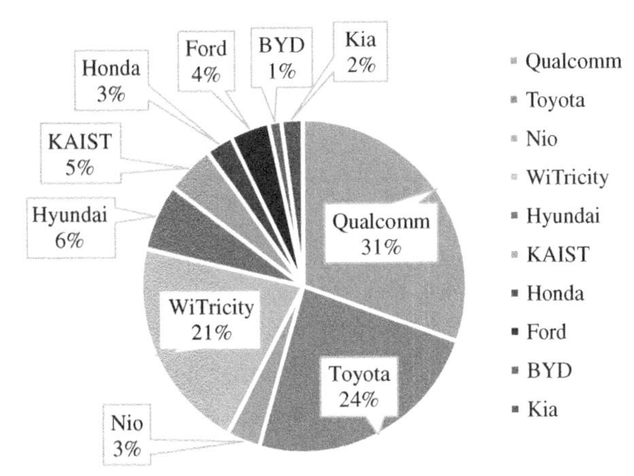

Figure 9.17 Proportion of patents among various enterprises.

9.3 Electromagnetic Field Reduction

The radiation of the EMF has already been one of the major pollutions after air pollution, water pollution, and noise pollution. As well known, the human body is a conductor of electricity, which means that the human tissues are inevitably affected by the surrounding electromagnetic environment. In order to ensure public health, the limits of the electric and the magnetic fields are necessary to be specified for various applications at different frequency bands, of course, including the EV wireless charging system. This section introduces the related standards and technical specifications and summarizes the elimination methods for the EMF radiation.

9.3.1 Standard

9.3.1.1 ICNIRP

International Commission on Non-Ionizing Radiation Protection (ICNIRP) aims to protect people and the environment against adverse effects of nonionizing radiation (NIR). To this end, ICNIRP develops and issues the science-based advice on limiting exposure to NIR. In 2010, *ICNIRP guidelines for limiting exposure to time-varying electric and magnetic fields (1 Hz to 100 kHz)* was published for the protection of humans exposed to electric and magnetic fields in the low-frequency range of the electromagnetic spectrum. It implies that human and animal bodies can significantly perturb the spatial distribution of a low-frequency electric field. The perturbed field lines external to the body are nearly perpendicular to the body surface. Since the body is a good conductor at the low-frequency band, oscillating charges can be induced on the surface of the exposed body, which can

Table 9.1 Basic restrictions for human exposure to time-varying electric and magnetic fields.

Exposure characteristic	Frequency range	Internal electric field (V/m)
Occupational exposure		
CNS tissue of the head	1–10 Hz	$0.5/f$
	10–25 Hz	0.05
	25–400 Hz	$2\times 10^{-3}f$
	400 Hz–3 kHz	0.8
	3 kHz–10 MHz	$2.7\times 10^{-4}f$
All tissues of head and body	1 Hz–3 kHz	0.8
	3 kHz–10 MHz	$2.7\times 10^{-4}f$
General public exposure		
CNS tissue of the head	1–10 Hz	$0.1/f$
	10–25 Hz	0.01
	25–400 Hz	$4\times 10^{-4}f$
	400 Hz–3 kHz	0.4
	3 kHz–10 MHz	$1.35\times 10^{-4}f$
All tissues of head and body	1 Hz–3 kHz	0.4
	3 kHz–10 MHz	$1.35\times 10^{-4}f$

produce currents inside the body. Limitations of exposure that are based on the physical quantity or quantities directly related to the established health effects are termed as basic restrictions. The basic restrictions are listed in Table 9.1. Note that the "f" in the table means the frequency, and the CNS for central nerve stimulation. For CNS tissue of the head, the restrictions give more detailed division and limitation.

As shown in Table 9.1, the guideline specifies the limit of the internal electric field at $1.35\times 10^{-4} f$ in view of the fact that the common operating frequency of the EV wireless charging system is 3 kHz to 10 MHz. Additionally, the reference levels of the electric and magnetic field strengths refer to the basic restrictions calculated by the mathematical modeling, which can be calculated for the condition of maximum coupling of the field to the exposed individual, as listed in Tables 9.2 and 9.3. For the general public, the electric field strength is limited to 83 V/m, the magnetic field strength is limited to 21 A/m, and the magnetic flux density should be controlled to lower than 2.7×10^{-5} T.

Table 9.2 Reference levels for occupational exposure to time-varying electric and magnetic fields (unperturbed rms values).

Frequency range	E-field strength, *E* (kV/m)	Magnetic field strength, *H* (A/m)	Magnetic flux density, *B* (T)
1–8 Hz	20	$1.63\times10^5/f^2$	$0.2/f^2$
8–25 Hz	20	$2\times10^4/f$	$2.5\times10^{-2}/f$
25–300 Hz	$5\times10^2/f$	8×10^2	1×10^{-3}
300 Hz–3 kHz	$5\times10^2/f$	$2.4\times10^5/f$	$0.3/f$
3 kHz–10 MHz	1.7×10^{-1}	80	1×10^{-4}

Table 9.3 Reference levels for general public exposure to time-varying electric and magnetic fields (unperturbed rms values).

Frequency range	E-field strength *E* (kVm^{-1})	Magnetic field strength, *H* (Am^{-1})	Magnetic flux density, *B* (T)
1–8 Hz	5	$3.2\times10^4/f^2$	$4\times10^{-2}/f^2$
8–25 Hz	5	$4\times10^3/f$	$5\times10^{-3}/f$
25–50 Hz	5	1.6×10^2	2×10^{-4}
50–400 Hz	$2.5\times10^2/f$	1.6×10^2	2×10^{-4}
400 Hz–3 kHz	$2.5\times10^2/f$	$6.4\times10^4/f$	$8\times10^{-2}/f$
3 kHz–10 MHz	8.3×10^{-2}	21	2.7×10^{-5}

9.3.1.2 IEC

The IEC Technical Committee issued the standard IEC 61980-1 *Electric Vehicle Wireless Power Transfer Systems – Part 1*, which gives the EMF measurement procedure demonstrative reference values. The field probe is utilized for the EMF measurement, which should be in accordance with the standard IEC 62233 *Measurement methods for electromagnetic fields of household appliances and similar apparatus with regard to human exposure* [75]. The measuring range includes the imaginary vertical plane with a distance of 20 cm from four vertical sides of the vehicle, as shown in Figure 9.18. The measurement should be carried out using the field probe to scan over the surface of the imaginary vertical plane for each side at least one time. The area with the maximum measured value of each side can be drawn using the maximum reading value. The values of all maximum reading positions can be recorded after measuring again for 10 seconds at least. All measured values should be in accordance with the requirements

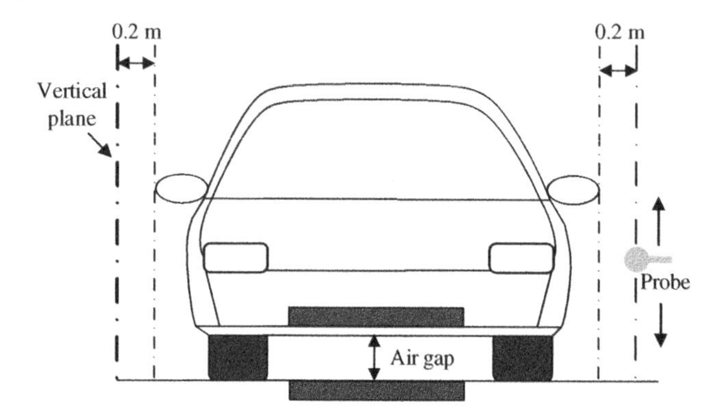

Figure 9.18 Front view of measurement.

of the local standards. In the absence of the corresponding local standards, the evaluation should refer to the requirements of *ICNIRP guidelines for limiting exposure to time-varying electric and magnetic fields* (1 Hz to 100 kHz).

9.3.1.3 SAE

SAE-drafted SAE J2954 *Wireless Power Transfer for Light-Duty Plug-In/Electric Vehicles and Alignment Methodology* and was revised in 2017 and 2019, respectively [27], where three physical regions are defined to facilitate the EMF safety management of the EV wireless charging systems. Figure 9.19 illustrates these three regions, that is Region 1 represents the entire area under the vehicle, Region 2 represents the region outside the periphery of the vehicle, and Region 3 represents the area inside the vehicle.

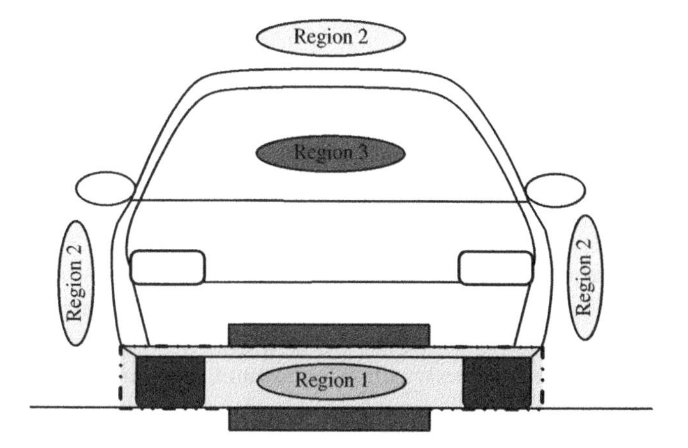

Figure 9.19 Front view of EMF regions.

Table 9.4 Reference levels in EMF exposure standard.

Quantity	ICNIRP 2010 general public reference level Regions 2 and 3
Magnetic field	27 μT or 21.5 A/m
Electric field	83 V/m

Source: Reference [27]/SAE International.

Table 9.5 Basic restriction levels in EMF exposure standard.

Quantity	ICNIRP 2010 general public basic restriction level Regions 2 and 3
Internal electric field	$1.35 \times 10^{-4} f = 10.7\,\text{V/m}$ at 79 kHz

Source: Reference [27]/SAE International.

The standard specifies the limits of the EMF reference levels and basic restriction levels with respect to different regions, which are consistent with the *ICNIRP guidelines for limiting exposure to time-varying electric and magnetic fields (1 Hz to 100 kHz)* as listed in Tables 9.4 and 9.5. The EMF of Region 2 and Region 3 should meet the reference levels listed in Table 9.4 or the basic restriction levels listed in Table 9.5. For Region 1, the reference and the basic restriction levels should be in accordance with both the tables simultaneously.

This standard also recommends the experimental procedure to find the maximum fields in Region 2 and Region 3 under different conditions of air gap and offset required by the standard. The operation process is similar to IEC 61980-1, while SAE J2954 gives more detailed procedure to assess the EMF radiation inside the vehicle as follows:

- Set up the offset and air gap conditions.
- Utilize the probe to find the maximum magnetic field from the occupant area in Region 3.
- Record the location and the value of the maximum magnetic field for each seating position. The measurement of the seating position should cover four areas, such as head, chest, seat cushion, and foot.

- Repeat the above steps to conduct the electric field measurement for nonmetallic vehicle floor pans. Meanwhile, for metallic floor pans, the assessment of electric field is not required in Region 3 except around floor openings.
- Repeat for each combination of various offset and gap conditions.

9.3.2 Mitigation Schemes

The adopting of some technical means to mitigate the electromagnetic radiation is the key to prevent from beyond limits required by the above standards for EV wireless charging systems. The EMF radiation issue is mainly caused by the magnetic leakage flux. Hence, in order to reduce the leakage flux, a commonly used method is to increase the magnetic coupling effect by optimizing both the primary and secondary coils or using ferrite materials, which is called the passive method. In some occasions, however, the passive methods may increase the power loss due to the usage of additional ferrites, especially for high-power WPT applications. Accordingly, the active intervention schemes are proposed to mitigate or even cancel out the EMF radiation. This section discusses the two main ways to realize the mitigation of EMF radiations.

9.3.2.1 Passive Methods

In [76], a three-phase power line structure is proposed to reduce the leakage in magnetic fields for OLEVs, which requires no additional ferrites apart from the ferrite components of power tracks. There are six overlaid power lines where each phase consists of two lines. Every three power lines from different phases are terminated to one point at their ends. Since the current is distributed into two power lines in each phase, and the overlaid part of different power lines can cancel the magnetic field to each other, the magnetic leakage can be dramatically reduced by 96% compared to the single-phase power line structure.

Another method to reduce the EMF radiation is shielding the electromagnetic waves emitted from the coupler with ferrite materials. The H-shaped core transformer is used as the coupler for EV wireless charging systems in [77]. The back of both the primary and secondary sides is covered with a layer of aluminum to shield the electromagnetic radiation outside the air gap of the coupler. Additionally, a ferrite plate is placed on the back of the aluminum sheet to reduce the EMF radiation. The measurement result of the leakage electric field is reduced by 7.2, 8.1, and 7.7 dB at 150, 250, and 350 kHz, respectively, while the coil-to-coil efficiency is increased from 90.1% to 92.7%.

Besides, the resonant reactive shield circuit is a passive compensation loop with an inductance coil and a capacitor for resonance, which can be used to reduce the EMF radiation [78]. Two compensation loops are mounted on both ends of the air gap to generate canceling magnetic fields to reduce the EMF radiation. Since no additional power source is required, the resonant reactive field also belongs to passive mitigation methods. The resonant reactive shield realizes a maximal reduction of 64% for magnetic field compared to that without the shield.

9.3.2.2 Active Methods

Being different from the passive method, the additional power source is also utilized to realize an active mitigation of the EMF radiation for EV wireless charging systems. For the OLEV with stretched power tracks, KAIST proposes generalized active EMF cancel methods [79]. By adding EMF canceling coils to both the transmitting coil and the pickup coil, the EMF radiation can be, respectively, reduced by the corresponding canceling coils. In addition, KAIST also applied the canceling coil for U-, W-, and novel I-type power tracks developed by themselves [57]. The measured results show that the mitigation scheme can ensure the radiation safety for these three types of coils in accordance with the basic restriction of the *ICNIRP guidelines for limiting exposure to time-varying electric and magnetic fields* (1 Hz to 100 kHz).

For the lumped track as mentioned in Section 9.2.3.1, the research team at The University of Auckland presents a ferrite-less pad for EV wireless charging systems, which adopts only one canceling coil connected in series with the transmitting coil but winded in the opposite direction. By adjusting the vertical distance and turns ratio between the transmitting coil and the canceling coil, it can effectively minimize the magnetic field in the areas under the pad and outside the charging area [80]. The validation results show that the proposed scheme can meet the requirement of 27-μT reference levels for general public exposure to time-varying electric and magnetic fields in the *ICNIRP guidelines for limiting exposure to time-varying electric and magnetic fields* (1 Hz to 100 kHz).

Furthermore, a novel design of the canceling coil is implemented for EV wireless charging systems, where the original circular canceling coil surrounding the transmitting coil is replaced by two separate semi-annular-shaped coils [81]. The two canceling coils can produce the magnetic field possessing the opposite phase with the magnetic flux outside the transmission channel and the same phase with the main flux used for the power transmission. Accordingly, it can reduce the EMF radiation and improve the coupling effect simultaneously.

9.4 Key Technologies

In the previous section, the history, principle, classification, typical works, and safety issues of wireless charging have been discussed for EVs, which implies the rapid development, especially during recent years as well as the promising application prospect. However, it should be noted that there are still critical technologies ahead; consequently, in this section, three typical technologies will be introduced to further elaborate the EV wireless charging systems.

9.4.1 Foreign Object Detection

Compared to other applications of the WPT techniques, the air gap of EV wireless charging systems is much wider, which increases the possibility of foreign objects impeding the transmission channel. The time-varying magnetic flux can induce the eddy current in metal material objects, thus leading to not only extra power loss but also heat dissipation in view of the higher power level. Accordingly, the FOD technique is extremely essential for EV wireless charging systems.

In [82], a power loss detection method is proposed and implemented to quantify the inherent power loss among the power transmitted to the load in WPT systems. By comparing the total input power, the corresponding difference value is the power loss generated by foreign objects. Consequently, the existence of this difference implies the presence of a foreign object in the area of the air gap. Then, the power transferring process should be shut down immediately. Although this method is only verified by low-power WPT applications, it really lights us with an idea to deal with the FOD for EV wireless charging systems.

Foreign objects not only cause power loss but also change the distribution of magnetic field. Such changes can also be utilized to detect the presence of foreign objects. In [83], a nonoverlapping coil set is implemented to realize the FOD scheme. The existence of metal objects is determined by the induced voltage difference of the nonoverlapping coil sets. The proposed coil set consists of two nonoverlapping symmetric coils. Due to the symmetric arrangement of the two coils, the corresponding induced voltages are the same when the magnetic field is generated only by the transmitting coil. However, when the metal object is located in the transmission channel, the magnetic flux is distorted by the eddy current induced in the metal object. As a result, the induced voltages are not the same anymore. Then, the FOD system can detect the existence of the metal object based on the voltage difference. Similarly, two-layer symmetrical coil sets are proposed to detect not only the presence but also the location of foreign objects [84]. Specifically, several coils are arranged on the first layer, while the second layer consists of other coils arranged in the vertical direction to the first layer. Using the constructed map of the magnetic field, the position of the foreign objects can be determined by the measurement results in the horizontal and vertical directions.

Being different from the induced-voltage-detection methods as aforementioned, an electromagnetic-model-based FOD (EM-FOD) is proposed by employing two open-circuited sensing coils to estimate the current in the primary circuit in [85]. The difference between the estimated and the measured current value is caused by foreign objects. As a significant advantage, the open-circuited sensing coils cause no change in the power loss and the output impedance of the whole system. In addition, the foreign object can also be detected based on the deviation of the resonance frequency [86, 87].

In addition to the detection of metal objects, live objects such as small animals are also required to handle EV wireless charging systems. The LOD is difficult to be realized by simply relying on the change in the power loss, the induced voltage, and the magnetic field as mentioned above. Accordingly, researchers attempt to use a deep learning algorithm known as a denoising convolutional autoencoder to identify the foreign object in real time from the image captured by a thermal camera [88].

9.4.2 Wireless Vehicle-to-Grid

The increasing number of EVs significantly increases the power demand, changes the load profile, and thus puts great pressure on the power grid. An appropriate management and control strategy can reduce the additional investment on power plants and grid infrastructures, wherein the vehicle-to-grid (V2G) technology has obtained increasing attentions in recent years, as depicted in Figure 9.20. The EV is taken as not only the transportation but also movable energy storage. By feeding the energy back to the grid, V2G can mitigate the load pressure for the grid during the peak power period. Interestingly, another similar concept of vehicle-to-home (V2H) is proposed to supply power from the EV battery to other household loads during power outage [89]. However, V2G can be realized only when the vehicle is parked in the lot due to the cable charging technique. If the V2G can also be conducted when the vehicle is running, this novel technology can keep ready for 24 hours 7 days. Accordingly, this section introduces the concept of V2G based on the wireless charging technologies.

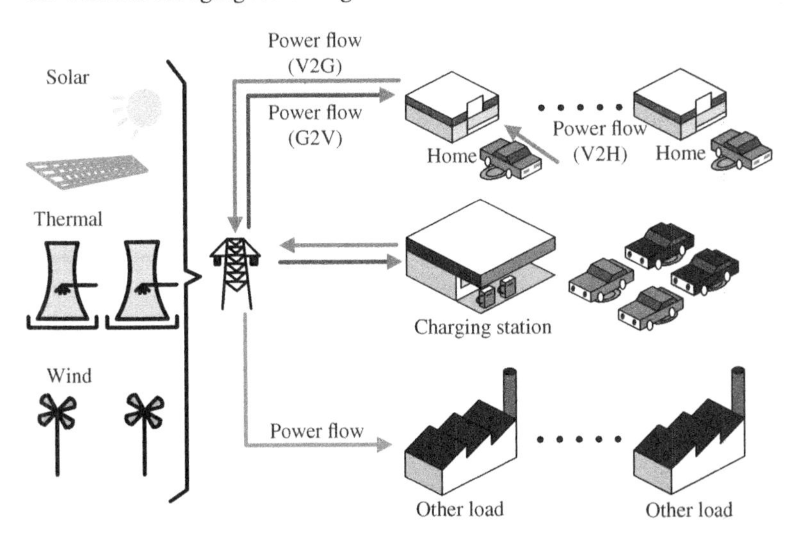

Figure 9.20 Overall configuration of V2G.

The realization of V2G technology requires a flexible, automatic reliable charging and discharging scheme for EVs. The design simplicity, high efficiency, low cost, flexibility, and automatic charging and discharging methods are highly emphasized for the research. As early as 2011, a current-source-based bidirectional WPT system was implemented in [90], which adopts parallel resonant circuits and controlled rectifiers on both primary and secondary sides to achieve a bidirectional power flow. The power and its flow direction are determined in two control modes: relative phase angle mode and the voltage amplitude mode. In the relative phase control mode, the power is regulated by varying the relative phase angles of the voltages generated by two rectifiers, while keeping the magnitude constant. In the voltage amplitude mode, each arm can be driven with a varying phase shift and relative phase angle equaling 90°. This model is verified using a 1.5-kW bidirectional WPT system [91]. To satisfy the high-power requirement of EV wireless charging system, a symmetric-architecture bidirectional wireless charging system is proposed in [92], which has been verified to achieve an output power of 5 kW with a maximum efficiency of 89.8%.

From the wireless charging model level to the whole grid level, an overall design concept of V2G is presented in [93], which consists of vehicle-mounted intelligent terminal (VMIT), stall intelligent terminal (SIT), subarea scheduling management system (SSM), and distribution network scheduling management system (DNSM). The VMIT is the interface between customers and vehicles, and the SIT is adopted to deal with the information from VMIT and to complete the charging operation. The SSM and the DNSM handle the information collecting and the power dispatching processes. Apart from the study on the system architecture, other technical solutions are also developed targeting specific issues. In [94], for instance a two-layer power flow controller for EV bidirectional WPT systems is implemented in which the first layer is responsible for gathering information from the driver, charging station, grid, and BMS as well as sending signals for charge or discharge, while the second layer receives the signal and generates the parameters for two resonant converters mounted on the vehicle and the grid sides to realize the power flow. Besides, a theoretical power flow model of a bidirectional wireless charging system for V2G application is presented in [95]. It provides accurate estimation for the active and the reactive power flow during V2G and grid-to-vehicle (G2V) modes; thus, the operators can predict and optimize the interaction between EVs and the grid.

9.4.3 Supercapacitor

The battery has been widely used for the energy storage of EVs by means of its relatively high energy density. However, its characteristic of low power density causes the difficulties of providing considerable power for the start-up or the fast acceleration of EVs. Although there are new types of batteries with high power density, the corresponding cost is higher than the commonly used batteries. More importantly, the instantaneous high-power output significantly deteriorates the battery life. In response to the abovementioned technical defects of batteries, the supercapacitor falls into our vision. Compared with the battery, the supercapacitor has high power density but lower energy density, thus being taken as the perfect compensation solution for batteries. The typical characteristics of battery cells and supercapacitor cells are listed in Table 9.6.

For example, the supercapacitor can be integrated with batteries to achieve overall performance with both instantaneous high-power output and large energy storage [97]. In other words, the utilization of supercapacitors can compensate the power fluctuation and improve the performance of the start-up or fast acceleration of EVs. In [98], a hybrid energy storage system (HESS) with the configuration of *battery plus supercapacitor* is proposed for EV wireless charging systems, where the wirelessly transferred power is acquired by the pickup coil mounted in the vehicle and then used to charge both the battery and the supercapacitor. The key of HESS is to develop a bidirectional dc–dc converter for the power management between the battery and the supercapacitor. For example, in order to maintain a relatively stable load profile for battery, the dc–dc converter is taken as a controllable switch to keep the voltage of the supercapacitor higher than that of the battery pack [96]. The battery provides the vehicle with the power only when the voltage of the supercapacitor drops below the value of the battery voltage. When the vehicle brakes, the energy is regenerated and fed back to the supercapacitor instead of the battery. In such way, the supercapacitors act as a protection umbrella for the battery to extend its life as long as possible. Additionally, in [99], a central controller is proposed to achieve the integration of the wireless charging system with the *battery plus supercapacitor* HESS, decouple the battery currents from high-frequency current, and

Table 9.6 Typical characteristics of battery cells and supercapacitor cells.

Types of cells	Nominal cell voltage (V)	Energy density (Wh/kg)	Power density (kW/kg)	Cycle life (Times)
Lead acid	2	30–40	0.18	Up to 800
Ni-Mh	1.2	55–80	0.4–1.2	Up to 1,000
Li-ion	3.6	80–170	0.8–2	Up to 1,200
Li-polymer	3.7	100–200	1–2.8	Up to 1,000
Li-iron phosphate	3.2	80–115	1.3–3.5	Up to 2,000
Supercapacitor	2.5/2.7	2–30	4–10	Over 1,000,000

Source: Adapted from Cao and Emadi [96].

Table 9.7 The cost of supercapacitors for EV application.

Brand	Model numbers	Nominal cell capacitance (F)	Energy density (Wh/kg)	Power density (kW/kg)	Cycle life (Times)
Maxwell	BMOD0165 P048 B01	3,000	3.9	6.8	Over 1,000,000
Maxwell	BMOD0063 P125 B04	3,000	2.3	3.6	Over 1,000,000
IOXUS	iRD3000K 270CT	2,880	6.6	9.6	Over 1,000,000
NINGBO CRRC	CRRC-M-0165-C1-0048	3,000	5.6	14.7	Over 1,000,000

manage the SOC of supercapacitors, which is particularly useful to offer instantaneous high output power for dynamic wireless charging systems.

Apart from the abovementioned hybrid energy storage, there are also a few EV wireless charging systems with only supercapacitor energy storage system while adopting no battery, which is applicable for short cruising mileage vehicles. In [100], a supercapacitor-based WPT system is proposed for a sightseeing vehicle, where the equivalent load resistance of supercapacitors is analyzed, and the CC charging is realized using a converter with proportional integral (PI) controller in the secondary side. The exemplified sightseeing vehicle can be fully charged in 200 seconds with an output power of 2.86 kW and the constant charging current equaling 31.5 A. For the product-level supercapacitors, the cost of some supercapacitors for EV application from several brands are listed as Table 9.7.

To sum up, both the HESS and the pure supercapacitor energy storage system illustrate that the supercapacitor is increasingly utilized for the EV energy storage system. Accordingly, its impedance characteristic should be taken into account, particularly with emphasis on the impact on the transmission performance of WPT systems.

9.5 Summary

This chapter discusses about the basic principle, typical works, regulations, and key technologies of wireless charging for EVs. Obviously, the research and development of EV wireless charging systems has become the hottest and typical application of WPT technologies. During the past years, a number of efforts have been made by many institutions all over the world, with no lacking landmark achievements. The WPT technology gives the development of EVs a promising future, although there are still technical issues and challenges ahead.

9.5.1 Improvement of the Charging Power

Nowadays, EV wireless charging systems are mainly realized at the power level of several kilowatts, whereas only a few enterprises can achieve tens to hundreds of kilowatts. For the static wireless charging system, the increasing charging power level means shorter awaiting time and more flexibility. In view of the dynamic wireless charging systems, the improvement in power can significantly reduce the length of power track, which can significantly reduce the construction cost of power tracks and increase the commercial value of the EV dynamic wireless charging system.

In response to the improvement in the charging power, the wide band gap (WBG) switching device first enters into our vision. By means of its salient advantages of high breakdown electric field intensity, improved thermal stability, and high carrier saturation mobility, the WBG is an ideal device-level solution to provide high-power WPT applications with high temperature resistance, high voltage resistance, high frequency performance, and low switching loss. From the perspective of the circuit topology, the multilevel converter is another expected solution for high-power EV wireless charging systems, since it can reduce the voltage stress on each power switching device while achieving a high-power output.

9.5.2 Enhancement of Misalignment Tolerance

The misalignment occurring in the charging process is another troublesome issue. Especially in the EV dynamic charging system, the lateral and the vertical misalignment is more considerable than the static wireless charging system, which inevitably deteriorates the charging power and the system efficiency. As a result, the vehicle cannot be charged at expected power level when it is running on the power track, thus giving end users a terrible use experience.

In response to the misalignment tolerance, the future study is suggested to focus on the optimization for the structure of stretched track as well as the design for the coil in the lumped track. More information can be obtained in Chapter 3. Apart from the study on the topology of power tracks and coils, another novel solution involves the application of metamaterials on WPT technologies. Its left-hand characteristics offer the negative permeability and the negative conductivity, which can confine the induced magnetic field within the expected area and thus improve the tolerance to misalignment during the charging process. However, the metamaterial commonly works at the MHz range, which limits its wide application on kHz high-power EV wireless charging systems. Accordingly, how to reduce the resonant frequency from the MHz to the kHz is the key in future studies.

9.5.3 Foreign Object Detection

Due to the time-varying magnetic field between the transmitting and the pickup coils, any foreign object falling into this region deteriorates the transmission performance of EV wireless charging systems. For metal foreign objects, the magnetic flux induces the eddy current, thus producing the power loss via the heat dissipation, which even leads to fire. For live foreign objects such as small animals, the radiation can be harmful for the body and even endanger their lives. Accordingly, the detection of foreign object is very important for EV wireless charging systems.

In response to avoid the influence of foreign objects, current studies mainly fall into indirect detection schemes, for example by monitoring the variation of power loss, the distribution of flux density, and the change of induced voltage. All the abovementioned schemes are feasible only for metal object. For live foreign objects, the detection is implemented based on the image captured by the thermal camera. Another detection scheme is to utilize the magnetoresistance sensor, where the core part is a special metal material whose resistance changes with the varying external magnetic field. As a consequence, the magnetoresistance sensor can be used to detect the change in magnetic flux caused by the existence of foreign objects.

9.5.4 Reduction of Cost

The construction cost is an inevitable issue which impedes the widespread commercialization of EV wireless charging systems. A great number of charging pads are required to be mounted for private apartments/houses, office buildings, parking lots, etc., which needs huge investment for the static wireless charging of EVs. Furthermore, the investment of EV dynamic wireless charging systems is much greater, for example the expenses in redesigning the road and laying large-scale power tracks as well as the daily maintenance cost. Consequently, how to reduce the construction cost becomes a troublesome issue that we are facing now.

In response to the issue of the construction cost, future studies should focus on the economic assessment of EV wireless charging system. Firstly, more installations of the EV wireless charging system, especially the dynamic charging, the lower requirement of batteries. Hence, the construction cost should be synthetically evaluated by considering the decrease in the battery cost, which is not a simple addition or subtraction. Secondly, the increasing charging power requires more investment on power devices, but it can reduce the length of the power track and the corresponding cost, which is another trade-off for the economic assessment of EV wireless charging systems. What's more, the planning and the location of power tracks is also an important technical concern, since traffic flow determines the construction scale and its cost. To sum up, the cost issue of EV wireless charging systems is complicated, which should be handled by synthetically considering various factors as mentioned above.

9.5.5 Impact on Power Grid

The huge charging demand from EV fleets can change the load profile and impact on the stability of power grid significantly. With the functionality of V2G, however, EVs also act as a flexible and movable energy storage to balance the power and the load for power grids. In other words, the penetration of EVs' impact on the grid not only originates from the perspective of the load but also in terms of the network topology and the control. Its influence is extensive, especially when the WPT technologies increase the interaction between the EVs and the grid.

In response to reduce the impact on the power gird caused by EV wireless charging systems, the daily load profile should be analyzed by considering the involvement of EV wireless charging systems. Meanwhile, the V2G technique needs to be further developed to release the load pressure for power grids. Besides, it should be noted that the bidirectional power flow of V2G systems may generate harmonics to deteriorate the power quality of the grid. The WPT technology increases the interaction between the EV and the grid, which means that it may further deteriorate the quality of power. Accordingly, the wireless V2G shows promising prospect, whereas the corresponding studies should particularly pay attention on the mitigation of harmonics.

9.5.6 Promotion of Its Commercialization

The commercialization is the final goal for most of the technical research, no exception for EV wireless charging technologies. Apart from the technical issues discussed above, the biggest barrier for the commercialization is how to coordinate and unify the standard of products among various automobile manufacturers, charging equipment producers, and constructors in accordance with limitations specified by other standards. It is not a pure technical issue, but a more complex work which we have to deal with. In response to the promotion of its commercialization, the most feasible way is to develop regulations or standards for EV wireless charging systems. The related work has been carried out as mentioned in Section 9.1.3, but it is not adequate. We have a lot of work ahead for addressing this issue.

References

1. Guarnieri, M. (2012). Looking back to electric cars. In: *Proceedings of the IEEE History of Electro Technology Conference*, Pavia, 1–6. IEEE.
2. Loeb, A.P. (2004). Steam versus electric versus internal combustion: choosing vehicle technology at the start of the automotive age. *Transportation Research Record* 1885 (1): 1–7.
3. Rim, C.T. and Mi, C. (2017). *Wireless Power Transfer for Electric Vehicles and Mobile Devices*. Wiley.

4 Patil, D., McDonough, M.K., Miller, J.M. et al. (2018). Wireless power transfer for vehicular applications: overview and challenges. *IEEE Transactions on Transportation Electrification* 4 (1): 3–37.

5 Dixon, J., Nakashima, I., Arcos, E.F., and Ortuzar, M. (2010). Electric vehicle using a combination of ultra capacitors and ZEBRA battery. *IEEE Transactions on Industrial Electronics* 57 (3): 943–949.

6 Ceraolo, M., di Donato, A., and Franceschi, G. (2008). A general approach to energy optimization of hybrid electric vehicles. *IEEE Transactions on Vehicular Technology* 57 (3): 1433–1441.

7 Mapelli, F.L., Tarsitano, D., and Mauri, M. (2010). Plug-in hybrid electric vehicle: Modeling, prototype, realization, and inverter losses reduction analysis. *IEEE Transactions on Industrial Electronics* 57 (2): 598–607.

8 Yilmaz, M. and Krein, P.T. (2013). Review of battery charger topologies, charging power levels, and infrastructure for plug-in electric and hybrid vehicles. *IEEE Transactions on Power Electronics* 28 (5): 2151–2169.

9 Kutkut, N.H., Divan, D.M., Novotny, D.W., and Marion, R.H. (1998). Design considerations and topology selection for a 120-kW IGBT converter for EV fast charging. *IEEE Transactions on Power Electronics* 13 (1): 169–178.

10 Praisuwanna, C. and Khomfoi, S. (2013). A quick charger station for EVs using a pulse frequency technique. In: *Proceedings of the IEEE Energy Conversion Congress and Exposition*, Denver, CO, 3595–3599. IEEE.

11 Lee, S., Huh, J., Park, C. et al. (2010). On-line electric vehicle using inductive power transfer system. In: *Proceedings of the IEEE Energy Conversion Congress Exposition*, Atlanta, GA, 1598–1601. IEEE.

12 Miller, J.M., Onar, O.C., and Jones, P.T. (2013). ORNL developments in stationary and dynamic wireless charging. In: *Proceedings of the IEEE Energy Conversion Congress Exposition*. IEEE.

13 D. M. Empey, E. H. Lechner, G. Wyess, et al. (1994). Roadway powered electric vehicle project track construction and testing program phase 3D. https://escholarship.org/content/qt1jr98590/qt1jr98590.pdf.

14 Shinohara, N. (2013). Wireless power transmission progress for electric vehicle in Japan. In: *Proceedings of the IEEE Radio and Wireless Symposium*, Austin, TX, 109–111. IEEE.

15 Qualcomm Halo, 2016. [Online]. Available: https://www.qualcomm.cn/news/blog-2016-12-29-3 (accessed April 2020).

16 WiTricity transfers electric energy or power over distance without wires, 2022. [Online]. Available: https://witricity.com/ (accessed April 2020).

17 INDUCTEV Wireless charging for electric vehicles, 2022. [Online]. Available: https://inductev.io/#wirelesscharging.

18 Bolger, J.G., Ng, L.S., Turner, D.B., and Wallace, R.I. (1979). Testing a prototype inductive power coupling for an electric highway system. In: *Proceedings of the IEEE Vehicular Technology Conference*, vol. 29, 48–56. IEEE.

19 Schladover, S.E. (1988). Systems engineering of the roadway powered electric vehicle technology. In: *Proceedings of the 9th International Electric Vehicle Symposium*.

20 Shin, J., Shin, S., Kim, Y. et al. (2014). Design and implementation of shaped magnetic-resonance-based wireless power transfer system for roadway-powered moving electric vehicles. *IEEE Transactions on Industrial Electronics* 61 (3): 1179–1192.

21 Onar, O.C., Miller, J.M., Campbell, S.L. et al. (2013). A novel wireless power transfer for in-motion EV/PHEV charging. In: *Proceedings of IEEE Applied Power Electronics Conference and Exposition*, Long Beach, CA, 3073–3080. IEEE.

22 Onar, O.C., Chinthavali, M., Campbell, S.L. et al. (2019). Vehicular integration of wireless power transfer systems and hardware interoperability case studies. *IEEE Transactions on Industry Applications* 55 (5): 5223–5234.

23 Sotiris, N., Yang, Y., and Georgiadis, A. (ed.) (2016). *Wireless Power Transfer Algorithms, Technologies and Applications in Ad Hoc Communication Networks*. Springer.

24 International Electrotechnical Commission IEC everywhere for a safer and more efficient world, 2022. [Online]. Available: https://www.iec.ch/ (accessed April 2020).

25 Electric vehicle wireless power transfer (WPT) systems – Part 2: Specific requirements for communication between electric road vehicle (EV) and infrastructure, 2022. [Online]. Available: https://webstore.iec.ch/publication/31050 (accessed April 2020).

26 Electric vehicle wireless power transfer (WPT) systems – Part 3: Specific requirements for the magnetic field wireless power transfer systems, 2022. [Online]. Available: https://webstore.iec.ch/publication/27435 (accessed April 2020).

27 SAE Wireless Power Transfer for Light-Duty Plug-In/Electric Vehicles and Alignment Methodology J2954_201711, 2017. [Online]. Available: https://www.sae.org/standards/content/j2954_201711/ (accessed April 2020).

28 SAE Communication for Wireless Power Transfer Between Light-Duty Plug-in Electric Vehicles and Wireless EV Charging Stations J2847/6_202009, 2020. [Online]. Available: http://standards.sae.org/j2847/6_201508/ (accessed April 2020).

29 SAE Electric vehicle wireless power transfer (WPT) systems – Part 3: Specific requirements for the magnetic field wireless power transfer systems

J2931/6_202208, 2022. [Online]. Available: http://standards.sae.org/j2931/6_201508/ (accessed April 2020).

30 Shoki, H. (2013). Issues and initiatives for practical deployment of wireless power transfer technologies in Japan. *Proceedings of the IEEE* 101 (6): 1312–1320.

31 Shinohara, N. (2018). *Wireless Power Transfer: Theory, Technology, and Applications*. Institution of Engineering & Technology.

32 Lu, F., Zhang, H., Hofmann, H., and Mi, C. (2015). A double-sided LCLC-compensated capacitive power transfer system for electric vehicle charging. *IEEE Transactions on Power Electronics* 30 (11): 6011–6014.

33 Luo, B., Mai, R., Guo, L. et al. (2019). LC–CLC compensation topology for capacitive power transfer system to improve misalignment performance. *IET Power Electronics* 12 (10): 2626–2633.

34 Dai, J. and Ludois, D.C. (2016). Capacitive power transfer through a conformal bumper for electric vehicle charging. *IEEE Journal of Emerging and Selected Topics in Power Electronics* 4 (3): 1015–1025.

35 Onar, O.C., Campbell, S.L., Seiber, L.E. et al. (2016). A high-power wireless charging system development and integration for a Toyota RAV4 electric vehicle. In: *Proceedings of the IEEE Transportation Electrification Conference and Expo*, Dearborn, MI, 1–8. IEEE.

36 Onar, O.C., Chinthavali, M., Campbell, S.L. et al. (2018). Modeling, simulation, and experimental verification of a 20-kW series–series wireless power transfer system for a Toyota RAV4 electric vehicle. In: *Proceedings of the IEEE Transportation Electrification Conference and Expo*, Long Beach, CA, 874–880. IEEE.

37 Qualcomm enables a world where everyone and everything can be intelligently connected, 2022. [Online]. Available: https://www.qualcomm.com/news/ (accessed April 2020).

38 WiTricity Acquires Qualcomm Halo, 2019. [Online]. Available: https://witricity.com/witricity-acquires-qualcomm-halo/ (accessed April 2020).

39 Kurs, A., Karalis, A., Moffatt, R. et al. (2007). Wireless power transfer via strongly coupled magnetic resonances. *Science* 317 (5834): 83–86.

40 PLUGLESS POWER wireless electric vehicle charging, 2022. [Online]. Available: https://www.pluglesspower.com/ (accessed April 2020).

41 INDUCTEV Wireless charging for electric vehicles, 2022. [Online]. Available: https://momentumdynamics.com/ (accessed April 2020).

42 Lu, F., Zhang, H., Hofmann, H., and Mi, C.C. (2016). A dynamic charging system with reduced output power pulsation for electric vehicles. *IEEE Transactions on Industrial Electronics* 63 (10): 6580–6590.

43 Nagendra, G.R., Covic, G.A., and Boys, J.T. (2014). Determining the physical size of inductive couplers for IPT EV systems. In: *Proceedings of the*

IEEE Applied Power Electronics Conference and Exposition, Fort Worth, TX, 3443–3450. IEEE.

44 Lee, K., Pantic, Z., and Lukic, S.M. (2014). Reflexive field containment in dynamic inductive power transfer systems. *IEEE Transactions on Power Electronics* 29 (9): 4592–4602.

45 Covic, G.A., Boys, J.T., Kissin, M., and Lu, H.G. (2007). A three-phase inductive power transfer system for roadway-powered vehicles. *IEEE Transactions on Industrial Electronics* 54 (6): 3370–3378.

46 Jang, Y.J., Suh, E.S., and Kim, J.W. (2016). System architecture and mathematical models of electric transit bus system utilizing wireless power transfer technology. *IEEE Systems Journal* 10 (2): 495–506.

47 Choi, S.Y., Huh, J., Lee, W.Y. et al. (2013). New cross-segmented power supply rails for roadway powered electric vehicles. *IEEE Transactions on Power Electronics* 28 (12): 5832–5841.

48 Zhang, W., Wong, S., Tse, C.K., and Chen, Q. (2014). An optimized track length in roadway inductive power transfer systems. *IEEE Journal of Emerging and Selected Topics in Power Electronics* 2 (3): 598–608.

49 Covic, G.A. and Boys, J.T. (2013). Inductive power transfer. *Proceedings of the IEEE* 101 (6): 1276–1289.

50 Cai, T., Duan, S., Feng, H. et al. (2016). A general design method of primary compensation network for dynamic WPT system maintaining stable transmission power. *IEEE Transactions on Power Electronics* 31 (12): 8343–8358.

51 Park, C., Lee, S., Jeong, S.Y. et al. (2015). Uniform power I-type inductive power transfer system with DQ-power supply rails for on-line electric vehicles. *IEEE Transactions on Power Electronics* 30 (11): 6446–6455.

52 Zheng, C., Ma, H., Lai, J.S., and Zhang, L. (2015). Design considerations to reduce gap variation and misalignment effects for the inductive power transfer system. *IEEE Transactions on Power Electronics* 30 (11): 6108–6119.

53 Huang, C.Y., Boys, J.T., and Covic, G.A. (2012). Resonant network design considerations for variable coupling lumped coil systems. In: *Proceedings of the IEEE Energy Conversion Congress and Exposition*, Raleigh, NC, 3841–3847. IEEE.

54 Esteban, B., Sid-Ahmed, M., and Kar, N.C. (2015). A comparative study of power supply architectures in wireless EV charging systems. *IEEE Transactions on Power Electronics* 30 (11): 6408–6422.

55 Deng, Q., Liu, J., Czarkowski, D. et al. (2017). Edge position detection of on-line charged vehicles with segmental wireless power supply. *IEEE Transactions on Vehicular Technology* 66 (5): 3610–3621.

56 Nagendra, G.R., Chen, L., Covic, G.A., and Boys, J.T. (2014). Detection of EVs on IPT highways. *IEEE Journal of Emerging and Selected Topics in Power Electronics* 2 (3): 584–597.

57 Huh, J., Lee, S.W., Lee, W.Y. et al. (2011). Narrow-width inductive power transfer system for online electrical vehicles. *IEEE Transactions on Power Electronics* 26 (12): 3666–3679.

58 Choi, S.Y., Gu, B.W., Jeong, S.Y., and Rim, C.T. (2015). Advances in wireless power transfer systems for roadway-powered electric vehicles. *IEEE Journal of Emerging and Selected Topics in Power Electronics* 3 (1): 18–36.

59 Choi, S.Y., Jeong, S.Y., Gu, B.W. et al. (2015). Ultraslim S-type power supply rails for roadway-powered electric vehicles. *IEEE Transactions on Power Electronics* 30 (11): 6456–6468.

60 Miller, J.M., Onar, O.C., White, C. et al. (2014). Demonstrating dynamic wireless charging of an electric vehicle: the benefit of electrochemical capacitor smoothing. *IEEE Power Electronics Magazine* 1 (1): 12–24.

61 Nagendra, G.R., Boys, J.T., Covic, G.A. et al. (2013). Design of a double coupled IPT EV highway. In: *Proceedings of the Annual Conference of the IEEE Industrial Electronics Society*, Vienna, 4606–4611. IEEE.

62 Chen, L., Nagendra, G.R., Boys, J.T., and Covic, G.A. (2015). Double-coupled systems for IPT roadway applications. *IEEE Journal of Emerging and Selected Topics in Power Electronics* 3 (1): 37–49.

63 Nagendra, G.R., Chen, L., Covic, G.A., and Boys, J.T. (2014). Detection of EVs on IPT highways. In: *Proceedings of the IEEE Applied Power Electronics Conference and Exposition*, Fort Worth, TX, 1604–1611. IEEE.

64 Lin, F., Covic, G.A., and Boys, J.T. (2018). A comparison of multi-coil pads in IPT systems for EV charging. In: *Proceedings of the IEEE Energy Conversion Congress and Exposition*, Portland, OR, 105–112. IEEE.

65 Pearce, M.G.S., Covic, G.A., and Boys, J.T. (2019). Passive reflection winding for ferrite-less double D topology for roadway IPT applications. In: *Proceedings of the IEEE Energy Conversion Congress and Exposition*, Baltimore, MD, USA, 1202–1209. IEEE.

66 Pearce, M.G.S., Covic, G.A., and Boys, J.T. (2019). Robust ferrite-less double D topology for roadway IPT applications. *IEEE Transactions on Power Electronics* 34 (7): 6062–6075.

67 Zaheer, A., Neath, M., Beh, H.Z.Z., and Covic, G.A. (2017). A dynamic EV charging system for slow moving traffic applications. *IEEE Transactions on Transportation Electrification* 3 (2): 354–369.

68 Bombardier's PRIMOVE Technology Enters Service on Scandinavia's First Inductively Charged Bus Line, 2022. [Online]. Available: http://primove.bombardier.com/ (accessed April 2020).

69 Olsson, O. (2014). "Project Report, Phase 1: Slide-in Electric Road System – Inductive project report," Viktoria Swedish ICT.

70 FeAsiBility analysis and development of on-Road chargIng solutions for future electric vehiCles, 2022. [Online]. Available: https://cordis.europa.eu/project/id/605405.

71 CEDR Website. Available: https://www.cedr.eu/download/other_public_files/2016_electric_road_system_workshop/20160314_CEDR-workshop-Electric-Road-Systems_03_Viktoria-ICT.pdf (accessed April 2020).

72 INSIDEEVs Tesla Model S Sales Hit 100,000 In U.S. Quicker Than Chevy Volt, Nissan LEAF, 2017. [Online]. Available: https://insideevs.com/news/333437/tesla-model-s-sales-hit-100000-in-us-quicker-than-chevy-volt-nissan-leaf/.

73 Quinn, J.C., Limb, B.J., Pantic, Z. et al. (2015). Feasibility of wireless power transfer for electrification of transportation: techno-economics and life cycle assessment. In: *Proceedings of the IEEE Conference on Technologies for Sustainability*, Ogden, UT, 245–249. IEEE.

74 Limb, B.J. et al. (2019). Economic viability and environmental impact of in-motion wireless power transfer. *IEEE Transactions on Transportation Electrification* 5 (1): 135–146.

75 Measurement methods for electromagnetic fields of household appliances and similar apparatus with regard to human exposure, 2022. [Online]. Available: https://webstore.iec.ch/publication/6618.

76 Kim, M., Kim, H., Kim, D. et al. (2015). A three-phase wireless-power-transfer system for online electric vehicles with reduction of leakage magnetic fields. *IEEE Transactions on Microwave Theory and Techniques* 63 (11): 3806–3813.

77 Jo, M., Sato, Y., Kaneko, Y., and Abe, S. (2014). Methods for reducing leakage electric field of a wireless power transfer system for electric vehicles. In: *Proceedings of the IEEE Energy Conversion Congress and Exposition*, Pittsburgh, PA, 1762–1769. IEEE.

78 Kim, S., Park, H., Kim, J. et al. (2014). Design and analysis of a resonant reactive shield for a wireless power electric vehicle. *IEEE Transactions on Microwave Theory and Techniques* 62 (4): 1057–1066.

79 Choi, S.Y., Gu, B.W., Lee, S.W. et al. (2014). Generalized active EMF cancel methods for wireless electric vehicles. *IEEE Transactions on Power Electronics* 29 (11): 5770–5783.

80 Tejeda, A., Carretero, C., Boys, J.T., and Covic, G.A. (2017). Ferrite-less circular pad with controlled flux cancelation for EV wireless charging. *IEEE Transactions on Power Electronics* 32 (11): 8349–8359.

81 Campi, T., Cruciani, S., Maradei, F., and Feliziani, M. (2019). Active coil system for magnetic field reduction in an automotive wireless power transfer system. In: *Proceedings of the. IEEE International Symposium on Electromagnetic Compatibility, Signal & Power Integrity*, New Orleans, LA, USA, 189–192. IEEE.

82 Kuyvenhoven, N., Dean, C., Melton, J. et al. (2011). Development of a foreign object detection and analysis method for wireless power systems. In: *Proceedings of the IEEE Symposium on Product Compliance Engineering Proceedings*, San Diego, CA, 1–6. IEEE.

83 Jeong, S.Y., Kwak, H.G., Jang, G.C. et al. (2018). Dual-purpose nonoverlapping coil sets as metal object and vehicle position detections for wireless stationary EV chargers. *IEEE Transactions on Power Electronics* 33 (9): 7387–7397.

84 Xiang, L., Zhu, Z., Tian, J., and Tian, Y. (2019). Foreign object detection in a wireless power transfer system using symmetrical coil sets. *IEEE Access* 7: 44622–44631.

85 Chu, S.Y. and Avestruz, A. (2019). Electromagnetic model-based foreign object detection for wireless power transfer. In: *Proceedings of the Workshop on Control and Modeling for Power Electronics*, Toronto, ON, Canada, 1–8. IEEE.

86 Jafari, H., Moghaddami, M., and Sarwat, A.I. (2019). Foreign object detection in inductive charging systems based on primary side measurements. *IEEE Transactions on Industry Applications* 55 (6): 6466–6475.

87 Moghaddami, M. and Sarwat, A.I. (2018). A sensorless conductive foreign object detection for inductive electric vehicle charging systems based on resonance frequency deviation. In: *Proceedings of the IEEE Industry Applications Society Annual Meeting*, Portland, OR, 1–6. IEEE.

88 Sonnenberg, T., Stevens, A., Dayerizadeh, A., and Lukic, S. (2019). Combined foreign object detection and live object protection in wireless power transfer systems via real-time thermal camera analysis. In: *Proceedings of the IEEE Applied Power Electronics Conference and Exposition*, Anaheim, CA, USA, 1547–1552. IEEE.

89 Kwon, M., Jung, S., and Choi, S. (2015). A high efficiency bi-directional EV charger with seamless mode transfer for V2G and V2H application. In: *Proceedings of the IEEE Energy Conversion Congress and Exposition*, Montreal, QC, 5394–5399. IEEE.

90 Madawala, U.K. and Thrimawithana, D.J. (2011). Current sourced bi-directional inductive power transfer system. *IET Power Electronics* 4 (4): 471–480.

91 Madawala, U.K. and Thrimawithana, D.J. (2011). A bidirectional inductive power interface for electric vehicles in V2G systems. *IEEE Transactions on Industrial Electronics* 58 (10): 4789–4796.

92 Matsumoto, H., Khan, N., and Trescases, O. (2019). A 5kW Bi-directional wireless charger for electric vehicles with electromagnetic coil based self-alignment. In: *Proceedings of the IEEE Applied Power Electronics Conference and Exposition*, Anaheim, CA, USA, 1508–1514.

93 Huang, X., Qiang, H., Huang, Z. et al. (2013). The interaction research of smart grid and EV based wireless charging. In: *Proceedings of the IEEE Vehicle Power and Propulsion Conference*, Beijing, 1–5. IEEE.

94 Mohamed, A.A.S. and Mohammed, O. (2018). Two-layer predictive controller for V2G and G2V services using on wireless power transfer technology. In: *Proceedings of the IEEE Industry Applications Society Annual Meeting*, Portland, OR, 1–8. IEEE.

95 Mohamed, A.A.S., Berzoy, A., and Mohammed, O.A. (2017). Experimental validation of comprehensive steady-state analytical model of bidirectional WPT system in EVs applications. *IEEE Transactions on Vehicular Technology* 66 (7): 5584–5594.

96 Cao, J. and Emadi, A. (2012). A new battery/ultracapacitor hybrid energy storage system for electric, hybrid, and plug-in hybrid electric vehicles. *IEEE Transactions on Power Electronics* 27 (1): 122–132.

97 Hanajiri, K., Hata, K., Imura, T., and Fujimoto, H. (2018). Maximum efficiency operation in wider output power range of wireless in-wheel motor with wheel-side supercapacitor. In: *Proceedings of the Annual Conference of the IEEE Industrial Electronics Society*, Washington, DC, 5177–5182.

98 Hiramatsu, T., Huang, X., and Hori, Y. (2014). Capacity design of supercapacitor battery hybrid energy storage system with repetitive charging via Wireless Power Transfer. In: *Proceedings of the International Power Electronics and Motion Control Conference and Exposition*, Antalya, 490–495. IEEE.

99 McDonough, M. (2015). Integration of inductively coupled power transfer and hybrid energy storage system: a multiport power electronics interface for battery-powered electric vehicles. *IEEE Transactions on Power Electronics* 30 (11): 6423–6433.

100 Li, Z., Zhu, C., Jiang, J. et al. (2017). A 3-kW wireless power transfer system for sightseeing car supercapacitor charge. *IEEE Transactions on Power Electronics* 32 (5): 3301–3316.

10

WPT for Low-Power Applications

By means of the remarkable characteristics of flexibility, position free, and movability, the wireless power transfer (WPT) is widely used not only in high-power applications represented by electric vehicles but also in various low-power applications, which is changing our lifestyle to some content. This chapter elaborates four typical low-power applications: portable consumer electronics, implantable medical devices (IMDs), drones, and underwater wireless charging.

10.1 Portable Consumer Electronics

10.1.1 Introduction

With the development of WPT for low-power applications and the increasing popularity of portable devices, the wireless charging for portable consumer electronics has become a hotspot in the mobile phone industry. Compared with the medical and underwater devices, wireless charging products for portable devices have appeared in large numbers, wherein the wireless charger for mobile phones is the most representative application. Hence, this chapter elaborates the technical challenges and opportunities of WPT technologies, with emphasis on portable consumer electronics.

Firstly, this section introduces the main organizations and alliances in the wireless charging industry, such as wireless power consortium (WPC), power matter alliance (PMA), and alliance for wireless power (A4WP). Secondly, this section introduces the wireless charging standard, with emphasis on the Qi standard. Thirdly, this section discusses the development process of wireless charging for mobile phones such as Xiaomi and Huawei. Lastly, the development prospects and challenges are both drawn for the wireless charging of portable consumer electronics.

Wireless Power Transfer: Principles and Applications, First Edition. Zhen Zhang and Hongliang Pang.

10.1.2 Wireless Charging Alliance

As the standard setters of the wireless charging industry, wireless charging alliances and organizations play an important role for not only the past decades but also for the future development. Hereby, three main alliances and organizations are introduced as follows.

10.1.2.1 Wireless Power Consortium

The WPC was established on 17 December 2008, which is a wireless charging organization composed of several independent companies. It is also the first standardization organization to promote wireless charging technology. This organization aims to create and promote the general wireless charging standards that are compatible with all rechargeable electronic devices. As a result, the intercommunication between various charging pads and portable consumer devices is realized.

So far, more than 250 members have participated in WPC, including Blackberry, Convenient Power, Royal Philips, Samsung, Sony, Texas Instruments, Toshiba, Verizon Wireless, and infrastructure providers. The most significant contribution of WPC in wireless charging field is the proposal of Qi wireless charging protocol, which is also called Qi standard. As the leading wireless charging standard in the world, Qi has brought more than 200 new wireless charging products to market, and these products have been widely applied in North America, South America, Asia Pacific, Europe, India, Africa, and Australia. Through the promotion of Qi wireless charging standard in the coming years, WPC hopes that a unified wireless charger with low energy consumption and high compatibility can be used in all kinds of portable consumer electronics such as mobile phone and laptop. The Qi wireless charging protocol will be introduced in detail in Section 10.3.1.2.

10.1.2.2 Power Matters Alliance

The PMA was initiated by Duracell Powermat and founded in 2012. It is a global nonprofit industry organization established by avant-grade industry leaders. It aims to create a better example of wireless charging technology. In recent years, PMA has grown into a diversified organization covering wireless communications, consumer products, transportation vehicles, railways, furniture, and other industries.

Through the development in recent years, the members of PMA are also expanding and enriching, including AT&T, Starbucks, DuPont, Qualcomm, HTC, Huawei, and Samsung. PMA is committed to create wireless power supply criterions for mobile phones and electronic devices in accordance with the IEEE association standards. For example, Duracell Powermat has launched a WiCC charging card that uses the PMA standard. The WiCC is a kind of card integrated with coils and electrodes for wireless charging. It is usually installed on the battery of mobile phone to realize wireless charging.

Both PMA and WPC mainly focus on inductive wireless charging systems, but there are still slight differences between them. Specifically, the Qi standard of WPC is more suitable for smart device, and it is supported by hundreds of mobile phones manufactures, while the PMA standard is more commonly used for charging devices such as coffee table and furniture.

10.1.2.3 Alliance for Wireless Power

The A4WP, which is also named Rezence, was established by Qualcomm, Samsung, and Powermat in 2012. It is also a global nonprofit industry organization. It aims to establish technical standards and industry dialogue mechanisms for wireless charging devices of electronic products including portable consumer electronics and electric vehicles. In addition to the founding companies and organizations, the members of A4WP also include Ever Win Industries, Gill Industries, Peiker Acustic, SK Telecom, and Intel.

Different from the magnetic induction wireless charging technique, which is emphasized in Qi and PMA standard, the A4WP standard pays more attention to magnetic resonance wireless charging technology. Compared with the magnetic induction wireless charging technology, the magnetic resonance wireless charging technology can achieve long-distance and multitarget charging, which undoubtedly increases the flexibility of wireless charging. A4WP hopes that wireless charging techniques will be popularized quickly so that it can be used anywhere almost. In other words, A4WP attempts to make wireless charging cheaper and increase the charging interface without increasing the volume of mobile phones, tablets, or laptops, which means that more and more manufacturers will choose wireless chargers as the default option.

In the Qi 1.2 version launched by WPC in 2014, the magnetic resonance technology was introduced for the first time. This version is compatible with the two transmission methods of magnetic induction and magnetic resonance. In order to further promote the unification of wireless charging technology standards, PMA and A4WP were merged in 2015 and renamed AirFuel. Such a merger means that the global wireless charging standards are moving toward the direction of unity.

10.1.2.4 Others

In addition to the three mainstream wireless charging alliances, there are some other alliances such as iNPOFi and Wi-Po. The iNPOFi wireless charging alliance was officially established in China on 19 November 2013. It is a nonprofit social organization voluntarily formed by enterprises and institutions that are actively involved in the nonradiative wireless charging industry. The whole name of iNPOFi is invisible power filed, which is a new wireless charging technology. Its wireless charging series products have the characteristics of no radiation, high-power conversion efficiency, and weak thermal effects. iNPOFi strives to promote and popularize its smart nonradiation wireless charging technology through the platform of the technology alliance. The Wi-Po magnetic resonance wireless charging independently developed by Weie Technology in China generates a resonant magnetic field of 6.78 MHz by magnetic generating device so as to achieve a longer transmission and has been granted more than 30 patent applications. The technology achieves the communication control through Bluetooth 4.0 and supports one-to-multiple synchronous communication. Wi-Po magnetic resonance wireless charging can be applied to various appliances such as mobile phones, computers, smart wearables, smart homes, medical equipment, and electric vehicles.

10.1.3 Wireless Charging Standard

10.1.3.1 Introduction

Before introducing the Qi wireless charging standard, let us first answer an important question: what is the wireless charging standard?

The wireless charging standard is a standardized requirement put forward by the wireless charging alliances or organizations in order to realize the unification and industrialization of wireless charging. Corresponding to the aforementioned three mainstream wireless charging alliances, the current wireless charging standards are Qi standard, PMA standard, and A4WP standard, wherein the Qi standard is the most widely prompted wireless charging standard, especially in portable consumer electronics such as mobile phones. Different standards determine their respective power transmission methods, scope of applications, related technical support, operating frequency, and foreign object handling. In addition to defining indicators, the most important role of wireless charging standard is to achieve communication between the transmitting and the receiving devices. In order to facilitate the understanding, we take the most widespread application of wireless charging technology as an example, namely the charging of mobile phone. In the entire process of wireless charging for mobile phones, there are three situations that should be paid attention particularly. The first situation is that the transmitting device needs to confirm whether the mobile phone has been placed above it. If there is no mobile phone, the transmitting device needs to be suspended to reduce power consumption. The second situation is that the transmitting device needs to determine whether there are metal objects such as keys placed on it. The metal objects inevitably produce heat, which adversely affects the charging safety and even results in fire. The third situation is that the current changes in different stages of charging process. For example, the current may be 1 A during normal charging, whereas it may drop to tens of mA when almost fully charged. Then, if the power of the transmitting device has been constant, the voltage will be very high, that is the reason why the load becomes smaller and the voltage rises.

Therefore, the power of the transmitting device needs to be adjusted when the load changes, just like the voltage feedback control in the switching power supply systems. All problems encountered in the three situations mentioned above need to be addressed through the communication mechanism of wireless charging standards.

In summary, the wireless charging standard is like a "pass certificate"; it identifies the receiving device that meets its standard requirements and performs charging operations in accordance with the established specifications. Moreover, it ensures the stability and compatibility of wireless charging applications and plays a decisive role in standardizing the wireless charging market.

10.1.3.2 Qi Wireless Charging Standard

As the first wireless charging standard, Qi standard is introduced by WPC, which has two characteristics of convenience and versatility. After years of development, Qi has become the most widely used standard in the global wireless charging market, especially in the applications of portable consumer electronics.

According to the specification documents issued by WPC, the low-power wireless charging system (power class 0 power transfer system) under the Qi standard is described from the aspects of principle, specifications, interface definitions, compliance testing, reference designs, and product registration.

The main features of Qi WPT system are as follows:

- The method of WPT is based on the principle of near-field magnetic induction.
- The actual charging power will not exceed the party with lower power tolerance between the transmitting and the receiving devices.
- The typical operating frequency is in the range of 87–205 kHz.
- The protocol communication and information synchronization are realized by the techniques of amplitude shift keying (ASK) and frequency shift keying (FSK), respectively.
- It supports two methods of placing the mobile on the surface of the transmitting devices, which are guided positioning and free positioning.
- It gives more than 30 kinds of transmitter design references.
- A very low stand-by power is achievable.

Qi also has its own certification standards. The power level and the coil design of wireless charging products determine the type of Qi certification. The wireless charging certification test of Qi standard includes the compliance test and the interoperability test. The compliance tests with respect to the power transmitter design, communication interface, and system performance are carried out by authorized test laboratory such as Microtest. In addition to the certification test, Qi also stipulates the certification level of the transmitter including BPP (baseline power profile), BPP + FOD (foreign object detection), and EPP (extended power profile). However, with a general increase in the power of wireless charging products, Qi simplified the certification level to BPP and EPP in 2018.

Nowadays, with the continuous development of wireless charging, the Qi standard is also advancing with the times. In order to meet the requirements of the wireless charging market, the power level of Qi and the coil design are also improved accordingly. The current Qi-based wireless charging systems typically use small coils that transmit power at a higher frequency, whereas the shortcomings of short transmission distance and multiple-target charging are also required to be addressed for Qi-based wireless charging systems.

10.1.4 Wireless Charging for Mobile Phones

Through the introduction of the wireless charging alliances and standards above, we understand the technical background and the industry specifications of wireless charging for portable consumer electronics. In this part, the mobile phone is chosen as an example to further demonstrate the application of wireless charging in portable consumer electronics.

With the acceleration of life rhythm especially for the development of the internet, the mobile phone has become an indispensable existence among numerous electronic devices. However, the limitation of the battery capacity greatly affects the usage experience because mobile phone needs to be charged from time to time. Although the capacity of mobile phone batteries is constantly increasing, it still cannot meet our daily requirements. Hence, the wireless charging for mobile phones has become a hotspot in the development of the mobile phone industry. Not only mobile phones, other portable consumer electronics such as smart watches, laptops, and headphones are also potential applications of wireless charging. On the other hand, the development of portable consumer electronics also provides platforms, challenges, and opportunities for the development of wireless charging technology in low-power applications.

Since WPC released the Qi wireless charging standard in 2008, the wireless charging for mobile phones has been constantly developing. In 2009, the first mobile phone-supporting wireless charging emerged, which was released by an American smartphone manufacturer Plam. This phone charges with the principle of magnetic induction, while it does not use the Qi standard. If the end user wants to experience wireless charging, he or she has to replace a special black shell called Touchstone which has integrated coils and control circuit inside. The charging power is 5 W, so it takes nearly 100 minutes to fully charge a battery of 1150 mAh. In addition to Plam, Powermat has also introduced a variety of portable and desktop wireless charging products to the market.

Two years later, the first mobile phone that supports the Qi standard in the world was invented in Japan in 2011, that is the Sharp SH-13C released by Sharp.

However, this phone is mainly for the Japanese market, so it has not stirred up the mobile phone wireless charging market world-widely. Not long after, the wireless charging for mobile phones began to enter public vision from two representative mobile phones, including Google Nexus 4 and Nokia Lumia 920. Both of them are charged with the power of 5 W based on Qi standard. However, the wireless charging of mobile phone fails to popularize due to the cognition limitation of people at that time. It was not until the Samsung Galaxy S6 series appeared in 2015 that the public began to realize the practicality of wireless charging. This situation can be explained that Samsung increased the power of wireless charging to 10 W, which is very important for mobile phones with 2000 mAh upward.

In 2017, Apple launched iPhone X which supports 7.5 W wireless charging. Unlike the traditional frequency conversion architecture, a fixed frequency and voltage regulation architecture is utilized in iPhone X for the first time, which greatly improves the performance of wireless charging. From then on, the mobile phone wireless charging market has been fully opened up. Subsequently, the Qi standard was updated to simplify the original certification level in 2018.

In the global mobile market, China gradually becomes the country with the largest mobile phone production, and the application of mobile phone wireless charging technology by Chinese mobile phone manufactures is also at the forefront of the world.

In the same year that iPhone X was released, the M7 Pulse was released by Gionee for the first mobile phone with wireless charging technology in China. In 2018, Huawei released a wireless charging board, which has increased the charging power to 15 W. Meanwhile, the launched 20 series mobile phone innovatively realized 5 W reverse charging. The release of Huawei 15 W wireless charging board opened the prelude to wireless charging applications for portable consumer electronics in China and played an indispensable role in the development of wireless charging industry around the world. A year later, Xiaomi 9 Pro once again raised the records of forward and reverse wireless charging to 30 and 10 W, respectively. Subsequently, many other manufactures have also launched their own wireless charging mobile phones and wireless charging devices. In 2020, Huawei raised the wireless charging power to 40 W, which enabled wireless superfast charging for mobile phones. Huawei 40 W wireless charger has also become the most powerful wireless charging product in the world.

Figure 10.1 shows several wireless charging devices from Huawei and Xiaomi. As shown in Figure 10.1, the mainstream wireless charging devices for mobile phones on the market are mainly divided into two types, including wireless charging boards and vertical wireless charging bases. Although the size of these wireless charging products is different, the overall design is similar. Figure 10.2 shows the disassembly diagram of Xiaomi 10 W wireless charging board. Although different products may have their own characteristics, all of them can be mainly divided

Figure 10.1 Wireless charging products. (a) Xiaomi 10W; (b) Huawei 15W; (c) Huawei 27W; (d) Huawei 27W for cars; (e) Xiaomi 30W; (f) Huawei 40W.

into three parts: the shell, the printed circuit board (PCB) of the control circuit, and the coil.

In addition to the appearance design and structural composition, the wireless charging principle is also an important part of the mobile phone wireless charger. At present, there are two wireless charging principles for mobile wireless chargers on the market. One is frequency conversion or fixed frequency modulation duty cycle architecture, and the other one is fixed frequency voltage regulation architecture. The working principle of the frequency conversion architecture is to adjust the transmission power by changing the operating frequency, which results in the interference to the mobile phone in certain transmission frequency bands. At present, the transmitting power with frequency conversion architecture on the market is limited between 5 and 10 W. Moreover, it has no quality factor detection so as to inevitably deteriorate the power transmission performance. This architecture has been passed only by BPP certification by now and is mainly proposed to support Apple 7.5 W wireless charging. However, it does not comply with the MFi specification and Qi standard.

Figure 10.2 Disassembly diagram of Xiaomi 10W wireless charging board.

The second type of architecture is the fixed frequency voltage regulation architecture, where the transmission power is controlled by adjusting the voltage. In this architecture, the operating frequency is fixed at about 127 kHz, and the duty cycle is constant at 50%. Then, it can effectively avoid the frequency band that interferes with the mobile phone so as to minimize the interference of wireless charging. With quality factor detection circuit, it can be more sensitive in FOD detection and meet the requirements of EPP. At present, the famous manufactures of fixed frequency voltage regulation schemes are NXP, IDT, and Convenient Power Systems (CPS).

After introducing the two solutions, the product features and development trends of wireless charging for mobile phone will be presented through several products of Xiaomi and Huawei. Figures 10.2 and 10.3 show the disassembly diagrams of wireless charging products of Xiaomi and Huawei. Additionally, Table 10.1 also gives a more intuitive comparison.

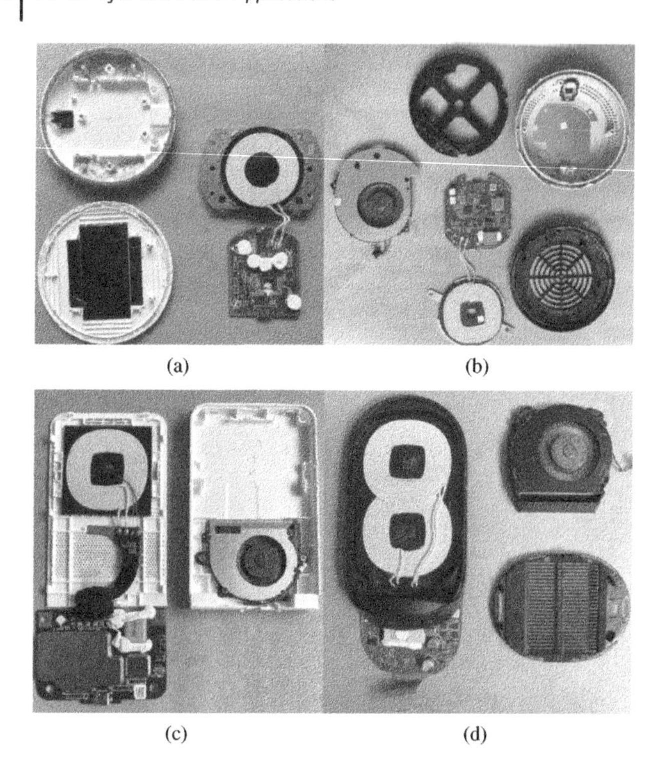

Figure 10.3 Disassembly diagram of wireless charging products. (a) Huawei 15W; (b) Huawei 27W; (c) Xiaomi 30W; (d) Huawei 40W.

Table 10.1 Comparison of wireless charging products from Xiaomi and Huawei.

Type	Year	Style	Cooling method	Number of coils
Xiaomi 10W	2018	Board	Passive	One
Huawei 15W	2018	Board	Passive	One
Huawei 27W	2019	Board	Initiative	One
Xiaomi 30W	2019	Vertical base	Initiative with air duct	One
Huawei 40W	2020	Vertical base	Initiative with air duct	Two

The development trends of wireless charging for mobile phones as well as the corresponding product features are mainly inflected in the following three aspects:

10.1.4.1 Transmission Performance

According to Table 10.1, it is a developing trend that the power is increasing continuously. Compared with the Huawei 15W wireless charging board, the charging efficiency of Huawei 27W wireless charging board is increased by 80%. Even further, Huawei 40W wireless charging device completes the wireless charging of Mate 20 Pro with a battery capacity of 4200 mAh in 68 minutes. Hence, the improvement in power alleviates the issue of low efficiency for wireless charging systems. What's more, a dual coil design is used to meet the power and misalignment requirement on the Huawei 40W. Similarly, with BP coil in electric vehicles, the design of the double coil in mobile phone will be increasingly used to achieve a better misalignment tolerance and greater flexibility.

10.1.4.2 Transmission Stability

In addition to the improvement in the charging performance, the stability is also an important aspect for wireless charging process, wherein the FOD is a vital indicator. In terms of the wireless charging safety, a lot of heat is generated by the eddy current in metal objects, which even results in fire hazard. Thus, the FOD is indispensable for wireless charging through quality factor detection which is mentioned earlier.

Apart from the FOD, the heat dissipation is also an indicator which is required to be considered for all wireless charging products. The increase in the charging power is accompanied with increasing heat. As a result, the heat dissipation has become an important aspect that needs to be continuously optimized. This can be reflected in the products of Xiaomi and Huawei, as shown in Figures 10.2 and 10.3a. A passive heat dissipation is used for both Xiaomi 10W and Huawei 15W. However, the traditional passive heat dissipation cannot meet the requirements as the power reaches 20 W. So, a number of products have added fans for initiative heat dissipation such as Huawei 27W wireless charger. In order to further improve the heat dissipation performance, the current 30 W power level wireless charging products began to adopt the air duct. On this basis, Huawei 40W is designed with a large area of heat dissipation holes under the charger base in cooperation with fan and air duct, as shown in Figure 10.4, which is somewhat similar to the heat dissipation design of laptops. In the future, the problem of heat dissipation will still be a knotty problem that troubles the development of wireless charging for mobile phones.

In addition to the FOD and heat dissipation, the current wireless charging products have also enriched the functions including overvoltage protection, overcurrent protection, overtemperature protection, undervoltage protection, and electrostatic protection to ensure the stability of wireless charging for mobile phones.

10.1.4.3 User Experience (Practicality)

Regardless of the transmission performance or stability, the ultimate objective of wireless charging products for mobile phones is to improve the convenience in

Figure 10.4 Cooling design of Huawei 40W wireless charger.

daily life. The structure of the vertical base used in Xiaomi 30W and Huawei 40W is designed not only for the air duct but also aiming to improve the user experience to a certain extent. Wireless charging technology releases mobile phone from the wire bondage and thus facilitates its normal use even while charging.

10.1.5 Discussion

As aforementioned, the products using wireless charging have attracted more and more attention, especially for mobile phone. However, there is still a long way to go for such an emerging technology, which has a great potential to spread to other areas. Apart from the extension of application scenarios, two other main development possibilities of wireless charging for portable consumer electronics are elaborated as follows:

Unification of wireless charging standards – Due to the difference in R&D strength, technology source, and market support institutions, the standards of wireless charging technology field are not uniform globally. Three major wireless charging alliances and corresponding standards have been discussed in the first part of this section. Looking at the global wireless charging market, the actual application situation has far fallen short of industry expectations although the technology prospect is huge. In addition to the immature technological development, another reason for this situation is that the wireless charging standards have not been unified. In other words, compatibility is one of the biggest problems encountered in the development of wireless charging. The disunity and incompatibility of wireless charging standards results in inconvenient purchase choices and poor experience for consumers. For the purpose of improving the consumer experience and accelerating market expansion, the unification of wireless charging standards is urgent, especially the interoperability of Qi and Rezence proposed by PMA and A4WP. This is also the premises of the popularity of wireless charging for portable consumer electronics in different regions.

Realization of long-distance wireless charging – The ultimate objective of wireless charging for portable consumer electronics is to be able to get rid of the limitations of the charging devices and places. At present, the wireless charging for portable consumer electronics is limited to a charging distance of a few millimeters. Hence, how to realize the long-distance wireless charging is very important. In response to this issue, Energous proposed the WattUp in 2015, which is a new type of wireless charging technology that can realize wireless charging operation within a range of 15 ft (about 4.6 m). It transmits power to other miniature receivers through a wireless hub. People can configure multiple receivers and plug them into sockets in different rooms such as bedrooms and living rooms. As a result, the truly wireless charging could be realized using a dedicated mobile phone. Unfortunately, this technology has not been popularized so far. In the view of the authors, there is really a long way to go for wireless charging technologies to be widely used like the way that was designed as Wi-Fi-style initially.

10.2 Implantable Medical Devices

10.2.1 Introduction

With the continuous improvement in technology and medical level, more and more technical innovations have been adopted for the treatment of various diseases and achieved better results. Among them, the research on IMDs by medical institutions, universities, and scientific research institution are becoming increasingly mature. The application of WPT technology in IMDs has become a new way of energy supply instead of battery pack, and it breaks through the application limit and service life of implantable devices [1].

The WPT-based IMDs avoid the suffering of patients caused by traditional wire connection and implantable battery. The medical devices are powered by the WPT technology, that is the electric energy is transmitted from external energy source to internal body and converted into stable DC electric energy through transcutaneous energy transfer. What's more, most of the IMDs are low-power devices, which are better for transcutaneous energy transfer [2]. After removing the battery, IMDs can not only reduce the volume greatly but also change the implantable mode and location of devices, which result in a better and broader treatment. Hence, the WPT technology provides favorable conditions for the development and improvement of IMDs.

10.2.2 Wireless Transfer for Implantable Medical Devices

Generally, there are four main schemes of WPT technology: inductive, capacitive, microwave, and ultrasonic [3], wherein the microwave power transmission cannot be directly used for implants as the power density is limited by human tissue exposed to electromagnetic (EM) field [4]. Hence, the other three WPT schemes will be discussed for IMDs in this chapter.

10.2.2.1 Inductive

As the most mature scheme, the inductive coupling has been proven in U.S. Food and Drug Administration (FDA)-approved implants. It is based on the EM induction, that is the transmitting coil adjacent to the skin generates a time-varying magnetic field which produces induced electromotive force by the pickup coil placed in the body. In the application of IMDs, the inductive coupling is generally used at the low-frequency range, since it can ensure the performance of the magnetic field penetrating through the body. In order to maximize the transmission efficiency, primary and secondary sides are tuned at the resonant frequency.

Even with extensive studies on inductive coupling, the WPT-based IMDs still face severe challenges. In addition to the restrictions of the WPT technology itself, such as transmission distance and misalignment, implantable applications also bring a series of challenges as listed below.

(1) The material of the implantable coil is flexible so as to ease the difficulties and injuries during the implantation. However, the real-time bending of the implantable coil will lead to serious detuning [5]. It indicates that the requirements of implantable application come at the expense of transmission efficiency.
(2) Due to the low resistivity of copper, the implantable copper coil shows excellent performance, whereas the copper is not biocompatible. In addition, the resistivity of most biocompatible conductors is 1 order of magnitude higher than that of copper.

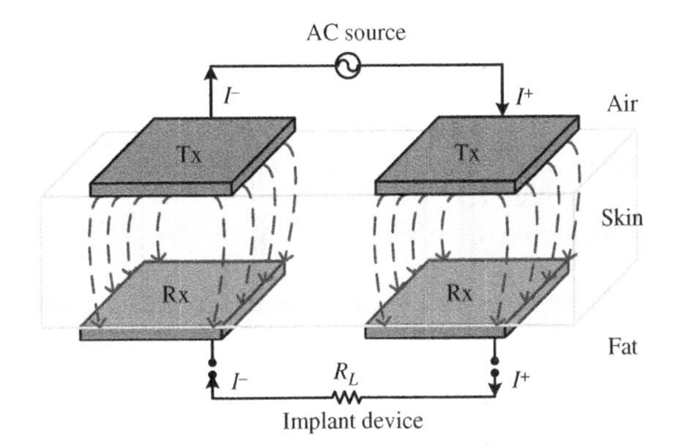

Figure 10.5 Schematic of the capacitive coupling method.

10.2.2.2 Capacitive

Capacitive coupling is a scheme corresponding to inductive coupling. It is based on the principle of electric coupling between two pairs of conductors just as shown in Figure 10.5. Energy transmission through human tissues is realized by displacement current without a physical medium. Extremely low current is generated due to the high mutual impedance of the primary conductor, whereas the mutual impedance will decrease as the pickup coil approaches, and then the secondary side will generate current to power-implantable devices.

Capacitive coupling is one of the recently proposed methods for driving subcutaneous implants, which shows great potential and prospective. However, its application in IMDs is still in the primary stage, which means that there are many challenges to be addressed:

(1) The tissue temperature increases with the capacitance plate, which brings discomfort to the patient.
(2) Capacitive coupling is sensitive to the separation between the transmitting and the pickup coils, unlike inductive coupling, which takes a larger separation (beyond 10 mm) to have a significant influence. Even a deviation of few millimeters can deteriorate the transmission efficiency. Hence, the power fluctuation of capacitive coupling is larger compared with the inductive coupling.

10.2.2.3 Ultrasonic

Ultrasonic power transmission uses ultrasound waves, and its frequency is commonly regulated over 20 kHz. This scheme requires medium to transmit energy, as shown in Figure 10.6, and the energy carried by ultrasound is transferred to implantable devices through the tissue where it is converted into electric energy by piezoelectric transducer [6]. The primary side is a kind of ultrasonic oscillator that generates surface vibration through electrical stimulation and the corresponding sound wave in the frequency range of 200 kHz to 1.2 MHz. The secondary side is an energy collector implanted inside the body.

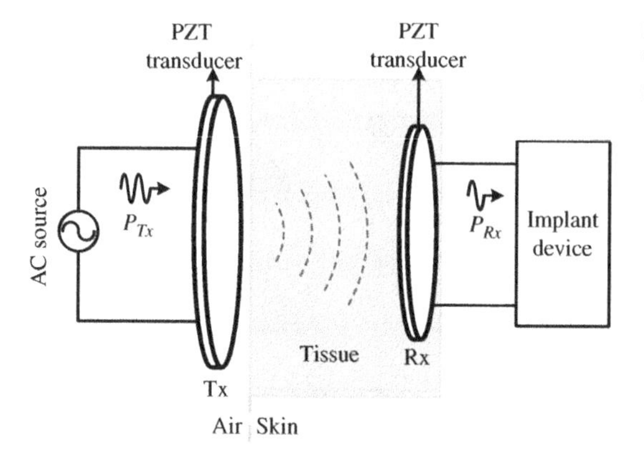

Figure 10.6 Schematic of the ultrasonic energy transfer method.

Ultrasonic transmission has the advantages of strong penetration, short wavelength, and no electromagnetic interference, whereas it is also required to deal with technical challenges as follows:

(1) Various organs of the body have different densities and acoustic impedances. The transmission of sound wave is greatly influenced by organs with high impedance, such as bones. This limits the placement of IMDs in the body.
(2) Vibration in the process of energy transmission may result in hidden danger to the patients.

10.2.3 Various Applications

10.2.3.1 Cochlear Implants

Cochlear implants, as a mature implantable technology, are widely used in hearing restoration of hearing impaired or deaf people [7]. At present, more than 20,000 patients around the world have recovered their hearing by cochlear implants. The transmitting part of the cochlear implants includes a microphone and an audio processor, as shown in Figure 10.7. By adopting the WPT technology, the electric signal converted by the audio processor is transmitted to the implantable units located beneath the skull behind the ear, and then the electric stimulates the auditory nerve fibers in the cochlea. Subsequently, the activity of the nerve fibers is transmitted to the brain where it is interpreted as auditory activity. It indicates that the working mechanism of the implanted cochlea is the same as that of normal hearing. Generally, the current pulse generated by the stimulation unit ranges from 10 μA to 2 mA, and the power of the cochlear implants vary between 20 and 40 mW. At present, the cochlear implant has been commercialized, and moreover, major manufacturers such as MED-EL and Advanced Bionics and Cochlear have been approved by FDA. Each of them is dedicated to developing new implantable technology to meet the needs of different patients.

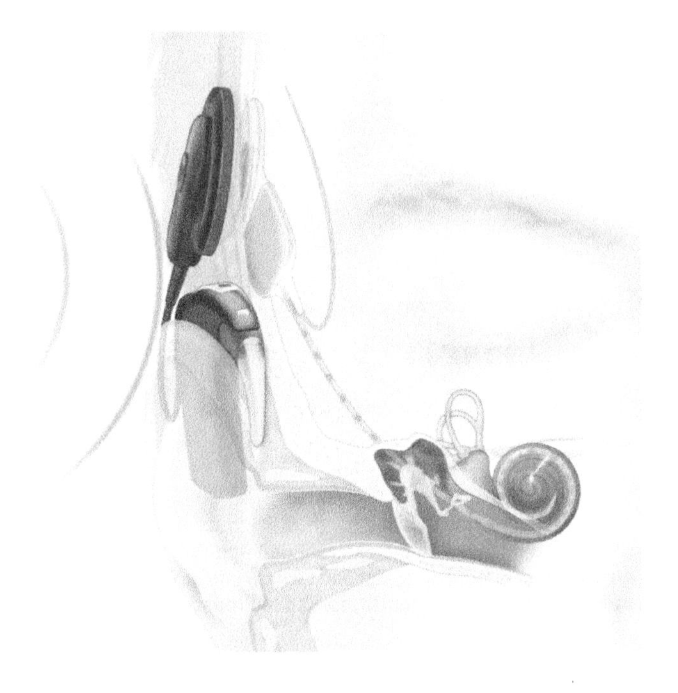

Figure 10.7 Schematic diagram of cochlear implant.

10.2.3.2 Retinal Implants

The electrical stimulation can help patients with vision loss or complete blindness caused by retinitis pigmentosa and age-related macular degeneration to achieve visual perception [8]. On this basis, the retinal implant has been developed and proved in function. For epiretinal implants, the pickup coil is implanted over the eye, while it is implanted beneath the scalp and attached to the retinal stimulator by a platinum wire for subretinal implants. A retinal prosthesis system is shown in Figure 10.8. The primary side includes microcamera, video processing unit (VPU), and a glass with transmitting coil. The electrical signal obtained by VPU is transmitted to the pickup coil installed on the side of the eye using the inductive coupling WPT scheme. Figure 10.8c shows the actual device produced by Second Sight Medical Products. There has been a marked improvement in the aspect of visual perception, motion recognition, and navigation ability by utilizing the retinal prosthesis system although it has not yet been able to help patients restore vision completely.

10.2.3.3 Cortical Implants

The appearance of high-density electrode array makes it possible for fully implantable recording device, which can be used to monitor the activity of the

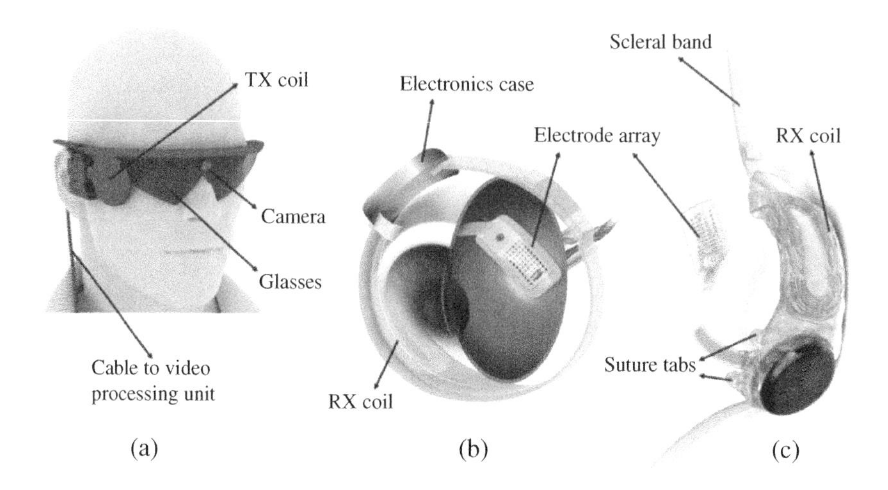

Figure 10.8 Retinal prosthesis system: (a) wearable external unit and (b) implantable unit on retina side. (c) an actual device produced by Second Sight Medical Products. Source: Heather Sheardown/Emily Anne Hicks/Aftab Tayab/Ben Muirhead/Elsevier B.V.

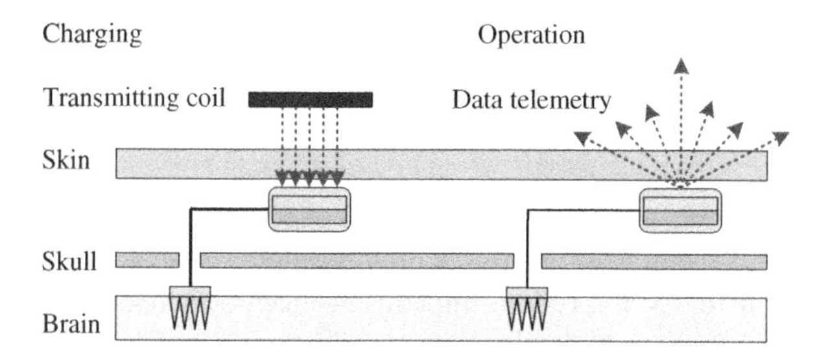

Figure 10.9 Structure of the cortical implant.

brain nerve and realize the brain–computer interface. However, how to energize the implant has become a major challenge for fully implantable recording device. The WPT technology provides an ideal solution, as shown in Figure 10.9. The neural interface uses the WPT technology for charging lithium battery, and for the transcutaneous link for the data for cortical implant applications with WPT technology, the inductive coupling is the preferred scheme.

10.2.3.4 Peripheral Nerve Implants

The peripheral nerve implants provide stimulation to restore motor impairment and sensory function of the limbs. They record the nerve signals and send them from the external stimulator to the implantable stimulator so as to bypass the damaged nerve area, as shown in Figure 10.10. In terms of muscle stimulation, they

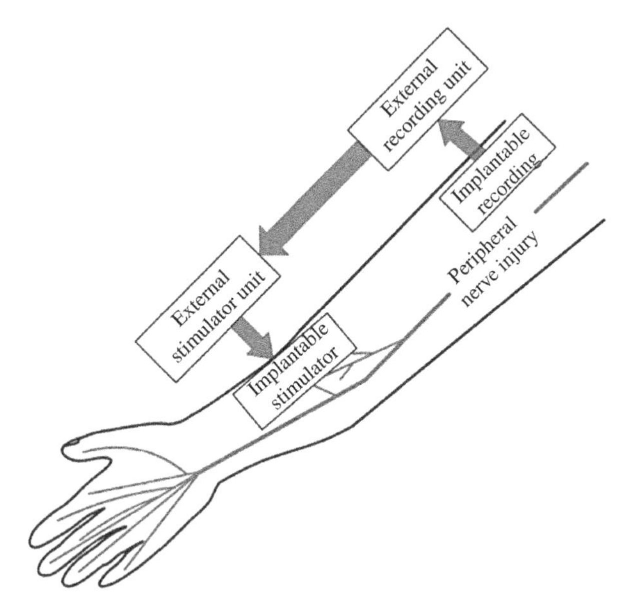

Figure 10.10 Schematic diagram of the peripheral nerve implant.

require current up to 10 mA, which corresponds to a 100 mW level stimulator. Considering the requirements of wireless power, the related works of the peripheral nerve implants are based on inductive coupling scheme. Nerve stimulation devices are either directly powered inductively coupled coils or wirelessly charged with rechargeable batteries integrated with pickup coils.

10.2.4 Safety Consideration

The safety consideration of IMDs is necessary because all implantable devices are in direct contact with the tissue. While ensuring the effective operation of implants, the potential safety hazard to the body should be minimized. Various hazards with respect to safety aspects will be elaborated in Sections 10.2.4.1–10.2.4.3.

10.2.4.1 EM Safety

IMDs transmit the energy and data through the EM field and thus inevitably results in electromagnetic compatibility (EMC) and electromagnetic interference (EMI) issues. As for EMC, it means that the EM energy produced by the IMDs is not allowed to produce intolerable interference to the human body. Otherwise, uncontrolled EM energy may cause unexpected tissue stimulation, which may even be fatal. Tissue burns are caused by electrode heating or the current generated when EM field penetrates the tissue. Tissue stimulation is due to the response of cells to external electric and magnetic fields. The main way to mitigate these effects is to reduce the magnetic field exposure. As for the aspect of EMI, it means that the IMDs in the patient may be affected due to strong EM field, especially in some special occasions, such as security check [9]. How to avoid EMI from outside should be fully considered in the design of implantable devices.

The International Electrotechnical Commission (IEC) regulates a series of standards for medical devices based on the safety considerations of implantable patients. IEC 60601-1-2 is particularly suitable for the basic safety consideration of medical equipment in the case of EMC and EMI.

10.2.4.2 Physical Safety

Physical safety refers to the physical structure of IMDs to ensure the safety of the patients. In order to reduce the damage caused by physical stress, the edge of device is usually designed to be curved. In addition, the volume of the device should be reduced as much as possible [10]. The flexibility of the device is also an important aspect; in nerve implants, the implant is encapsulated in semiflexible silicone. All of the considerations are aimed at alleviating the harm of the body when the patient is constantly exercising. However, it also increases the complexity for system designing since more factors need to be taken into consideration.

10.2.4.3 Cyber Safety

Currently, the information interaction of IMDs is realized using wireless devices, which also leads to two major threats to data access. Firstly, information of IMDs may be stolen or even tampered with, which may result in misleading and working abnormally. Secondly, IMDs may be illegally controlled so that some vital functions cannot be operated. Hence, IMDs need to take these cyber level issues into account. Data encryption and security software are some ways to deal with the threats at the cost of complexity and power dissipation. Unfortunately, FDA does not have a clear regulatory system for the safety of IMDs.

10.2.5 Future Challenges

IMD is a new application field of WPT in recent years, which is still in the developing stage. Compared with other applications such as electric vehicles and portable consumer electronics, IMD-based WPT systems need higher requirements and technical specifications. There is still a lot of work ahead, both in theory and practice.

First of all, the transmission of WPT systems needs to be improved, especially in transcutaneous energy transmission. With the increase in the device implantation depth, the transmission distance also needs to increase accordingly. What's more, whether it is deep tissue or subcutaneous implantation, the body may be affected by power-transmitting equipment. Then, how to reduce the leakage flux in the process of energy transmission is also a technical issue which should be addressed.

IMD is also closely related to the development of other industries. At the moment, electronic devices have been miniaturized to millimeter level and can be directly implanted into nerve fibers. With the development of semiconductor industry, is can be expected that IMD will be more and more miniaturized. In addition, the emergence of new materials (metamaterials, flexible materials, and biocompatible materials) can also adapt well to the IMDs.

10.3 Drones

This section aims to introduce the WPT system of drones including its challenges and technical issues. The first part discusses about the flight endurance of electric-driven drones as well as its market prospective. Then, the challenges of drone WPT systems as well as other wireless charging methods of drone are discussed. As a typical wireless charging mode of drone, the wireless in-flight charging of drone is elaborated in detail including the analysis, design, and challenge. Lastly, the opportunities of drone WPT systems will be drawn.

10.3.1 Introduction

As a pretty valuable technology in the twenty-first century, as well as one of the typical WPT applications, drones have been applied to extensive fields including commerce, industry, and military to perform tasks that are difficult for humans to achieve by themselves, such as aerial photography, aerial monitoring, and material rescue in no man's land. Under this high-speed trend, the global market of drones is forecast to grow by at least a half in the next decade due to the advance of technology and expanding demand for drones [11]. However, the technology of drones faces with a realistic barrier in battery capacity, which greatly limits the distance and duration of their flight. Two main methods are presented to solve the problem mentioned above in the previous studies. The first is to increase the capacity of the batteries by means of the current advanced battery materials; however, this may lead to add the weight of batteries so as to reduce the flight time of drones and increase the cost against commercial utilization. Another attractive alternative is to charge drones by installing battery-charging platforms, which can be classified into the wired and the wireless charging ways for drones. Compared with the traditional charging method using cables or connectors, the wireless charging technique provides a more convenient, intelligent, and reliable way obviously [12].

This section focuses on the WPT technology of drones, which can ensure the rapid charging of drones to increase its continuous flight time. In addition, two important indicators of WPT systems, namely transmission power and transmission efficiency, are sensitive to coupling coefficient. A number of control strategies and design methods have been applied to WPT systems in order to improve the coupling between the transmitting coil and the pickup coil so as to transmit energy stably and efficiently [13, 14]. Nevertheless, the application scenarios of drones bring great challenges to the design of WPT systems, which will be introduced in Section 10.3.2.

10.3.2 Challenges

In general, the WPT system of drones involves two main forms: landing on a charging pad (land and charge) to recharge and the wireless in-flight charging (flight and charge). First of all, the WPT system requires minimal or no additional auxiliary control circuits in the secondary side to reduce the volume and the weight of the energy pickup device as much as possible, which aims to reduce the power loss when drones fly in the air. In addition, the coupling effect between the transmitting and the pickup coils continuously fluctuates, especially for wireless in-flight charging of drones. Since the application scenario of drones is generally outdoors, it is inevitably affected by the weather condition. For example, the drone may land on other unexpected positions or swing in the air by the wind, which deteriorates the power transmission performance of WPT systems significantly. Then, how to deal with such continuous disturbance is the key technical issue to offer a stable and efficient wireless charging for electric-driven drones.

For the land-and-charge mode, an autonomous charging base station for drones is presented [15], where drones can land autonomously and charge the battery wirelessly without human intervention. As shown in Figure 10.11, the pickup coil and the corresponding circuitry of the WPT system are mounted on the landing

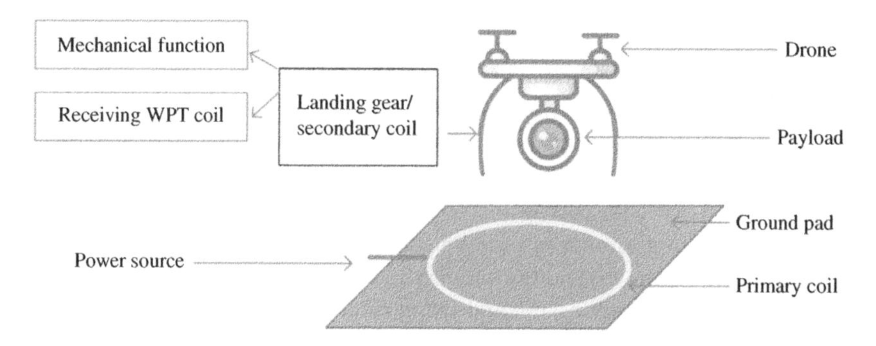

Figure 10.11 Application of WPT for electric-driven drones.

gear, namely a 3-D aluminum tube of appropriate shape for both the mechanism and the electrical function. Such a design adds less weight to the drone and has little impact on the aerodynamics of the drone. After determining the pickup coil, the transmitting coil on the ground base station is optimized to achieve high power and high efficiency as well as showing insensitivity to coil misalignment. The corresponding design criteria for the transmitting and the pickup coils can be referred in Ref. [15]. However, in the land-and-charge mode, landing the drone in the specific place smoothly and safely is the biggest challenge, which is different from the flight-and-charge mode.

To avoid the system malfunction problem caused by the failure of the drone to land, the flight-and-charge mode is popular in the academic and business fields. In the mode, the drone can hover over the charging platform and wirelessly receive the energy from the transmitter. As shown in Figure 10.12, the drone hovers above the energy-transmitting coil during charging stage and keeps its position relative to the transmitting coil constant by dynamically adjusting its altitude [16]. Actually, in severe weather such as strong winds, the hovering position of the drone will still fluctuate continuously, which will lead the continuous change in the mutual inductance in the WPT system. To stabilize the output power performance under the flight-and-charge mode, a nonlinear parity-time (PT) symmetric model was proposed with the analytical expressions of the output power and the transmission efficiency [16]. In the PT symmetrical region, the constant transmission efficiency and transmission power against the change in mutual inductance can be achieved, which has also been demonstrated by the experimental verification with a 100 W commercial drone.

In comparison with other WPT applications, the flight-and-charge mode faces the larger challenges due to the continuous change in mutual inductance, the variation in the desired charging power, and the carrying weight limits of

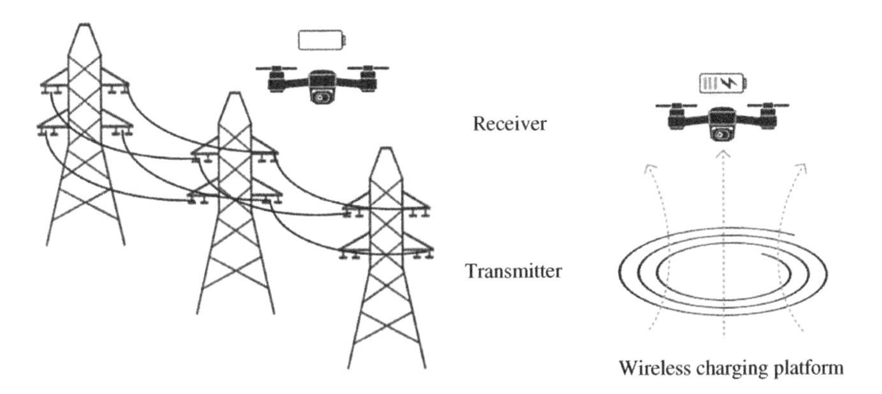

Figure 10.12 Schematic of wireless in-flight charging for drones.

drones. For further description, the fluctuation in mutual inductance between the transmitting and pickup coils will be affected by the position deviation of the flight control system and the harsh application environment such as windy weather. Accordingly, the drone wireless in-flight charging should ensure the stable charging power under continuous changes in mutual inductance caused by the transmission distance and angle changes. Besides, due to the difference in in-flight drone power losses under different working conditions, the drone needs to flexibly adjust the charging power to meet the rapid energy supply. Thus, the constant power control should respond quickly to changes in different control demands, especially for the desired charging power. In addition, the constant power control scheme of the flight-and-charge mode should pay attention to the lightweight design of the drone side since the additional control device at the pickup will increase the power consumption of drones. Thus, with the novel flight-and-charge mode, the aforementioned three key issues should be addressed, otherwise the stability of the drone output power cannot be achieved. Above all, the key to achieve constant output power of drone wireless in-flight charging is to improve the stability under drone-specific issues, namely the continuous change in mutual inductance, the variation in the desired charging power, and the carrying weight limits of drones.

10.3.3 Wireless In-flight Charging of Drones

As mentioned above, the drone wireless in-flight charging is an ideal way to realize the unmanned and flexible aerial energy supply, which can avoid the closing of drone flight control system compared with the wireless charging after landing on the platform. The continuous operation of the flight control system in the drone wireless in-flight charging system can ensure the stability and safety of the energy supply, especially under continuous environmental disturbances (such as strong winds). In addition, taking advantage of its large shooting distance and angle of view, in-flight drone can simultaneously realize the hovering monitoring and rapid energy supply, which improves the work efficiency of intelligent power lines inspection with drones. Considering the challenges of the drone wireless in-flight charging in Section 10.3.2, the current researches are mainly concentrated in the following aspects: the constant current control method, the lightweight design scheme of the transmitting coil, and the high-efficiency charging.

Firstly, for the usual lithium-ion battery-powered drones, the constant current control is a necessary technology because of its benefits to the battery lifetime and safe operation. In Ref. [17], a mutual-inductance-dynamic-predicted constant current control is proposed to realize the secondary-feedback-free output current adjustment for drone wireless in-flight charging systems. In practical systems, the challenge is to keep a constant current output for drones under the continuous fluctuations of mutual inductance, the variation in the desired charging current, and the carrying weight limits of drones. Accordingly, the novel mutual inductance prediction scheme combined with optimized phase-shift control is introduce to maintain the desired constant charging current, which can be implemented at the transmitting side to address the impact of the above challenges. The simulated and experimental results are both given to verify the feasibility of the proposed control scheme, wherein the prediction accuracy is above 92.5%, the current control accuracy is within 5%, and the average response time is less than 320 ms. It shows that the proposed dynamic-predicted CC control scheme has the improved real-time capability and enhanced robustness, which is an ideal technical means for drone wireless in-flight charging.

Besides, the lightweight design of coils is critical for the wireless in-flight charging of drones, which can reduce the energy consumption of the in-flight drone. With the development in electronic technology and the high-frequency technology of WPT system, the coils and electronic circuits of the system can achieve high power density with limited weight, which provides the possibility for the lightweight realization of drones. In Ref. [18], a multi-megahertz WPT system with the features of its light weight and energy saving is introduced. A hollow pickup coil fabricated by the copper-plated plastic structure is designed, as shown in Figure 10.13 [18], which is shaped like the propeller shroud on the drone itself and is mounted on the body of the drone to replace the position of the batteries. The coil structure, which does not require a ferrite core, minimizes the aerodynamic impact of the coil on the drone and provides support for high coupling over a wide area. The transmitting coil is composed of two inner and outer circular magnetic conducting materials with an outer ring diameter of 20 cm. Accordingly, sufficient coupling can be achieved between the transmitting and the pickup coils. Meanwhile, the proposed system can meet the power

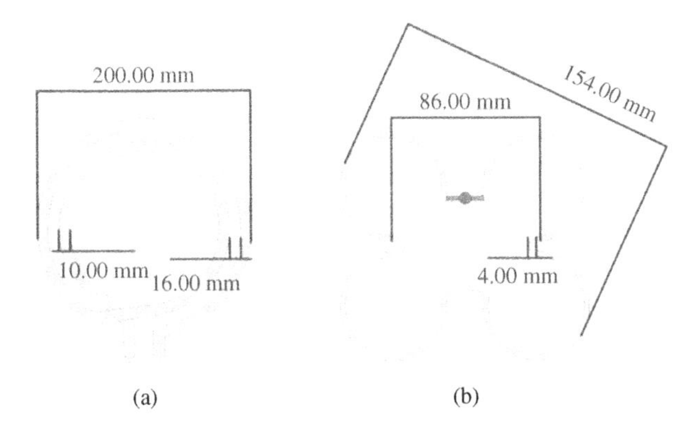

Figure 10.13 WPT coils: (a) transmitting coil and (b) receiving coil.

demand of the dynamic load change and cope with the dynamic changes in the altitude of in-flight drones.

In addition, the high efficiency charging is also challenging for the drone wireless in-flight charging system. Fortunately, the application of semiconductors in wide band-gap devices such as silicon carbide (SiC) and gallium nitride (GaN) has provided important support for the design of more high-efficiency MHz circuit. These materials are applied in the soft-switching resonant inverter and rectifier to provide power support for the WPT systems, which can also help to improve the voltage and current regulation ability for the drone wireless charging systems with large load variation and power variation. In Ref. [19], a load-independent Class EF inverter is used at the transmitting coil, which can provide a stable current amplitude for the transmitting coil and realize zero-voltage switching for various operation situation. At the pickup coil, a hybrid Class E rectifier is used to adjust for various changes in both the coupling and the power demand. Through the overall system design, the 15-W mini-battery-free drone is efficiently charged when hovering near the charging area. Furthermore, considering the magnetic field safety of the system, specific absorption rate (SAR) simulation shows that the magnetic field exposed to human body in the system is far lower than 1000 times of the magnetic field limit of the human's head and trunk proposed by relevant standards [20].

10.3.4 Discussion

The drone application of the WPT has attracted global attention as a replaceable scheme to charge batteries in recent years, which has the potential to solve the problem of the drone flight time. This section introduces the challenges of the drone WPT systems including the land and charge and the flight and charge. As an ideal way to realize the unmanned and flexible aerial energy supply, the technology of the drone wireless in-flight charging is introduced in detail, including the constant current control method, the lightweight design scheme of the transmitting coil, and the high-efficiency charging and EMI suppression method.

10.4 Underwater Wireless Charging

This section aims to introduce the underwater wireless power transfer (UWPT) and its relative applications. First, the challenges facing UWPT systems are elaborated, with emphasis on the impact on the parameters and performance of UWPT systems caused by the characteristic of seawater. The corresponding technical solutions are the basis to develop the UWPT. Then, the relative applications are given to illustrate the feasibility and current research outputs of UWPT. Lastly, the opportunities of UWPT systems will be discussed in detail.

10.4.1 Introduction

With the vigorous development of technology and the desire of human for survival resources, the ocean technology has obtained increasing attentions and development in recent years, for example the autonomous underwater vehicles (AUVs). The ocean technology relies on a number of underwater devices, which mainly depend on the electric power supply. Accordingly, how to deliver power reliably for such electric-driven underwater devices under seawater has become an important issue.

The traditional solution relies on the metal wire to transmit power from the power supply to the device. However, if adopting the traditional power transmission under seawater, it requires to ensure the tightness of the devices, thus inevitably increasing additional friction when the device is docked and detached, which is inconvenient for implementation and daily maintenance and even results in abrasion. Besides, the leakage accident is easy to happen. Accordingly, underwater devices are expecting for a brand-new scheme of power transfer all the time.

With the development of power electronic technologies in recent years, the WPT has been increasingly studied and widely applied in practical applications. The transmitter and pickup of WPT systems are completely isolated and independently packaged, which effectively addresses the technical limit of traditional power transfer schemes for underwater devices. In particular, WPT technologies play an important role in extreme environments. For example, in the marine environment, it improves the security, flexibility, and concealing of underwater devices, whereas the studies on UWPT are still in a primary developing stage since the seawater is quite different from the air. The seawater is utilized as a special electromagnetic medium to transmit power in UWPT systems. However, the characteristics of seawater, such as conductive nature, enormous pressure in deep sea, and force of waves, have significant impact on the performance of WPT systems. To apply WPT systems in the marine environment, these technical issues should be considered and addressed successively.

10.4.2 Analysis of UWPT

As aforementioned, the seawater is utilized as the special electromagnetic medium to transmit power in UWPT systems. It can be seen that the characteristic of seawater is the key to UWPT technologies. Accordingly, this part mainly discusses the challenges facing UWPT systems.

10.4.2.1 Challenges

The UWPT systems will be immersed in the seawater, which means that the parameters of the UWPT systems are inevitably affected by the marine environment, such as temperature fluctuation and corrosive salts. Hence, the tightness is the first concern which should be ensured in the design of UWPT systems to prevent the seawater from contacting the underwater devices. In addition, due to using the seawater as the transmission medium, other challenges facing UWPT systems are summarized as follows:

(1) As an inherent characteristic of seawater, the conductance increases difficulties to transmit electromagnetic waves, since the high-frequency electromagnetic field results in eddy currents to produce the opposing magnetic field and thus reduce the original field strength. Thus, the loss of eddy currents deteriorates the power-transmitting performance of the UWPT systems.
(2) The force of waves inevitably changes the relative position of the transmitting and pickup coils. Accordingly, the coupling coefficient of UWPT systems fluctuates continuously, which means that the UWPT system cannot be stabilized at the optimal operating point if simply immigrating the WPT technology from the land to the sea. In other words, the transmitting efficiency is difficult to be ensured for UWPT systems.
(3) The study by Kraichman [21] shows that the conductive characteristic of seawater adds a resistive component to UWPT systems. The corresponding resistive component increases rapidly when the frequency is above 200 kHz. Accordingly, such a resistive component limits the range of the operating frequency for UWPT systems. As the most direct consequence, the transmitting distance is inevitably limited under seawater, accordingly.
(4) The huge pressure under deep sea conditions causes the "piezomagnetic effect." Then, the magnetic permeability of ferrite decreases, which causes the parameters jump of the UWPT systems and deteriorates the coupling strength between the transmitting and the pickup coils.

10.4.2.2 Analysis

The resistance of a coil consists of three parts including the DC resistance, AC resistance, and the radiation resistance. The DC resistance relies on the conductor size, while the AC resistance relies on its skin depth. Since the values of DC resistance and AC resistance only rely on the quality factor of the wire and the coil configuration, the value of resistance is irregulated to the environment, that is the values of DC and AC resistance in seawater are equal to that in air. However, resulting from the conductive nature of seawater, the current is induced by the magnetic field around the coils. Accordingly, the radiation resistance in seawater is obviously different from that in air. Hereby, the radiation resistance of coil in the seawater, derived by Kraichman, is given by

$$R_{rad}^{sea} = \omega\mu a\left[\frac{4}{3}(\beta a)^2 - \frac{\pi}{3}(\beta a)^3 + \frac{2\pi}{15}(\beta a)^5 - \cdots\right] \tag{10.1}$$

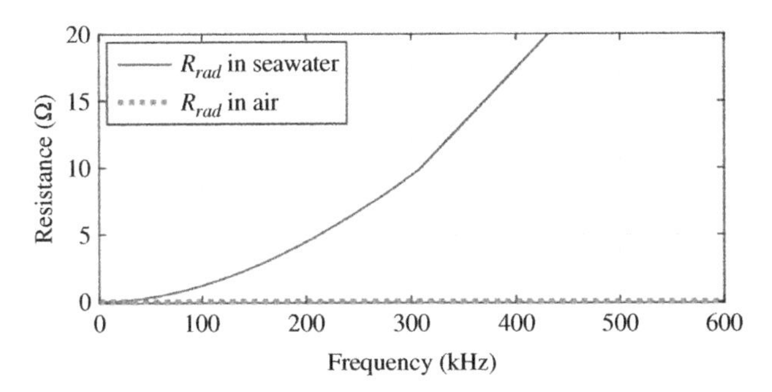

Figure 10.14 Simulation result of radiation resistance in seawater and air. Source: Adapted from Kim et al. [22].

Correspondingly, the radiation resistance of the coil in the air is given by

$$R_{rad}^{air} = \frac{\pi}{6} \frac{\omega^4 \mu a^4}{c^3} \tag{10.2}$$

where ω is the frequency, μ is the permeability of the medium, a is the radius of coils in meters, $\beta = (\mu\omega\sigma/2)^{1/2}$, and σ is the conductivity of the medium. Using (10.1) and (10.2), the simulation result is given in Figure 10.14. It can be obtained that the radiation resistance in air is almost zero, whereas the radiation resistance in seawater varies with the frequency.

As a lossy medium, the seawater affects the mutual inductance between the transmitting and the pickup coils. Accordingly, the mutual inductance is shown as

$$M = \frac{\mu}{4\pi} N_1 N_2 \oint \oint \frac{e^{-\gamma |r_2 - r_1|}}{|r_2 - r_1|} dl_1 dl_2 \tag{10.3}$$

where $\gamma = \sqrt{j\omega\mu\sigma}$, and N_1 and N_2 are the number of turns of the transmitting and pickup coils. In addition, the circular integrals are over the path of a single turn of each coil.

Figure 10.15 depicts the equivalent circuit of the exemplified UWPT system, which shows that there are two obvious differences by comparing with classic

Figure 10.15 Equivalent circuit of UWPT systems.

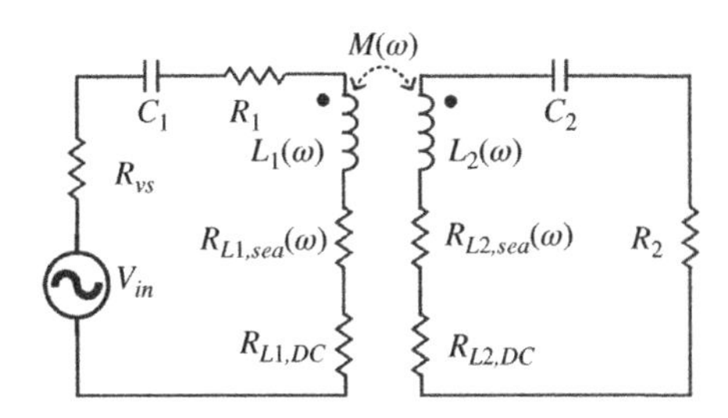

WPT systems. First, the conductive characteristic of seawater adds a resistive component to each coil of UWPT systems, namely ${R_{L1}}^{sea}$ and ${R_{L2}}^{sea}$, which is different from the DC wire resistance. Besides, the conductive characteristic of seawater causes the additional resistive component and the coil inductances related to the frequency. Second, there is an imaginary component in the mutual inductance, which may also produce power loss.

10.4.3 Applications

So far, the UWPT system has three main applications in the underwater including AUV as depicted in Figure 10.16, ocean buoy as depicted in Figure 10.17, and ocean observation network. This section reviews the development of UWPT applications chronologically.

The underwater wireless charging for AUVs was first proposed by Feezor, Sorrel, and others in 2001. The underwater inductive power transfer (IPT) system is utilized to transmit power from the subsea base station to AUV. In the 2000 m underwater, the transmission power of the proposed system is up to 200 W, and the transmission efficiency is up to 79%.

Under the force of waves, the relative position of the coils has inevitable offset for AUVs normally. Since the traditional E-shape core is sensitive to the horizontal offset and thus unsuitable for AUV, the tapered and pot cores are generally utilized for AUVs. In 2004, Tohoku University and Nippon Electronics company jointly proposed an underwater IPT system, which adopts a special shape of ferrite and

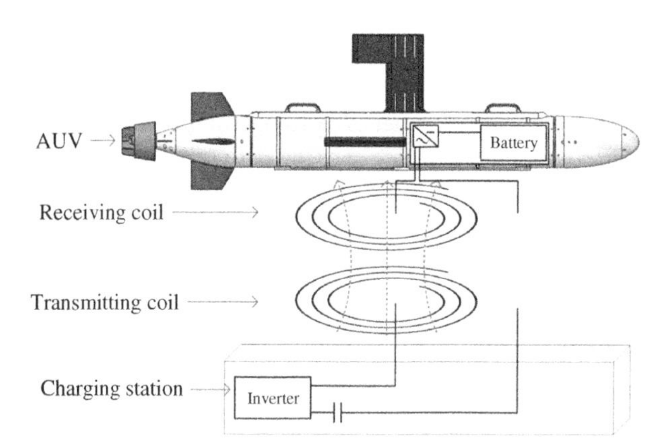

Figure 10.16 Underwater wireless charging for AUV. Source: Adapted from Orekan et al. [23].

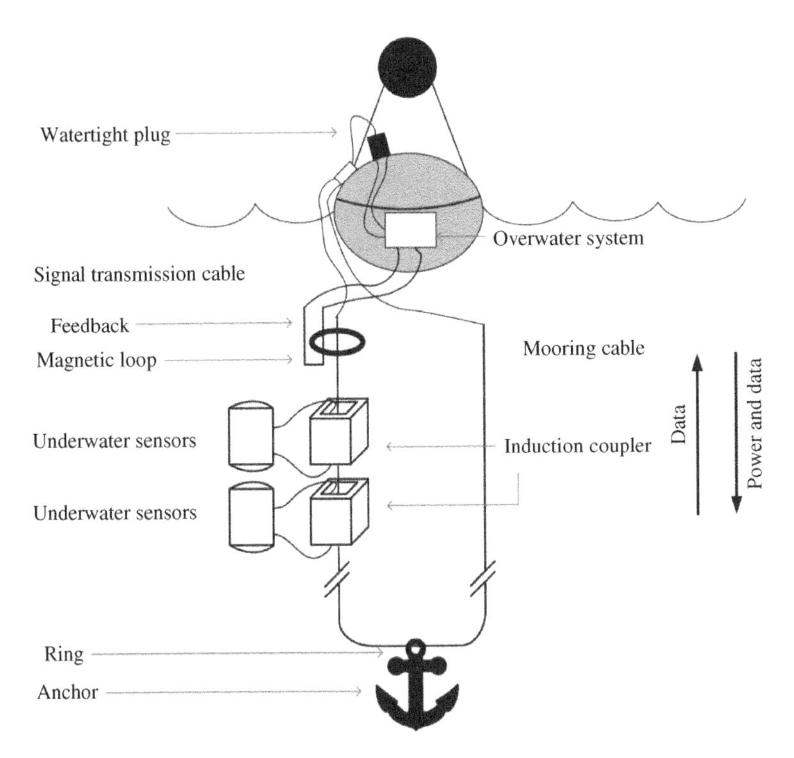

Figure 10.17 Underwater wireless charging for ocean buoy.

tapered coil for the wireless charging of AUVs. The corresponding transmission power is up to 500 W, and the transmission efficiency is up to 90%.

In 2007, based on the ALOHA-MARS submarine observation system, McGinni of Washington University designed an inductive charging system for the anchor profile observer. Its transmission power and efficiency are up to 250 W and 70%, respectively. In Japan, coincidentally, Sojo University and the Institute of Marine and Earth Sciences also jointly developed a similar anchoring system. In 2010, Zhejiang University studied on the electromagnetic coupler of WPT systems for deep-sea applications. The experimental results show that the transmission power and the efficiency are up to 450 W and above 85%, respectively.

Since it is difficult to supply power for traditional buoy systems in the seawater, the battery has been taken as the most common technical solution for the energization. However, batteries normally have a large volume and weight, thus limiting the mounting space of main electronic equipment in the buoy. Accordingly, a research team from Tianjin University proposed an IPT system for charging ocean buoy in 2011 where a solar panel is fixed on the buoy to realize energy self-supported.

In 2014, Northwestern Polytechnical University studied on the transmission mechanism of power in the seawater by utilizing the principle of magnetic coupling resonance. Then, an annular ferrite electromagnetic coupler is developed to transmit power to AUVs. The transmission power is up to 500 W. However, this annular ferrite electromagnetic coupler is required to personalize the structure according to various AUVs, thus significantly deteriorating the generality.

10.4.4 Discussion

Technically, UWPT is derived by classic WPT systems, namely the transmission principle and the system architecture are both consistent when WPT technologies are applied in air or in seawater. However, due to the specific characteristic of the medium and underwater devices, there are still a number of research blanks for further improvements in the UWPT systems.

First, UWPT systems have power loss caused by the eddy current; especially, the power loss significantly increases in the high-frequency range. Hence, the eddy current loss is necessary to be taken into consideration for UWPT systems in future. The theoretical analysis and the mathematical expression of eddy current loss are given in [24], where the eddy current loss is only considered between the transmitter and the pickup coils. In future studies, the eddy current loss should be analyzed in all dimensions, which will provide a theoretical basis for improving the transmission efficiency of the UWPT systems.

Second, the characteristic of seawater results in the change in the system parameters and transmitting performance of UWPT systems. Currently, the studies on the disturbance of environmental variables mainly focus on a single variable, such as temperature, salinity, horizontal offset, and coupling angle. However, the impact of multiple environmental variables has been nearly unexplored in previous studies, which meanwhile specifies the future research directions for UWPT technologies.

Lastly, the mechanism of WPT systems determines the transmission distance; for example, the transmission distance of IPT system is generally in millimeters level. In addition, the complex environment of seawater further limits the transmission distance. According to the rule of IPT systems that the transmission distance cannot be greater than the diameter of the transmitting coil, the size of the transmitting coil should be increased to realize a relatively long power transmission. However, the design of the underwater devices is generally compact, which limits the size of the coils. Accordingly, how to further increase the transmission distance without affecting the volume of underwater devices needs to be taken into account in the practical applications for UWPT systems.

References

1 Hauser, R.G. (2005). The growing mismatch between patient longevity and the service life of implantable cardioverter-defibrillators. *Journal of the American College of Cardiology* 45 (12): 2022–2025.
2 Li, X., Tsui, C.Y., and Ki, W.H. (2014, 2015). A 13.56 MHz wireless power transfer system with reconfigurable resonant regulating rectifier and wireless power control for implantable medical devices. *IEEE Journal of Solid-State Circuits* 50 (4): 978–989.
3 Olivo, J., Carrara, S., and De Micheli, G. (2011). Energy harvesting and remote powering for implantable biosensors. *IEEE Sensors Journal* 11 (7): 1573–1586.
4 Agarwal, K., Jegadeesan, R., Guo, Y., and Thakor, N.V. (2017). Wireless power transfer strategies for implantable bioelectronics. *IEEE Reviews in Biomedical Engineering* 10: 136–161.
5 Qusba, A., RamRakhyani, A.K., So, J. et al. (2014). On the design of microfluidic implant coil for flexible telemetry system. *IEEE Sensors Journal* 14 (4): 1074–1080.
6 Ozeri, S. and Shmilovitz, D. (2010). Ultrasonic transcutaneous energy transfer for powering implanted devices. *Ultrasonics* 50 (6): 556–566.
7 Zeng, F., Rebscher, S., Harrison, W. et al. (2008). Cochlear implants: system design, integration, and evaluation. *IEEE Reviews in Biomedical Engineering* 1: 115–142.
8 Bashirullah, R. (2010). Wireless implants. *IEEE Microwave Magazine* 11 (7): S14–S23.
9 Niehaus, M. and Tebbenjohanns, J. (2001). Electromagnetic interference in patients with implanted pacemakers or cardioverter-defibrillators. *Heart* 86 (3): 246–248.
10 Patrick, J.F., Busby, P.A., and Gibson, P.J. (2006). The development of the nucleus freedom cochlear implant system. *Trends in Amplification* 10 (4): 175–200.
11 Lu, M., Bagheri, M., James, A.P., and Phung, T. (2018). Wireless charging techniques for UAVs: a review, reconceptualization, and extension. *IEEE Access* 6: 29865–29884.
12 Hui, R., Zhong, W., and Lee, C.K. (2014). A critical review of recent progress in mid-range wireless power transfer. *IEEE Transactions on Power Electronics* 29 (9): 4500–4511.
13 Tavakoli, R. and Pantic, Z. (2018). Analysis, design, and demonstration of a 25-kW dynamic wireless charging system for roadway electric vehicles. *IEEE Journal of Emerging and Selected Topics in Power Electronics* 6 (3): 1378–1393.

14 Sample, A.P., Meyer, D.A., and Smith, J.R. (2011). Analysis, experimental results, and range adaptation of magnetically coupled resonators for wireless power transfer. *IEEE Transactions on Industrial Electronics* 58 (2): 544–554.

15 Campi, T., Cruciani, S., Maradei, F., and Feliziani, M. (2019). Innovative design of drone landing gear used as a receiving coil in wireless charging application. *Energies* 12 (18): 3483.

16 Zhou, J., Zhang, B., Xiao, W. et al. (2019). Nonlinear parity-time-symmetric model for constant efficiency wireless power transfer: application to a drone-in-flight wireless charging platform. *IEEE Transactions on Industrial Electronics* 66 (5): 4097–4107.

17 Gu, Y., Wang, J., Liang, Z., and Zhang, Z. (2022). Mutual-inductance-dynamic-predicted constant current control of LCC-P compensation network for drone wireless in-flight charging. *IEEE Transactions on Industrial Electronics* https://doi.org/10.1109/TIE.2022.3142427.

18 Arteaga, J.M., Aldhaher, S., Kkelis, G. et al. (2019). Dynamic capabilities of multi-MHz inductive power transfer systems demonstrated with batteryless drones. *IEEE Transactions on Power Electronics* 34 (6): 5093–5104.

19 Aldhaher, S., Yates, D.C., and Mitcheson, P.D. (2018). Load-independent Class E/EF inverters and rectifiers for MHz-switching applications. *IEEE Transactions on Power Electronics* 33 (10): 8270–8287.

20 Christ, A., Douglas, M.G., Roman, J.M. et al. (2013). Evaluation of wireless resonant power transfer systems with human electromagnetic exposure limits. *IEEE Transactions on Electromagnetic Compatibility* 55 (2): 265–274.

21 Kraichman, M.B. (1962). Impedance of a circular loop in an infinite conducting medium. *Journal of Research of the National Bureau of Standards, Section D: Radio Propagation* 499.

22 Kim, J., Kim, K., Kim, H. et al. (2019). An efficient modeling for underwater wireless power transfer using Z-parameters. *IEEE Transactions on Electromagnetic Compatibility* 61 (6): 2006–2014.

23 Orekan, T., Zhang, P., and Shih, C. (2018). Analysis, design, and maximum power-efficiency tracking for undersea wireless power transfer. *IEEE Journal of Emerging and Selected Topics in Power Electronics* 6 (2): 843–854.

24 Yan, Z., Zhang, Y., Kan, T. et al. (2018). Eddy current loss analysis of underwater wireless power transfer system. *AIP Advances* 8 (10): 8s81–884.

Index

a

acoustic wireless power transfer 8
active rectifier 149, 158–161, 169
air gap 28, 32, 60, 64–68, 79–83, 85, 89, 279–280, 282–284, 287–295, 297, 299, 306–308, 310
aluminum shielding 68–69, 80
amplitude shift keying (ASK) 331
asymmetric cipher 216
asymmetric coil 292

b

band-pass filter 237
battery management system (BMS) 282–283, 290–291, 312
bifurcation phenomenon 151–152, 218–219
bipolar pad 86–88, 297

c

capacitive coupling 3, 6–7, 284, 341
capacitor array 111, 114, 122, 153, 220–224, 228, 230, 232
chaotic cryptography 217
chaotic system 217–218, 232
coil optimization 70, 90
compensation network 12, 14, 22–23, 27, 32, 34, 37, 95–96, 100, 103, 109–112, 193, 196–200, 202, 282–283, 292
conduction losses 125
conductive WPT 280
constant current (CC) 34, 45, 103–106, 108–110, 149, 285, 350, 352
constant voltage (CV) 33, 50, 103–104, 106–107, 109–110, 149, 179, 205
converter 15, 20, 22, 24–25, 31, 123, 148–149, 155–158, 160–162, 164, 167, 169–170, 172–173, 186, 196, 204, 219, 279, 282, 290–291, 312–315
coupling coefficient 20, 28, 39, 42, 45–46, 49, 57–58, 60–62, 64–68, 79, 81, 83, 85, 88, 103–106, 108, 144–146, 149–150, 160, 163–165, 168–172, 205, 211, 215, 263–264, 292, 348, 354
cross-coupling 37–39, 42, 45, 48–49, 51, 111, 121–122, 179, 183, 202–203, 205

d

DC–DC 24, 148–149, 155–158, 160–162, 164, 167, 169–170, 204, 282–285, 290–291, 313
DDQ pad 62, 86–89
dead zone 129, 131, 189, 268–269
decryption 216, 220
degrees of freedom (DoF) 260–261, 271
double-D pad (DD pad) 84–85, 297
dual-frequency 180, 191–193, 196–199
dynamic charging 57, 104, 277, 279–280, 295, 297, 300, 315–316

e

efficiency 6, 9, 12, 14–15, 21–23, 27, 31, 34, 36–37, 41–42, 44–45, 48–51, 57, 60–61, 64, 67–70, 78–80, 90, 98, 100–101, 103–104, 108, 111, 120–122, 124, 132, 136, 143–149, 151–155, 157, 160–173, 179, 183, 185, 189, 202–205, 210, 212, 235–237, 239–240, 244–246, 251, 254–255, 262–264, 270–272, 279, 281, 283–285, 287, 291, 293–297, 299, 308, 312, 315, 329, 337, 340–341, 348–350, 352, 354, 356–358
electric vehicles (EVs) 13, 27, 57, 104, 124, 136, 179, 209, 277, 281, 294, 327, 329–330, 337, 346
electromagnetic fields (EMFs) 27, 209, 281, 305
electromagnetic induction 266, 283
electromagnetic interference (EMI) 279, 342, 345
energy encryption 25, 209–210, 215–217, 220–224, 226, 229–232
equivalent impedance 33, 40, 43, 45, 47–50, 63, 97, 99–102, 115, 117, 132–133, 144, 163, 170, 202–203

f

far-field transmission 6
ferrite core 60–61, 64, 82, 267–269, 290–294, 351
finite element method (FEM) 75
foreign object detection (FOD) 281, 310, 316, 331
Fourier series analysis 186
frequency sensitivity 135, 210–211, 215, 217–218, 222, 229, 232
frequency shift keying (FSK) 331
frequency splitting 95, 135, 150, 152–153, 209–212, 215
fundamental harmonic analysis (FHA) 184

g

gallium nitride (GaN) 124, 352
grid-to-vehicle (G2V) 312

h

high-order harmonics 191, 194–196
hybrid electric vehicles (HEVs) 277
hybrid energy storage system (HESS) 313
hybrid topologies 104, 109

i

impedance matching 12, 15, 95–96, 111, 114, 119, 122, 147–149, 153–154, 179, 196
impedance mismatching 95
induced electromagnetic field 25–27, 37, 51–52, 180, 184, 209–212
inductive power transfer (IPT) 12, 19–28, 32–39, 42, 45–46, 48–52, 180, 203, 279–280, 285, 356–358

inductor-capacitor-capacitor (LCC) 107–110, 205
in-flight charging 347–352
interoperability 13, 280–281, 331, 339
inverter 14–15, 20, 22, 24–27, 34, 37, 57–58, 95, 115, 120, 123, 125–136, 148–149, 153, 157, 160–161, 170, 180, 186, 191–192, 195–196, 204, 209, 219, 261, 279, 282, 284–285, 290–291, 295, 352, 356

k

Kirchhoff's voltage law (KVL) 34, 39, 43, 46, 50, 181, 184, 197, 199, 203

l

leakage inductance 20, 22, 28, 32, 109, 150
leakage magnetic flux 28
live object detection (LOD) 287, 311
load detection 12, 235, 255, 269–271
logistic map 217–219, 221, 229, 232

m

magnetic field 13, 20, 23–28, 37, 65, 68, 70–71, 73, 77, 80, 83–86, 96, 235–238, 244–248, 250–262, 264–265, 268–271, 280, 283, 291, 293, 303–311, 315–316, 330, 340, 346, 352, 354
magnetic flux 20, 28, 31, 52, 61, 68–69, 80, 82–83, 85–86, 153, 211, 229, 245–247, 256–257, 261, 263–266, 268–270, 283, 291, 293–295, 304–305, 309–310, 316
magnetic resonant coupling (MRC) 11–12, 19, 232, 237
magnetic shielding 68
maximum efficiency 15, 36, 42, 45, 60–61, 143, 147, 149, 155, 157, 160–173, 179, 205, 239, 246, 270, 297, 312
maximum power 51, 148, 152, 160, 245–246, 291, 296
microcontroller unit (MCU) 188
microwave wireless power transfer 9
multifrequency excitation 25–26, 37, 51, 179–180, 185, 189, 195
multifrequency simultaneous transmission 179, 184, 204
multifrequency superposition methodology 27, 180, 191
multiple-input multiple-output (MIMO) 205
multiple-pickup 15, 111, 114–115, 121, 179, 196, 198–199, 204–205, 209
multitransmitter 179, 205, 236
mutual inductance estimation 143, 164–165, 170

n

near-field transmission 6
Newton–Raphson iteration algorithm 188
Nikola Tesla 3–5
nonionizing radiation (NIR) 303
n-to-n transmission 21

o

omnidirectional WPT 14, 235–239, 244, 246–248, 255, 260, 262–264, 266, 269–271
1-to-n transmission 14–15, 19, 24–27, 37, 50, 179–180
1-to-1 transmission 14, 19, 21, 28–31, 36, 50, 179–180, 205
on-line electric vehicles (OLEVs) 294–295, 301, 308–309
optical WPT 6

p

phase-locked loop (PLL) 150–152
pickup coils 10, 20, 22–23, 28–29, 32, 35–36, 38, 42, 45, 51, 58, 60–62, 64, 68, 79–81, 95, 120, 145, 179, 192, 213, 218, 222–224, 230–231, 235–237, 248, 253, 256, 261–267, 282–283, 292–293, 295, 316, 341, 345, 348–351, 354–355, 358
pickups 12, 14, 19, 21, 24–28, 37–52, 117, 120–122, 180, 182–183, 191, 195, 202, 204–205, 209–210, 223–224, 226, 228–229, 235–238, 242, 244–247, 260, 270, 293
plug-in hybrid electric vehicles (PHEVs) 277, 279, 300–301
power allocation 15, 179, 204
power distribution 3, 25–26, 37, 41, 44–45, 47–48, 50–51, 179, 204–205, 223–224, 232, 270
power electronic 3, 20, 124, 189, 291, 301, 353
printed spiral coil (PSC) 74–75, 77
proximity effect 70, 74–75, 77
pulse-density modulation (PDM) 158–159
pulse-width modulation (PWM) 14, 26, 123, 127–129, 132, 135, 150–152, 158, 180, 185–190
pure electric vehicles (PEVs) 277, 300–301

q

quality factors (Q) 19–20, 60–61, 146–147

r

radio-frequency 3, 6
random sequence 217
reactive power 31–32, 96, 199, 218, 312
rectifier 9–10, 14, 22, 24–26, 37, 58, 132, 139, 148–149, 155, 158–161, 167–170, 175–176, 192, 282, 284–285, 290–291, 295, 312, 352, 359–360
resonate frequency 179–180, 182
reverse recovery current 132
rotating magnetic field 236–238, 247–248, 260–261, 271

s

selective harmonic elimination pulse width modulation (PWM) 180, 189
selective WPT 180, 232
self-resonance coil 58, 90
semiconductor 3, 14, 23, 123–126, 131–132, 135, 148, 347, 352
series–series (SS) 14, 34–35, 37, 39–40, 45, 49–50, 57, 59–60, 96–98, 103–104, 143, 145, 210
signature algorithm 216
silicon carbide (SiC) 124–126, 167, 352
single-frequency excitation 25–26, 37, 179
single-frequency simultaneous transmission 180
single-frequency time-sharing transmission 179–180
soft switching 23, 34, 131–132, 148–149
state of charge (SOC) 285, 314
supercapacitor 312–314
switching losses 131–132, 134
symmetric cipher 216

t

3-dimensional (3D) WPT 235, 247
3D rotating magnetic field 260

total harmonic distortion (THD) 22, 188
transformer-based superposition 180, 191–192
transformer-free superposition 180, 192, 195
transmitter 3, 10–11, 21, 24, 27–28, 37, 39, 86, 95, 154, 179, 195, 205, 209, 222–224, 226, 228, 230, 236–238, 242, 244–247, 251, 253–254, 256–260, 270, 290–291, 331, 349, 353, 358
triple-frequency 199–202

u
ultracapacitors 296
unauthorized pickup 52, 205, 223–224, 226, 228–229, 231–232
underwater wireless power transfer (UWPT) 14, 352–356, 358
unique equilibrium point 118

v
vehicle-to-grid (V2G) 311–312, 317
virtual capacitor 114
voltage–ampere (VA) 22, 31, 96, 150, 196, 218, 221

w
wide-bandgap semiconductor device 123–125
wireless power consortium (WPC) 12, 123, 327–328
wireless power transfer (WPT) 3, 5–6, 9–15, 19–20, 25, 57–62, 64–65, 67–70, 74, 78–80, 86, 88–90, 95–97, 106, 109, 111–112, 114–115, 120–123, 126, 135, 143–148, 150, 153–155, 158–160, 163, 165, 179–182, 184, 186, 188, 190, 192, 194–196, 198–200, 202, 204–213, 215–224, 226, 228–229, 232, 235–240, 244–248, 250, 255–256, 258, 260, 262–264, 266, 269–271, 277–278, 280–286, 288, 290, 292, 294–298, 300, 302, 304, 306, 308, 310, 312, 314–317, 327–328, 330–331, 339–340, 342–344, 346–358

z
zero current switching (ZCS) 14, 22, 132–133
zero phase angle (ZPA) 150–153, 201
zero voltage switching (ZVS) 14, 22

IEEE Press Series on Power and Energy Systems

Series Editor: Ganesh Kumar Venayagamoorthy, Clemson University, Clemson, South Carolina, USA.

The mission of the IEEE Press Series on Power and Energy Systems is to publish leading-edge books that cover a broad spectrum of current and forward-looking technologies in the fast-moving area of power and energy systems including smart grid, renewable energy systems, electric vehicles and related areas. Our target audience includes power and energy systems professionals from academia, industry and government who are interested in enhancing their knowledge and perspectives in their areas of interest.

1. *Electric Power Systems: Design and Analysis, Revised Printing*
   Mohamed E. El-Hawary
2. *Power System Stability*
   Edward W. Kimbark
3. *Analysis of Faulted Power Systems*
   Paul M. Anderson
4. *Inspection of Large Synchronous Machines: Checklists, Failure Identification, and Troubleshooting*
   Isidor Kerszenbaum
5. *Electric Power Applications of Fuzzy Systems*
   Mohamed E. El-Hawary
6. *Power System Protection*
   Paul M. Anderson
7. *Subsynchronous Resonance in Power Systems*
   Paul M. Anderson, B.L. Agrawal, and J.E. Van Ness
8. *Understanding Power Quality Problems: Voltage Sags and Interruptions*
   Math H. Bollen
9. *Analysis of Electric Machinery*
   Paul C. Krause, Oleg Wasynczuk, and S.D. Sudhoff
10. *Power System Control and Stability, Revised Printing*
    Paul M. Anderson and A.A. Fouad

11. *Principles of Electric Machines with Power Electronic Applications,* Second Edition
Mohamed E. El-Hawary

12. *Pulse Width Modulation for Power Converters: Principles and Practice*
D. Grahame Holmes and Thomas Lipo

13. *Analysis of Electric Machinery and Drive Systems*, Second Edition
Paul C. Krause, Oleg Wasynczuk, and S.D. Sudhoff

14. *Risk Assessment for Power Systems: Models, Methods, and Applications*
Wenyuan Li

15. *Optimization Principles: Practical Applications to the Operations of Markets of the Electric Power Industry*
Narayan S. Rau

16. *Electric Economics: Regulation and Deregulation*
Geoffrey Rothwell and Tomas Gomez

17. *Electric Power Systems: Analysis and Control*
Fabio Saccomanno

18. *Electrical Insulation for Rotating Machines: Design, Evaluation, Aging, Testing, and Repair*
Greg C. Stone, Edward A. Boulter, Ian Culbert, and Hussein Dhirani

19. *Signal Processing of Power Quality Disturbances*
Math H. J. Bollen and Irene Y. H. Gu

20. *Instantaneous Power Theory and Applications to Power Conditioning*
Hirofumi Akagi, Edson H. Watanabe, and Mauricio Aredes

21. *Maintaining Mission Critical Systems in a 24/7 Environment*
Peter M. Curtis

22. *Elements of Tidal-Electric Engineering*
Robert H. Clark

23. *Handbook of Large Turbo-Generator Operation and Maintenance,* Second Edition
Geoff Klempner and Isidor Kerszenbaum

24. *Introduction to Electrical Power Systems*
Mohamed E. El-Hawary

25. *Modeling and Control of Fuel Cells: Distributed Generation Applications*
M. Hashem Nehrir and Caisheng Wang

26. *Power Distribution System Reliability: Practical Methods and Applications*
    Ali A. Chowdhury and Don O. Koval

27. *Introduction to FACTS Controllers: Theory, Modeling, and Applications*
    Kalyan K. Sen and Mey Ling Sen

28. *Economic Market Design and Planning for Electric Power Systems*
    James Momoh and Lamine Mili

29. *Operation and Control of Electric Energy Processing Systems*
    James Momoh and Lamine Mili

30. *Restructured Electric Power Systems: Analysis of Electricity Markets with Equilibrium Models*
    Xiao-Ping Zhang

31. *An Introduction to Wavelet Modulated Inverters*
    S.A. Saleh and M.A. Rahman

32. *Control of Electric Machine Drive Systems*
    Seung-Ki Sul

33. *Probabilistic Transmission System Planning*
    Wenyuan Li

34. *Electricity Power Generation: The Changing Dimensions*
    Digambar M. Tagare

35. *Electric Distribution Systems*
    Abdelhay A. Sallam and Om P. Malik

36. *Practical Lighting Design with LEDs*
    Ron Lenk and Carol Lenk

37. *High Voltage and Electrical Insulation Engineering*
    Ravindra Arora and Wolfgang Mosch

38. *Maintaining Mission Critical Systems in a 24/7 Environment*, Second Edition
    Peter Curtis

39. *Power Conversion and Control of Wind Energy Systems*
    Bin Wu, Yongqiang Lang, Navid Zargari, and Samir Kouro

40. *Integration of Distributed Generation in the Power System*
    Math H. Bollen and Fainan Hassan

41. *Doubly Fed Induction Machine: Modeling and Control for Wind Energy Generation Applications*
    Gonzalo Abad, Jesús López, Miguel Rodrigues, Luis Marroyo, and Grzegorz Iwanski

42. *High Voltage Protection for Telecommunications*
    Steven W. Blume

43. *Smart Grid: Fundamentals of Design and Analysis*
    James Momoh

44. *Electromechanical Motion Devices*, Second Edition
    Paul Krause, Oleg Wasynczuk, Steven Pekarek

45. *Electrical Energy Conversion and Transport: An Interactive Computer-Based Approach*, Second Edition
    George G. Karady and Keith E. Holbert

46. *ARC Flash Hazard and Analysis and Mitigation*
    J.C. Das

47. *Handbook of Electrical Power System Dynamics: Modeling, Stability, and Control*
    Mircea Eremia and Mohammad Shahidehpour

48. *Analysis of Electric Machinery and Drive Systems*, Third Edition
    Paul C. Krause, Oleg Wasynczuk, S.D. Sudhoff, and Steven D. Pekarek

49. *Extruded Cables for High-Voltage Direct-Current Transmission: Advances in Research and Development*
    Giovanni Mazzanti and Massimo Marzinotto

50. *Power Magnetic Devices: A Multi-Objective Design Approach*
    S.D. Sudhoff

51. *Risk Assessment of Power Systems: Models, Methods, and Applications*, Second Edition
    Wenyuan Li

52. *Practical Power System Operation*
    Ebrahim Vaahedi

53. *The Selection Process of Biomass Materials for the Production of Bio-Fuels and Co-Firing*
    Najib Altawell

54. *Electrical Insulation for Rotating Machines: Design, Evaluation, Aging, Testing, and Repair*, Second Edition
    Greg C. Stone, Ian Culbert, Edward A. Boulter, and Hussein Dhirani

55. *Principles of Electrical Safety*
    Peter E. Sutherland

56. *Advanced Power Electronics Converters: PWM Converters Processing AC Voltages*
Euzeli Cipriano dos Santos Jr. and Edison Roberto Cabral da Silva

57. *Optimization of Power System Operation*, Second Edition
Jizhong Zhu

58. *Power System Harmonics and Passive Filter Designs*
J.C. Das

59. *Digital Control of High-Frequency Switched-Mode Power Converters*
Luca Corradini, Dragan Maksimoviæ, Paolo Mattavelli, and Regan Zane

60. *Industrial Power Distribution*, Second Edition
Ralph E. Fehr, III

61. *HVDC Grids: For Offshore and Supergrid of the Future*
Dirk Van Hertem, Oriol Gomis-Bellmunt, and Jun Liang

62. *Advanced Solutions in Power Systems: HVDC, FACTS, and Artificial Intelligence*
Mircea Eremia, Chen-Ching Liu, and Abdel-Aty Edris

63. *Operation and Maintenance of Large Turbo-Generators*
Geoff Klempner and Isidor Kerszenbaum

64. *Electrical Energy Conversion and Transport: An Interactive Computer-Based Approach*
George G. Karady and Keith E. Holbert

65. *Modeling and High-Performance Control of Electric Machines*
John Chiasson

66. *Rating of Electric Power Cables in Unfavorable Thermal Environment*
George J. Anders

67. *Electric Power System Basics for the Nonelectrical Professional*
Steven W. Blume

68. *Modern Heuristic Optimization Techniques: Theory and Applications to Power Systems*
Kwang Y. Lee and Mohamed A. El-Sharkawi

69. *Real-Time Stability Assessment in Modern Power System Control Centers*
Savu C. Savulescu

70. *Optimization of Power System Operation*
Jizhong Zhu

71. *Insulators for Icing and Polluted Environments*
Masoud Farzaneh and William A. Chisholm

72. *PID and Predictive Control of Electric Devices and Power Converters Using MATLAB®/Simulink®*
Liuping Wang, Shan Chai, Dae Yoo, and Lu Gan, Ki Ng

73. *Power Grid Operation in a Market Environment: Economic Efficiency and Risk Mitigation*
Hong Chen

74. *Electric Power System Basics for Nonelectrical Professional*, Second Edition
Steven W. Blume

75. *Energy Production Systems Engineering*
Thomas Howard Blair

76. *Model Predictive Control of Wind Energy Conversion Systems*
Venkata Yaramasu and Bin Wu

77. *Understanding Symmetrical Components for Power System Modeling*
J.C. Das

78. *High-Power Converters and AC Drives*, Second Edition
Bin Wu and Mehdi Narimani

79. *Current Signature Analysis for Condition Monitoring of Cage Induction Motors: Industrial Application and Case Histories*
William T. Thomson and Ian Culbert

80. *Introduction to Electric Power and Drive Systems*
Paul Krause, Oleg Wasynczuk, Timothy O'Connell, and Maher Hasan

81. *Instantaneous Power Theory and Applications to Power Conditioning*, Second Edition
Hirofumi, Edson Hirokazu Watanabe, and Mauricio Aredes

82. *Practical Lighting Design with LEDs*, Second Edition
Ron Lenk and Carol Lenk

83. *Introduction to AC Machine Design*
Thomas A. Lipo

84. *Advances in Electric Power and Energy Systems: Load and Price Forecasting*
Mohamed E. El-Hawary

85. *Electricity Markets: Theories and Applications*
Jeremy Lin and Jernando H. Magnago

86. *Multiphysics Simulation by Design for Electrical Machines, Power Electronics and Drives*
Marius Rosu, Ping Zhou, Dingsheng Lin, Dan M. Ionel, Mircea Popescu, Frede Blaabjerg, Vandana Rallabandi, and David Staton

87. *Modular Multilevel Converters: Analysis, Control, and Applications*
Sixing Du, Apparao Dekka, Bin Wu, and Navid Zargari

88. *Electrical Railway Transportation Systems*
Morris Brenna, Federica Foiadelli, and Dario Zaninelli

89. *Energy Processing and Smart Grid*
James A. Momoh

90. *Handbook of Large Turbo-Generator Operation and Maintenance,* Third Edition
Geoff Klempner and Isidor Kerszenbaum

91. *Advanced Control of Doubly Fed Induction Generator for Wind Power Systems*
Dehong Xu, Dr. Frede Blaabjerg, Wenjie Chen, and Nan Zhu

92. *Electric Distribution Systems*, Second Edition
Abdelhay A. Sallam and Om P. Malik

93. *Power Electronics in Renewable Energy Systems and Smart Grid: Technology and Applications*
Bimal K. Bose

94. *Distributed Fiber Optic Sensing and Dynamic Rating of Power Cables*
Sudhakar Cherukupalli and George J Anders

95. *Power System and Control and Stability,* Third Edition
Vijay Vittal, James D. McCalley, Paul M. Anderson, and A. A. Fouad

96. *Electromechanical Motion Devices: Rotating Magnetic Field-Based Analysis and Online Animations,* Third Edition
Paul Krause, Oleg Wasynczuk, Steven D. Pekarek, and Timothy O'Connell

97. *Applications of Modern Heuristic Optimization Methods in Power and Energy Systems*
Kwang Y. Lee and Zita A. Vale

98. *Handbook of Large Hydro Generators: Operation and Maintenance*
Glenn Mottershead, Stefano Bomben, Isidor Kerszenbaum, and Geoff Klempner

99. *Advances in Electric Power and Energy: Static State Estimation*
Mohamed E. El-hawary

100. *Arc Flash Hazard Analysis and Mitigation*, Second Edition
J.C. Das

101. *Maintaining Mission Critical Systems in a 24/7 Environment,* Third Edition
Peter M. Curtis

102. *Real-Time Electromagnetic Transient Simulation of AC-DC Networks*
Venkata Dinavahi and Ning Lin

103. *Probabilistic Power System Expansion Planning with Renewable Energy Resources and Energy Storage Systems*
Jaeseok Choi and Kwang Y. Lee

104. *Power Magnetic Devices: A Multi-Objective Design Approach*, Second Edition
Scott D. Sudhoff

105. *Optimal Coordination of Power Protective Devices with Illustrative Examples*
Ali R. Al-Roomi

106. *Resilient Control Architectures and Power Systems*
Craig Rieger, Ronald Boring, Brian Johnson, and Timothy McJunkin

107. *Alternative Liquid Dielectrics for High Voltage Transformer Insulation Systems: Performance Analysis and Applications*
Edited by U. Mohan Rao, I. Fofana, and R. Sarathi

108. *Introduction to the Analysis of Electromechanical Systems*
Paul C. Krause, OlegWasynczuk, and Timothy O'Connell

109. *Power Flow Control Solutions for a Modern Grid using SMART Power Flow Controllers*
Kalyan K. Sen and Mey Ling Sen

110. *Power System Protection: Fundamentals and Applications*
John Ciufo and Aaron Cooperberg

111. *Soft-Switching Technology for Three-phase Power Electronics Converters*
Dehong Xu, Rui Li, Ning He, Jinyi Deng, and YuyingWu

112. *Power System Protection*, Second Edition
Paul M. Anderson, Charles Henville, Rasheek Rifaat, Brian Johnson, and Sakis Meliopoulos

113. *High Voltage and Electrical Insulation Engineering*, Second Edition
Ravindra Arora and Wolfgang Mosch

114. *Modeling and Control of Modern Electrical Energy Systems*
Masoud Karimi-Ghartemani

115. *Control of Power Electronic Converters with Microgrid Applications*
Arindam Ghosh and Firuz Zare

116. *Coordinated Operation and Planning of Modern Heat and Electricity Incorporated Networks*
Mohammadreza Daneshvar, Behnam Mohammadi-Ivatloo, and Kazem Zare

117. *Smart Energy for Transportation and Health in a Smart City*
Chun Sing Lai, Loi Lei Lai, and Qi Hong Lai

118. *Wireless Power Transfer: Principles and Applications*
Zhen Zhang and Hongliang Pang

Printed and bound by CPI Group (UK) Ltd, Croydon, CR0 4YY

16/04/2025

14658584-0004